T0263103

The Role of Animals in Emerging Viral Diseases

The Role of Animals in Emerging Viral Diseases

Edited by

Nicholas Johnson

AMSTERDAM • BOSTON • HEIDELBERG • LONDON
NEW YORK • OXFORD • PARIS • SAN DIEGO
SAN FRANCISCO • SINGAPORE • SYDNEY • TOKYO
Academic Press is an Imprint of Elsevier

Academic Press is an imprint of Elsevier
525 B Street, Suite 1900, San Diego, CA 92101-4495, USA
32 Jamestown Road, London NW1 7BY, UK
225 Wyman Street, Waltham, MA 02451, USA

Notice

British Library Cataloguing-in-Publication Data
A catalog record for this book is available from the British Library.

Library of Congress Cataloging-in-Publication Data
A catalog record for this book is available from the Library of Congress.

ISBN: 978-0-12-405191-1

For information on all Academic Press publications
visit our website at elsevierdirect.com

Typeset by TNQ Books and Journals
www.tnq.co.in

Working together
to grow libraries in
developing countries

www.elsevier.com • www.bookaid.org

This book is dedicated to Clive
and Jean for a lifetime of support

Contents

1. A Short Introduction to Disease Emergence

Nicholas Johnson

2. Patterns of Foot-and-Mouth Disease Virus Distribution in Africa

Miriam B. Casey, Tiziana Lembo, Nick J. Knowles, Robert Fyumagwa, Fredrick Kivaria, Honori Maliti, Christopher Kasanga, Raphael Sallu, Richard Reeve, Satya Parida, Donald P. King and Sarah Cleaveland

3.   Parvoviruses of Carnivores

Andrew B. Allison and Colin R. Parrish

4.   Rabies

Conrad Freuling, Ad Vos, Nicholas Johnson, Ralf-Udo Mühle
and Thomas Müller

10. Hantavirus Emergence in Rodents, Insectivores and Bats

Mathias Schlegel, Jens Jacob, Detlev H. Krüger, Andreas Rang and Rainer G. Ulrich

11. Nipah Virus

David T.S. Hayman and Nicholas Johnson

12. Synthesis

Philip R. Wakeley, Sarah North and Nicholas Johnson

Contributors

Andrew B. Allison   Baker Institute for Animal Health, Department of Microbiology and Immunology, College of Veterinary Medicine, Cornell University, USA

Eric J. Arts   Division of Infectious Disease, Case Western Reserve University, Cleveland, Ohio, USA

Miriam B. Casey   Institute of Biodiversity, Animal Health and Comparative Medicine, College of Medical, Veterinary and Life Sciences, University of Glasgow, United Kingdom; The Pirbright Institute, United Kingdom

Sarah Cleaveland   Institute of Biodiversity, Animal Health and Comparative Medicine, College of Medical, Veterinary and Life Sciences, University of Glasgow, United Kingdom

Elisabeth Fichet-Calvet   Bernhard-Nocht Institute of Tropical Medicine, Hamburg, Germany

Conrad Freuling   Institute of Molecular Biology, Friedrich-Loeffler-Institut, Federal Research Institute for Animal Health, Greifswald-Insel Riems, Germany

Robert Fyumagwa   Tanzania Wildlife Research Institute, Tanzania

Paul Gale   Animal Health and Veterinary Laboratories Agency, Surrey, United Kingdom

David T.S. Hayman   Department of Biology, Colorado State University, Fort Collins, Colorado, USA; Department of Biology, University of Florida, Gainesville, FL, USA

Jens Jacob   Julius Kühn-Institute, Federal Research Center for Cultivated Plants, Institute for Plant Protection in Horticulture and Forests, Vertebrate Research, Münster, Germany

Petrus Jansen van Vuren   Center for Emerging and Zoonotic Diseases, National Institute for Communicable Diseases of the National Health Laboratory Service, South Africa

Nicholas Johnson   Animal Health and Veterinary Laboratories Agency, Surrey, United Kingdom

Christopher Kasanga   Sokoine University of Agriculture, Tanzania

Donald P. King   The Pirbright Institute, United Kingdom

Fredrick Kivaria   Ministry of Livestock and Fisheries Development, Tanzania

Nick J. Knowles   The Pirbright Institute, United Kingdom

Detlev H. Krüger   Institute of Medical Virology, Helmut-Ruska-Haus, Charité Medical School, Berlin, Germany

Tiziana Lembo   Institute of Biodiversity, Animal Health and Comparative Medicine, College of Medical, Veterinary and Life Sciences, University of Glasgow, United Kingdom

Honori Maliti   Tanzania Wildlife Research Institute, Tanzania

Glenn A. Marsh   CSIRO Australian Animal Health Laboratory, Geelong, Australia

Ralf-Udo Mühle   Institute for Biochemistry and Biology, University of Potsdam, Potsdam, Germany

Thomas Müller   Institute of Molecular Biology, Friedrich-Loeffler-Institut, Federal Research Institute for Animal Health, Greifswald-Insel Riems, Germany

Sarah North   Animal Health and Veterinary Laboratories Agency, Surrey, United Kingdom

Satya Parida   The Pirbright Institute, United Kingdom

Colin R. Parrish   Baker Institute for Animal Health, Department of Microbiology and Immunology, College of Veterinary Medicine, Cornell University, USA

Janusz T. Paweska   Center for Emerging and Zoonotic Diseases, National Institute for Communicable Diseases of the National Health Laboratory Service, South Africa; Division Virology and Communicable Diseases Surveillance, School of Pathology, University of the Witwatersrand, Johannesburg, South Africa

Andreas Rang   Institute of Medical Virology, Helmut-Ruska-Haus, Charité Medical School, Berlin, Germany

Richard Reeve   Institute of Biodiversity, Animal Health and Comparative Medicine, College of Medical, Veterinary and Life Sciences, University of Glasgow, United Kingdom

Raphael Sallu   Tanzania Veterinary Laboratory Agency, Tanzania

Mathias Schlegel   Friedrich-Loeffler-Institut, Federal Research Institute for Animal Health, Institute for Novel and Emerging Infectious Diseases, Greifswald - Insel Riems, Germany

Denis M. Tebit   Division of Infectious Disease, Case Western Reserve University, Cleveland, Ohio, USA; Myles H. Thaler Center for AIDS and Human Retrovirus Research, Department of Microbiology, University of Virginia, Charlottesville, Virginia, USA

Rainer G. Ulrich   Friedrich-Loeffler-Institut, Federal Research Institute for Animal Health, Institute for Novel and Emerging Infectious Diseases, Greifswald - Insel Riems, Germany

Ad Vos   IDT Biologika GmbH, Dessau-Rosslau, Germany

Philip R. Wakeley   Animal Health and Veterinary Laboratories Agency, Surrey, United Kingdom

Lin-Fa Wang   CSIRO Australian Animal Health Laboratory, Geelong, Australia; Duke-NUS Graduate Medical School, Singapore

Foreword

A definition of an emerging disease is one that has newly appeared in a population or that has been known for some time but is rapidly increasing in incidence or geographic range. These events are commonly described as outbreaks and they impact on human populations through disease and death with great cost to human health and the economy. A report commissioned by The World Bank has calculated that six disease outbreaks alone between 1997 and 2009 resulted in economic losses of US$80 billion.[1] Five of these disease outbreaks were caused by viruses (Nipah virus, West Nile virus, SARS coronavirus, High Pathogenic Avian Influenza and Rift Valley fever virus), of which three are considered in this book. All of these viruses originated in animal reservoirs and were spread through complex interactions with particular animal species. The purpose of this book is to explore these interactions with an underlying message that a One Health approach to disease emergence is needed uniting a range of disciplines to detect emerging viruses and implement effective control measures before they impact on human populations.

On embarking on the preparation of this book, each contributor was asked to consider the role of animals in the emergence of a particular virus. The choice of viruses selected for inclusion in this book is not significant other than to illustrate the role of animals in the processes that lead to emergence. It is certainly not a complete or exhaustive list of the range of emerging viruses that challenge human health in the 21st century. However, I think that they do reflect many of the different interactions and interfaces between humans, animals and viruses. Likewise, the order of the chapters in this book has no particular significance as each describes the emergence of a particular virus, and while there will be common themes that will be discussed in the final chapter, each also has unique features. Each chapter has been written by authors experienced in the study of that virus and is a stand-alone piece. An introductory chapter provides an overview of many aspects of disease emergence, and includes the anthropogenic factors that drive disease emergence. While the majority of the viruses discussed in this book are clearly defined as zoonotic, the first two examples have been included that have a greater impact on livestock and companion animals. These are foot-and-mouth disease (FMD) virus and canine parvovirus type-2 (CPV-2)

---

1. *People, Pathogens and Our Planet.* Report commissioned by The World Bank. Report Number 69145-GLB. http://un-influenza.org/files/PeoplePathogensandOur%20Planet.pdf

respectively. Current knowledge suggests that these viruses do not cause significant disease in humans. FMD has been included because it has the capacity to suddenly emerge and is a constant challenge to those areas that are free of disease. Also it is now recognized as one of the major threats to food supplies in the developing world and in turn is a major driver of poverty. Control of this disease would make a real difference to the lives of millions of subsistence farmers throughout Africa and Asia. The inclusion of CPV-2 is merited by the rapid global spread observed in the mid-1970s that occurred despite stringent control of dog movements between many countries. Understanding how this virus switched host and spread so rapidly could reveal how future disease might emerge and spread in companion animal populations.

These chapters are followed by one on the varied reservoirs of rabies virus, a virus that has been with us since antiquity. Rodents have a long association as a reservoir of human disease and are represented by chapters on hantaviruses and Lassa fever virus. Primates by contrast have a relatively recent, but devastating, history as a source of human disease and the chapter on the emergence of simian immunodeficiency virus (SIV) provides a thorough review of the primate species associated with the emergence of both SIV and human immunodeficiency virus.

In recent years bats have emerged in their own right as a source of zoonotic viruses and it is timely to include two chapters on the paramyxoviruses, Nipah virus and Hendra virus, with a very recent history of disease outbreaks in humans with very high levels of human mortality. West Nile virus (WNV) and Rift Valley fever virus (RVFV) are examples of disease transmission by mosquitoes. In the case of WNV, birds also play a key role in disease dispersal and could be responsible for the worldwide spread of this virus. By contrast, RVFV has been restricted to Africa, with a well-documented outbreak in the Arabian Peninsula. However, its maintenance within the environment in between explosive outbreaks remains unresolved.

The purpose of the final synthesis chapter is to bring together common themes identified by the preceding chapters and to consider the impact of new technologies on our ability to detect and control viruses. Detection of animal-associated viruses is revolutionizing the way we view the current pathogens affecting the human race and identify new viruses that could be the pathogens of the future.

There are a number of viruses that some readers might consider should have been included. Obvious examples include influenza virus, severe acute respiratory syndrome (SARS virus), hepatitis E and Ebola virus. In each case animals act as a reservoir, for translocating the virus, acting as a potential mixing vessel (influenza virus) and transmitting virus to humans. In each case I would have to agree that they could have been included. Beyond viruses there are emerging bacteria such as *Anaplasma phagocytophilum*, the causative agent of granulocytic anaplasmosis, and emerging fungi, such as white nose syndrome affecting bats in North America. So, in the end, the focus for this book is the relationship between the virus and its animal reservoirs and not an exhaustive list, and the examples within the book illustrate this thoroughly.

I would like to express my thanks to all the contributors who have worked to create a series of chapters that illustrate viruses that have emerged and spread from, and through interactions with, animals. I hope these will prove thought-provoking to both the expert and nonexpert alike. I would also like to thank Denise Penrose for encouraging me to embark on the process of preparing a book, and Halima Williams from Elsevier who has supported me in bringing this project together. Without them, I probably would have consigned this to the file marked good ideas but not worth pursuing.

Nicholas Johnson

Chapter 1

A Short Introduction to Disease Emergence

Nicholas Johnson

Animal Health and Veterinary Laboratories Agency, Surrey, United Kingdom

INTRODUCTION

Sudden disease emergences, now often considered events, have been reported since antiquity. Perhaps one of the earliest reported was described by Thucydides, the Athenian, who in his *History of the Peloponnesian War* (411 BC) described an as-yet unidentified plague that weakened the population of Athens in 430 BC at a critical point during its siege by the armies of Sparta. The Roman Empire was repeatedly beset by plague epidemics during its long dominance over Europe, North Africa and the Near East. The numerous transfers of viruses from the Old to the New World following the rediscovery of the Americas by Columbus in 1492 caused devastation of the native populations, particularly through the introduction of smallpox, reducing them to less than 10 percent of preconquest levels. In the 17th century, Daniel Defoe described in detail the impact of the Black Death on London in 1665 in his fictional book *A Journal of a Plague Year*. In modern times we have had the devastating pandemics of influenza, cholera and human immunodeficiency virus (HIV). Each of these events provokes similar questions. Where did it come from? How did it get here and why? The processes that underlie these events continue to this day and are collectively termed emerging and re-emerging diseases (Morse, 1995). The definition of an emerging disease is one that has newly appeared in a population or that has been known for some time but is rapidly increasing in incidence or geographic range. A number of recurring features of emerging diseases have been noted. Most are zoonotic, thus can infect humans and other animals. Many are viruses, particularly those with RNA genomes (Taylor et al., 2001).

The main purpose of this book is to explore the role of animals in the emergence of viruses. Figure 1.1 provides a schematic of the interfaces between humans, animals and viruses. It is these interfaces that are critical in understanding how and potentially when new diseases will emerge. However, many factors influence the emergence of diseases from animals and these are considered in the following sections.

The Role of Animals in Emerging Viral Diseases. http://dx.doi.org/10.1016/B978-0-12-405191-1.00001-6

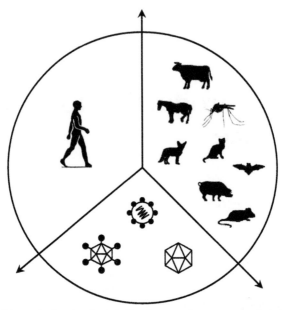

FIGURE 1.1   Schematic showing the interface between humans, animals (wildlife, livestock, companion animals) and viruses.

When considering the factors that lead to the emergence of viral diseases it becomes very apparent that there is a high degree of interrelatedness. For example, human encroachment in a virgin ecosystem provides the opportunity for a previously unencountered virus to infect human beings or their domestic animals. Viral mutability could enable a virus to adapt to the new host. Subsequent trade and migration could allow spread of the new virus to susceptible individuals and the large populations found in urban centers could ensure that the virus is maintained. The interplay of all these factors provides a rational framework and explanation for virus emergence. However, to gain an appreciation of the components of this process it is necessary to focus on each in itself before linking them together.

HUMAN FACTORS THAT INFLUENCE EMERGENCE

Perhaps above all else, human activities are the most influential factors driving emergence events and it would be remiss to ignore them, despite the focus of this book being primarily on the role of animals. The following sections outline a range of human factors that have shaped, are shaping and will continue to shape disease emergence.

Animal domestication

One of the defining acts that led to the transition of humans from hunter/gatherers to urban dwelling populations was the domestication of plants and

TABLE 1.1 A Summary of Key Domestication Events

Species	Likely geographical origin	Years from present
Dog	East Asia	15,000
Goat/sheep	Asia	8,000
Pigs	Middle East / China	7,000
Horse	Eurasia	4,000
Cattle	Asia / North Africa	6,000
Camel	Asia	4,000
Donkey	Egypt	4,000

animals (Diamond, 2002). This act has caused two of the most important activities that challenge the world today and is a major factor driving disease emergence. The first is the relentless modification of the environment. This ranges from major destruction of the environment through activities such as mining and waste disposal to relatively minor modifications such as provision of access for tourism. However, the most common modification usually begins with the clearance of vegetation for livestock pasture and crop cultivation. This may involve deforestation resulting in the catastrophic disruption of the existing ecosystem, displacing or destroying many species within it.

The second has been the exponential growth of the global human population (see next section). The main benefits resulting from animal domestication are twofold: the provision of food, particularly from species such as cattle, sheep, pigs and poultry; and provision of function, examples of which include the domestication of the horse for transport or the dog to assist in hunting. More recently, domestic animals have served a further function, that of companionship, particularly the dog and cat.

The domestication of the dog provides an example of the process. The grey wolf (*Canis lupus*) is considered the progenitor of the domestic dog, which is now classified as a separate subspecies (*Canis lupus familiaris*) (Mech and Boitani, 2006). Phylogenetic analysis of mitochondrial DNA derived from 654 domestic dogs from around the world suggests that there were relatively few domestication events, all taking place in southeast Asia (Savoilainen et al., 2002). This then led to the spread of dogs throughout the Old and New Worlds (Leonard et al., 2002). Documented remains of domestic dogs date back to 12,000 years ago in the Middle East (Davis and Valla, 1978) and this, together with molecular clock analysis of mitochondrial DNA, suggests a likely date of domestication around 15,000 years ago (Table 1.1). Subsequent selective breeding has led to the plethora of varieties that are now apparent today but all are derived from the wolf. Domestic dogs now provide a variety of functions that support human lives but have also brought their pathogens into close proximity to humans and their dwellings. The act of domesticating animals has been considered one of the major drivers in the emergence of human infectious diseases (Wolfe et al., 2007).

Human population expansion

This factor above all others is driving the emergence of new diseases. Data collected by the United Nations suggested that the human population was estimated to be just short of seven billion in 2010 (anon, 2010) and continues to grow exponentially. Two countries, China and India, have populations of over one billion. This in turn drives all other human activities that might influence disease emergence. It increases the demand for food, which in turn increases the area needed for arable farming and the number of livestock to meet this demand. It has also changed the way in which livestock is reared, leading to increased intensive farming practices. More land is needed to accommodate agricultural production, which leads to further environmental encroachment that leads to displaced or altered behaviors in wildlife species.

Population increase is fueling increased urbanization and increased population density. Singapore is now the most densely populated place on earth with five million humans contained within 707 km^2. In less-developed regions of the world, this places stress on sanitation and clean water infrastructures if present, and creates a breeding ground for fecal–oral infections if not. The increase in both numbers and densities of humans provides larger populations in close proximity to one another, supporting the maintenance and spread of infectious diseases that are discussed later in this chapter.

Changes in human behavior

From the emergence of humans in Africa, the development of the species has been closely associated with migration throughout the world. This continues today as people move because of deteriorating local conditions (drought, crop failure, conflict) or in search of improved economic circumstances. A prominent feature of migration over recent centuries has been the movement of populations from rural to urban centers, to a point where the majority of humans now reside in cities. This process has been driven by increases in the human population, developments in the intensification of agriculture reducing the need for human labor, and industrialization of societies that create more jobs in urban areas. This constant movement of populations provides one of the main vehicles for disease translocation.

International travel, commerce and conflict

Historically, human travel has been the principal and effective method for disease spread and introduction. There are numerous examples, particularly the movement of armies in antiquity (plague) to the Korean War (Hantaan virus) that have been associated with the emergence and spread of disease. Probably the most dramatic was the introduction of smallpox to the Americas following the Spanish conquest of Mesoamerica. The combination of disease introduction and abuse of the indigenous people led to the rapid depopulation across the continent.

Commerce and travel go hand in hand and have also played a part in the movement of disease. Yellow fever was introduced into the Americas, along with its vector, the mosquito *Aedes aegypti*, by the mass movement of slaves from Africa (Bryant et al., 2007). The subsequent trading network between Africa, the Americas and Europe led to sudden outbreaks of the disease in European ports during the 19th century. One of the best documented disease outbreaks occurred in 1865 in Swansea, a port city in southern Wales, following the embarkation of a ship from Cuba delivering copper ore (Meers, 1986). A combination of the introduction of infected mosquitoes from the ship and a particularly warm period in October led to a local outbreak at this unusually high latitude. This highlights the role of invasive species. Fortunately for South Wales, its climate is too cold in winter, although it certainly receives enough precipitation, to support over-wintering *Ae. aegypti* mosquitoes and it is assumed that the imported mosquitoes died out, causing a cessation in cases of yellow fever. However, many mosquito species have proven more robust and with a little human help have managed to move between continents. The trade in used tires by container ships has been instrumental in dispersing the aggressive biting mosquito, *Aedes albopictus*. Originally from Asia, it has spread to the Americas and then Europe (Benedict et al., 2007). The presence of this species in Italy contributed to an outbreak of Chikungunya fever following the return of viremic individuals returning to the country having visited islands in the Indian Ocean that were experiencing outbreaks of the disease. Other species such as *Aedes japonicus* have also emerged in Central Europe (Schaffner et al., 2009).

The increasing speed of international travel has had a dramatic effect on the emergence and spread of pathogens. For the emergence of SARS coronavirus and avian influenza, the movement of infected individuals by air resulted in the rapid dispersal of viruses away from the likely point of origin to new countries around the globe.

Technology and industry

One of the unintended consequences of human technological development has been the emergence of disease. The intensification of agriculture over the past 200 years has led to the concentration of susceptible animals. Imposition of these in close proximity to wildlife reservoirs has resulted in species jumps and outbreaks such as that of Nipah virus. The emergence of Nipah virus in Malaysia in 1998 was triggered by the transfer of virus from roosting fruit bats of the genus *Pteropus*, to intensively farmed pigs (Eaton et al., 2006). This led to a rapid amplification of virus within the pig farm and provided the opportunity for infection of farm workers (Mohd Nor et al., 2000). This case is remarkable in that it resulted from the rapid jump of a virus across two species barriers within a short period of time (see Chapter 11).

Changes to food processing led to the emergence of bovine spongiform encephalopathy (McKintosh et al., 2003). Contamination of food remains a constant problem in the world. The scale and distribution of food networks can lead to rapid dissemination of a pathogen to many people. The complexity of these networks, often together with criminal activity, creates a challenge to identify the source of an outbreak and prevent further infections (Schimmer et al., 2008).

Medical developments from the hypodermic needle to organ transplantation, that have saved many thousands of lives, have also inadvertently been highly efficient at transmitting viruses (Simonsen et al., 1999; Razonable, 2011). Blood transfusion in particular has been associated with transmission of many viruses including human immunodeficiency virus, human T cell lymphotropic virus, hepatitis B virus and West Nile virus (Bihl et al., 2007). While it is possible to reduce the risk for known viruses, it is impossible to screen blood supplies for unknown viruses.

VIRUS FACTORS THAT INFLUENCE EMERGENCE

Virus structure

Viruses come in many different forms but have a number of common features. Firstly, they all have a nucleic acid genome. However, they show a highly diverse range in genomic size and composition. Virus genomes can consist of ribonucleic acid (RNA) or deoxyribonucleic acid (DNA), they can be single- or double-stranded, and they range in size from a thousand base pairs coding for a small number of proteins to tens of thousands of base pairs coding for hundreds of proteins. Table 1.2 provides details of a number of viruses that cause severe disease in livestock to illustrate this variety.

The variety continues in the structure of viruses. Some are enveloped with a host-derived lipid bilayer into which virus proteins are inserted, whereas others are not and have a particularly tough protein capsid structure that makes the virus highly resistant to desiccation, such as norovirus.

A critical feature of all viruses that is relevant to transmission is the surface proteins that project from the surface of the virion, either through a lipid envelope or as a component of the capsid. The primary function of this protein or proteins is to engage with a host receptor and initiate entry of a virus into the target cell. It is assumed that for most viruses evolutionary pressure has led to a degree of specificity of this interaction at the level of host and even the cell type that the virus infects. This in turn will influence the mode of transmission of the virus and restrict its target cell range, its cellular tropism. This aspect of transmission forms the concept of the species barrier. However, other interactions between the virus and the host cell, such as the ability to subvert the innate immune response, utilization of cellular components and modification of cellular activities also provide barriers to virus replication. Simplistically, the species barrier restricts a particular virus to bind to a particular receptor, of a particular cell type of a particular host species. The ability to deviate away from this apparent

TABLE 1.2 Structural Details of the Genomes of Common Virus Pathogens of Livestock

Family	Genus	Example	Enveloped	Genome	Genome size (kilobase pairs)
Circoviridae	Circovirus	Porcine circovirus	−	Single-stranded DNA	1
Rhabdoviridae	Vesiculovirus	Vesicular stomatitis virus	+	Single-stranded negative sense RNA	11.2
Picornaviridae	Aphthovirus	Foot-and-mouth disease virus	−	Single-stranded positive sense RNA	8
Reoviridae	Orbivirus	Bluetongue virus	−	Double-stranded RNA (10 segments)	19.2
Coronaviridae	Coronavirus	Infectious bronchitis virus	+	Single-stranded positive sense RNA	30
Bunyaviridae	Orthobunyavirus	Akabane virus	+	Single-stranded negative sense RNA (3 segments)	12
Poxviridae	Capripoxvirus	Sheep pox virus	+	Double-stranded DNA	150

host restriction forms the basis of cross-species transmission (CST) and is influenced by nonviral factors, such as the opportunity to be exposed to a new host, or by changes in the virus, as discussed below.

Transmission of viruses

Viruses can be transmitted between susceptible individuals by a range of routes as outlined in Table 1.3. Transmission of viruses forms a critical component of how viruses emerge and this may change as a disease epidemic evolves. HIV is an example of this. The initial events that caused HIV to jump from chimpanzees to humans cannot be known with any certainty although it may have occurred during butchery of bush meat, perhaps resulting from infection of an open wound, and is thus an example of mechanical transmission (see Chapter 9). However, there is no doubt that the resulting pandemic was spread primarily by sexual contact and to a lesser degree by contaminated needles and medical blood products. This example also returns to the role of human behavior in the rate of transmission. A combination of promiscuity and air travel promoted the rapid spread of a sexually transmitted disease.

TABLE 1.3 Mechanisms of Virus Transmission

Method of transmission	Comments	Virus examples
Airborne transmission	This can result from expelling virus by coughing or sneezing but can result from deposition of virus on surfaces and subsequent contact by a susceptible individual.	Rhinoviruses Foot-and-mouth disease virus
Fecal–oral	Shedding of virus in diarrhea and contamination of food or water leads to onward infection. Particularly prevalent where sanitation is poor.	Rotavirus
Mechanical transmission	This could involve direct deposition of virus through contact or by transmission by a vector such as mosquito or tick.	Yellow fever (mosquito-borne) African swine fever virus (tick-borne) Toscana virus (sandfly-borne)
Sexually transmitted	Transmission through intimate contact.	Human immunodeficiency virus
Iatrogenic	Medical activities lead to transmission of virus such as blood transfusion or use of blood products and organ transplant.	Hepatitis C

Another human activity, congregating in large numbers, promotes transmission of airborne viruses. Activities such as attendance of schools, universities or sporting events are all high risk activities for contracting airborne diseases. This brings susceptible individuals into close contact with those that are infected. This has been the foundation for modeling of disease spread and persistence, as pioneered by Maurice Bartlett (Bartlett, 1957), who in the 1950s published a number of studies on the persistence of measles virus (Box 1.1). From these

Box 1.1 Measles Virus

The Virus

Measles virus is a nonsegmented negative strand RNA virus classified within the family *Paramyxoviridae*, genus *Morbillivirus*. Related viruses include canine distemper virus, rinderpest virus and peste de petit ruminant virus (Barrett, 1999).

The Disease

Measles is transmitted by aerosols with infection initiating in the respiratory tract. The virus spreads to secondary lymphoid tissue followed by viremia 7 days following infection. This disseminates the virus to epithelial tissue in various organs and the skin (Griffin, 2010). The infected individual develops a range of symptoms including fever, cough and a characteristic maculopapular rash. The virus is highly contagious but death is rare. In a small proportion of cases, virus persists in neurological tissue in a condition called subacute sclerosing pan encephalitis (SSPE), which can lead to destruction of the neurons and death.

Eradication

Measles is one of the few pathogens that meet the criteria for disease eradication, the state where the agent of disease no longer exists in nature (Bremen et al., 2011). These are: 1) humans are the only reservoir for the pathogen, 2) diagnostic tests exist, and 3) an effective intervention, principally a vaccine, is available (Dowdle, 1999). Elimination, the interruption of transmission of a pathogen to a point where disease incidence becomes zero in a population within a defined geographical area, has been achieved for measles in the Americas (Moss and Strebel, 2011). The closely related rinderpest virus has recently been eradicated (Roeder, 2011).

Vaccine

A live attenuated vaccine is available for measles that is given up to 18 months of age. It is safe and effective, providing life-long immunity. Vaccination programs aim to reach as many susceptible individuals as possible to reduce the ability of the virus to persist in the population. However, in recent years a decline in vaccine coverage has led to the re-emergence of measles virus in countries in which it had been virtually eliminated, such as the United Kingdom.

Virus Mutability

Reports suggest that measles virus mutates at comparable rates to other RNA viruses in the range of 10^{-4} to 10^{-3} mutations per nucleotide per year (Kuhne et al., 2006), but this does not appear to have reduced the protective immunity conferred by vaccination.

came the concept of Critical Community Size (CCS), the size a population needed to support persistence of a particular pathogen. For measles virus this was estimated to vary between 250,000 and 500,000 individuals. Below this, further transmission of virus would be reduced to a level that would cause the epidemic to fade out. Viral factors such as the persistence of the virus in the host, the duration of immunity to the virus and virus variation in response to host immunity will affect the CCS. A further factor that influences this is the birth rate of the host population (Conlan and Grenfell, 2007). Those populations with a higher birth rate will replenish the subpopulation of susceptible individuals and overall have a lower CCS, enhancing the ability of the infection to persist.

Other factors that affect transmission are the period over which the infected host sheds virus. This is often divided into acute and chronic disease. An example of acute is the common cold, which results from rhinovirus infection, where the host sheds virus for a short period that can be as little as two days, but the infection resolves rapidly and the host is immune to further infections. In more extreme virus infections, disease does not resolve and the host dies. The converse of this is the development of a chronic infection where the host is capable of infecting susceptible individuals for a considerable period, often years. Viral infections such as hepatitis B virus and HIV are examples of this scenario. An intermediate condition exists where the host is chronically infected but sheds virus intermittently. A clear example of this is herpes simplex virus infections, where the virus remains dormant within infected dorsal root ganglia but emerges intermittently to cause a cold sore from which virus is shed. These are often triggered following periods of stress but resolve through immune control.

One potential factor that may prove useful in predicting virus emergence is that of force of infection. This term is usually defined as the number of new infections divided by the number of susceptible individuals exposed, multiplied by the average duration of exposure. In practice these are difficult parameters to define although within human medicine, scenarios such as transmission of bloodborne infections such as hepatitis B and C within drug users have been studied where the period of use is known. This has enabled the calculation of the force of infection of a particular virus within a defined population. This approach has also been applied to transmission of vector-borne infections such as malaria and Dengue virus where the period of vector activity can be predicted. A similar approach has been used to estimate the force of infection of rabies virus transmitted by vampire bats based on the number of biting incidents per member of the population (Schneider et al., 1996). If force of infection can be estimated it could help in predicting the risk of host jumps by viruses and thus the likelihood of disease emergence.

Impact of infection

Simplistically there are three outcomes for the host following infection with a virus: 1) the host dies as a result of the infection; 2) the host suffers a period of morbidity but produces an effective immune response and eliminates the virus from the body; 3) the host becomes persistently infected with intermittent virus shedding, fluctuating levels of virus shedding or constant shedding of high levels of virus. Virus infection can trigger oncogenesis and cancer in the host (Liao, 2006). Examples of this include hepatitis B virus as a causative agent of hepatocellular carcinoma (Tan, 2011); Epstein-Barr virus infection can lead to both Burkitt (Brady et al., 2007) and Hodgkin (Kapatai and Murray, 2007) lymphoma; infection with certain strains of human papillomavirus cause almost 100% of cases of cervical cancer (Gravitt, 2011).

Virus mutation

The key feature of viruses that enhances their ability to infect new hosts under different environmental conditions is the rapidity with which their genome can mutate. This takes many forms, outlined below, but critically, changes in the genome, ranging from single nucleic acid base changes to wholesale reassortment of segmented genomes, can lead to changes in the infectious properties of the progeny virus and thus the ability to adapt to new or different circumstances.

Point mutations

All virus genomes encode a polymerase that copies new genomes from the infecting virus nucleic acid. Both DNA and RNA virus polymerases have limited proofreading ability and so can introduce single base mutations to progeny genomes at rates considerably higher than that found in prokaryotic and eukaryotic cells. This is expressed as substitutions (s) per nucleotide position (n) per cell infection (c) (Sanjuán et al., 2010). This in turn could potentially affect the amino acid composition of the proteins encoded by the genome. Many mutations will be silent, having no effect on the properties of the next generation of viruses, many will be deleterious leading to defective virions, but some may enhance the ability of the progeny virus to infect a new host species or replicate more efficiently in an existing one.

RNA viruses have a greater mutation rate than DNA viruses, and there is a clear relationship between genome length and mutations rate, suggesting that beyond a certain length of genome, the mutation rate is sufficiently high that each new genome contains a deleterious mutation. The high rate of mutation within RNA viruses has led to the development of quasispecies theory (Eigen, 1993). This theory attempts to explain some of the properties of RNA viruses through the existence of a highly variant virus population (see Box 1.2).

Insertions/deletions

A further means of genome mutation that could lead to a change of virus properties is that of insertions or deletions, sometimes referred to as in/dels. The mechanism of this form of mutation is unclear but is also likely to be caused by polymerase errors. The insertion of charged residues to the hemagglutinin

Box 1.2 Quasispecies Theory

Quasispecies theory has developed through the application of mathematics to explain the effects of higher mutation rates on RNA virus behavior. Some key definitions are (Domingo et al., 2005):

Quasispecies
A weighted distribution of mutants centered around one master sequence.

Mutation rate
The frequency of occurrence of a mutational event during genome replication.

Mutant spectrum
The ensemble of mutant genomes that compose a quasispecies.

Fitness
A parameter that quantifies the adaptation of an organism or a virus to a given environment.

Consensus sequence
The sequence resulting from taking for each position the most frequent residue found at the corresponding position in the homologous set of aligned sequences.

Sequence space
A theoretical representation of all possible variants of a genomic sequence for a single-stranded RNA virus.

The study of quasispecies has led to theories on virus fitness, persistence and mutation in the face of antivirus treatment (Lauring and Andino, 2010). It has also led to new concepts in treating RNA virus infections through increasing mutation rates that in turn increase deleterious mutations that cause virus population extinction (Ojosnegros et al., 2011). Some authors have argued against the existence of quasispecies (Holmes and Moya, 2002) and conceptually it is difficult to understand where quasispecies exist in time and space outside of an experimental setting.

protein of influenza can dramatically change its susceptibility to cleavage by host proteases and in turn increase its virulence in avian and mammalian hosts (Webster and Rott, 1987; Horimoto and Kawaoka, 1995). In/dels have also been observed within the genome of lyssaviruses, although in this case the effect on virus phenotype was not clear as it was reported to occur within an intergenic region (Johnson et al., 2007).

Reassortment

One of the most dramatic ways in which viruses can alter their genomes is through reassortment of genome segments. This is restricted to those viruses that have segmented genomes but has led to the emergence of viruses with clearly different properties such as increased virulence for humans. Bunyaviruses have a tripartite genome (see Chapters 8 and 10) and so are capable of reassortment. Although rare, a recent case of reassortment of an Orthobunyavirus isolated from a human with febrile illness has been reported (Aquilar et al., 2011). Orthomyxoviruses have eight genome segments and reassortment has led to antigenic shift of the virus surface proteins and emergence of influenza viruses to which there is no prior immunity. Finally Orbiviruses have 10 segments and reassortment is suspected to have led to the emergence of bluetongue virus variants in Europe (Batten et al., 2008).

Recombination

Certain viruses use recombination as part of their replicative cycle, i.e., lentiviruses (see Chapter 9). Recombination in other viruses is more controversial. There is strong evidence that Western equine encephalitis virus emerged as a recombination between Eastern equine encephalitis virus and a Sindbis-like virus (Hahn et al., 1988). Further analysis has shown that many New World *Alphaviruses* have emerged following recombination events (Weaver et al., 1997). There is also growing evidence that sections of RNA virus genomes have been converted into double-stranded DNA and recombined within the genomes of certain host species. For example, sequences homologous to flavivirus RNA have been identified in the mosquito genome (Crochu et al., 2004) and ancestral fragments of Ebola virus have been identified in bat genomes (Taylor et al., 2011). It is not clear how this could have occurred with such viruses replicating exclusively in the cytoplasm of the cell.

In summary, mutational changes within the virus genome can lead to emergence of a virus phenotype that has increased replication fitness within its environment. This could take the form of an antigenic change to its surface protein that has not previously been encountered by the host, an increase in virulence for a host or hosts, an ability to infect a different host cell or the ability to infect a new host species. All can lead to the emergence of disease.

Virus subversion of the host innate immune response

The innate immune response is considered the first line of defense against viruses and consists of a variety of mechanisms that first detect infectious agents or more particularly structures that are associated with pathogens, referred to as microbe-associated molecular patterns or MAMPS. These are recognized by pattern recognition receptors (PRRs), which fall into three categories: the Toll-like receptors, the Rig-I receptors and the Nod-like receptors (Gerlier and Lyles, 2011). Activation of PPRs leads to transcriptional activation of type-1 interferon, principally interferon beta, which in turn activates further interferons and other proteins that actively inhibit viral replication and prevent further viral spread. Unsurprisingly, viruses have evolved many mechanisms to inhibit interferon activation by blocking signaling pathways that stimulate gene transcription (reviewed extensively by Randall and Goodbourn, 2008). The virus proteins from each of the viruses discussed in this book will be identified in the following chapters.

ANIMAL FACTORS THAT INFLUENCE EMERGENCE

As discussed previously—and indeed it is the basic premise of this book—many emerging viruses have their origin within animal populations. Each animal on earth hosts its own spectrum of pathogens and it is only now with the advent of sensitive gene amplification and mass sequencing that researchers can fully reveal the extent of this (Drexler et al., 2012a). Following the emergence of the SARS-coronavirus, bats have come under intense scrutiny as the potential origin of the SARS outbreak and a source of zoonotic viruses. Surveillance in bats from around the world has shown that they are infected with a diverse range of coronaviruses. Even European bats have been shown to host a range of SARS-related coronaviruses (Drexler et al., 2012b) although there is no evidence to suspect that these viruses are capable of jumping the species barrier to humans at the present time. A recent study screening illegally imported products of wildlife origin simultaneously demonstrated the species from which the product originated and detected the presence of retroviruses and herpes viruses within it (Smith et al., 2012). In some instances, the original species were shown to be nonhuman primates and thus the viruses that infected these animals were theoretically adapted to primates and thus have a shorter "species jump" to humans and present a higher threat to the human population. A number of features of animals enable them to play a role in virus emergence and these are discussed in the following sections.

Avian migration

The activities that animals of all species undertake greatly influences the emergence of zoonotic diseases. Migration is often cited as a behavior that leads to the translocation of diseases. Birds are associated with a range of pathogenic organisms, extensively listed by the Czech biologist Zdenek Hubálek (2004), including viruses belonging to the families *Bunyviridae, Flaviviridae, Togaviridae, Orthomyxoviridae* and *Paramyxoviridae*. Key examples of the spread of zoonotic viruses by migrating birds are the westward spread of H5N1 avian influenza from Asia to Europe and the movement of West Nile virus from Africa to Western Europe following well-traveled migration routes across the Mediterranean Sea (see Chapter 7). Migration therefore provides a direct vehicle for long-distance translocation of pathogens and thus the opportunity to emerge in new locations.

Feeding behavior

Feeding habits can also be a critical point at which animals have the opportunity to transmit viruses to other species. The common vampire bat (*Desmodus rotundus*), found throughout Latin America, is adapted to blood-feeding as virtually its only source of nutrition (Greenhall, 1988). The species has various adaptations to achieve this, including self-sharpening incisor teeth, anticoagulants in its saliva and grooves in its tongue to enhance blood lapping. As an unintended consequence of these adaptations, the vampire bat is a highly efficient transmission vector of rabies virus and is responsible for large numbers of livestock deaths (Belotto et al., 2005) and occasional transmission to humans in Latin America (Barbosa et al., 2008). Blood-feeding arthropods, especially mosquitoes, are also highly efficient at transmitting viruses from animals to humans and between humans. Examples of these are given in Table 1.4.

Dispersal

A final example of behavior that has also been influential in the transmission of rabies virus has been that of dispersal, particularly dispersal of juvenile foxes. The red fox (*Vulpes vulpes*) is one of the major wildlife reservoirs for rabies in Europe and North America. This has only been fully revealed through modeling disease movements (Thulke et al., 1999) and corroborated with historical data from the expansion of fox rabies through the second half of the 20th century (Bourhy et al., 1999). Dispersal can also result from human activities that lead to the displacement of wildlife (see Chapter 4).

TABLE 1.4 Zoonotic Viruses Transmitted by Biting Arthropods

Virus family	Species	Arthropod vector
Bunyaviridae	Crimean-Congo Haemorrhagic fever virus	Ticks (*Hyalomma* spp.)
	La Crosse virus	Mosquitoes (*Aedes* spp.)
	Oropouche virus	Biting midges (*Culicoides* spp.)
	Rift Valley fever virus	Mosquitoes (*Aedes* spp.)
	Sandfly fever virus	Sandfly (*Phlebotomus* spp.)
Flaviviridae	Dengue virus	Mosquitoes (*Aedes* spp.)
	Japanese encephalitis virus	Mosquitoes (*Culex* spp.)
	St. Louis encephalitis virus	Mosquitoes (*Culex* spp.)
	Tick-borne encephalitis virus	Ticks (*Ixodes* spp.)
	West Nile virus	Mosquitoes (*Culex* spp.)
	Yellow fever virus	Mosquitoes (*Aedes* spp.)
Reoviridae	Colorado tick fever virus	Ticks (*Dermocentor* spp.)
Togaviridae	Chikungunya virus	Mosquitoes (*Aedes* spp.)
	Equine encephalitidies (Eastern, Western, Venezuelan)	Mosquitoes

While all animal species will be infected by a range of viruses, some groups of animals are of particular interest either because they share close genetic links to humans, such as the primates, or are associated with known zoonotic viruses with the potential to transmit disease to human populations. Both avian and rodent species have long been known to act as reservoirs for zoonotic diseases. In recent years, bats in particular have been identified as a new source of emerging viruses (Calisher et al., 2006; Halpin et al., 2007; Kuzmin et al., 2011). This has been based on a long-standing association with rabies and the emergence of SARS, Hendra virus, Nipah virus and Ebola virus.

EMERGING VIRUSES

The aim of this book is to show how animals in all their forms have contributed to the emergence of viruses. In order to do this, examples of emerging viruses have been selected and reviewed in depth. In a comprehensive review of human pathogens Taylor and co-workers (2001) identified over 1,400 disease-causing agents, over half of which were zoonotic. In addition, 177 were considered emerging or re-emerging and of these, the majority were viruses. By its nature, the emergence of unknown diseases is impossible to predict (Tesh, 1994). However, by studying those diseases that have emerged we may better understand why they emerged, when they did and what measures in future may prevent or control the impact of such emergent pathogens.

REFERENCES

Anon. (2010). http://esa.un.org/wpp/excel-Data/population.htm. Accessed March, 2013.

Aquilar, P. V., Barrett, A. D., Saeed, M. F., Watts, D. M., Russell, K., Guevara, C., Ampuero, J. S., Suarez, L., Cespedes, M., Montgomery, J. M., Halsey, E. S., & Kochel, T. J. (2011). Iquitos virus: a novel reassortant Orthobunyavirus associated with human illness in Peru. *PloS Neglected Tropical Diseases*, 5, e1315.

Barbosa, T. F. S., Mederiros, D. B. D., da Rosa, E. S. T., Casseb, L. M. N., Medeiros, R., Pereira, A. D., Vallinoto, A. C. R., Vallinoto, M., Begot, A. L., Lima, R. F. D., Vasconcelos, P. F. D., & Nunes, M. R. T. (2008). Molecular epidemiology of rabies virus isolated from different sources during a bat-transmitted human outbreak occurring in Augusto Correa municipality, Brazilian Amazon. *Virology*, *370*, 228–236.

Barrett, T. (1999). Morbillivirus infections, with special emphasis on morbilliviruses of carnivores. *Vet. Microbiol.*, *69*, 3–13.

Bartlett, M. S. (1957). Measles periodicity and community size. *Journal of the Royal Statistical Society*, *A120*, 48–70.

Batten, C. A., Maan, S., Shaw, A. E., Maan, N. S., & Mertens, P. P. (2008). A European field strain of bluetongue virus derived from two parenteral vaccine strains by genome segment reassortment. *Virus Research*, *137*, 56–63.

Belotto, A., Leanes, L. F., Schneider, M. C., Tamayo, H., & Correa, E. (2005). Overview of rabies in the Americas. *Virus Research*, *111*, 5–12.

Benedict, M. Q., Levine, R. S., Hawley, W. A., & Lounibos, L. P. (2007). Spread of the Tiger: Global risk of invasion by the mosquito *Aedes albopictus*. *Vector Borne and Zoonotic Disease*, *7*, 76–85.

Bihl, F., Castelli, D., Marincola, F., Dodd, R. Y., & Brander, C. (2007). Transfusion-transmitted infections. *Journal of Translational Medicine*, *5*, 25.

Bourhy, H., Kissi, B., Audry, L., Smereczak, M., Sadkowska-Todys, M., Kulonen, K., Tordo, N., Zmudzinski, J. F., & Holmes, E. C. (1999). Ecology and evolution of rabies virus in Europe. *J. Gen. Virol*, *80*, 2545–2557.

Brady, G., MacArthur, G. J., & Farrell, P. J. (2007). Epstein-Barr virus and Burkitt lymphoma. *J. Clin. Path.*, *60*, 1397–1402.

Bremen, J. G., de Quadros, C. A., Dowdle, W. R., Foege, W. H., Henderson, D. A., John, T. J., & Levine, M. M. (2011). The role of research in viral disease eradication and elimination programs: Lessons for malaria eradication. *PLoS Medicine*, *8*, e1000405.

Bryant, J. E., Holmes, E. C., & Barrett, A. D. T. (2007). Out of Africa: A molecular perspective on the introduction of yellow fever virus into the Americas. *PLoS Pathogens*, *3*, e75.

Calisher, C. H., Childs, J. E., Field, H. E., Holmes, K. V., & Schountz, T. (2006). Bats: Important reservoir hosts of emerging viruses. *Clin. Microbiol. Rev.*, *19*, 531–545.

Conlan, A. J., & Grenfell, B. T. (2007). Seasonality and the persistence and invasion of measles. *Proc. Biol. Sci.*, *274*, 1133–1141.

Crochu, S., Cook, S., Attoui, H., Charrel, R. N., De Cheese, R., Belhouchet, M., Lemasson, J. J., de Micco, P., & de Lamballerie, X. (2004). Sequences of flavivirus-related RNA viruses persist in DNA from integrated in the genome of Aedes spp. Mosquitoes. *J. Gen. Virol.*, *85*, 1971–1980.

Davis, S., & Valla, F. (1978). Evidence for domestication of the dog 12,000 years ago in the Natufian of Israel. *Nature*, *276*, 608–610.

Diamond, J. (2002). Evolution, consequences and future of plant and animal domestication. *Nature*, *418*, 700–707.

Domingo, E., Escarmís, C., Lázaro, E., & Manrubia, S. C. (2005). Quasispecies dynamics and RNA virus extinction. *Virus Research*, *107*, 129–139.

Dowdle, W. (1999). The principles of disease elimination and eradication. *Morbidity and Mortality Weekly Report*, *48*(SU01), 23–27.

Drexler, J. F., Corman, V. M., Müller, M. A., Magangam, G. D., Vallo, P., Binger, T., Gloza-Rausch, F., Rasche, A., Yordanov, S., Seebens, A., Oppong, S., Sarkodie, Y. A., Pongombo, C., Lukashev, A. N., Schmidt-Chanasit, J., Stöcker, A., Carniero, A. J. B., Erbar, S., Maisner, A., Fronhoffs, F., Beuttner, R., Kalko, E. K. V., Kruppa, T., Franke, C. R., Kallies, R., Yandoko, E. R. N., Herrler, G., Reusken, C., Hassanin, A., Krüger, D. H., Matthee, S., Ulrich, R., Leroy, E. M., & Drosten, C. (2012a). Bats host major mammalian paramyxoviruses. *Nature Commun.*, *3*, 796.

Drexler, J. F., Gloza-Rausch, F., Glende, J., Corman, V. M., Muth, D., Goettsche, M., Seebens, A., Niedrig, M., Pfefferle, S., Yordanov, S., Zhelyazkov, L., Hermanns, U., Vallo, P., Luakshev, A., Müller, M. A., Deng, H., Herrler, G., & Drosten, C. (2012b). Genomic characterization of severe acute respiratory syndrome-related coronavirus in European bats and classification of coronaviruses based on partial RNA-dependent RNA polymerase gene sequences. *J. Virol.*, *84*, 11336–11349.

Eaton, B. T., Broder, C. C., Middleton, D., & Wang, L.-F. (2006). Hendra and Nipah viruses: different and dangerous. *Nat. Rev. Microbiol.*, *4*, 23–35.

Eigen, M. (1993). Viral quasispecies. *Scientific American*, *269*, 42–49.

Gerlier, D., & Lyles, D. S. (2011). Interplay between innate immunity and negative-strand RNA viruses: towards a rational model. *Microbiology and Molecular Biology Reviews*, *75*, 468–490.

Gravitt, P. E. (2011). The unknowns of HPV natural history. *J. Clin. Inv.*, *121*, 4593–4599.

Greenhall, A. M. (1988). Feeding behaviour. In A. M. Greenhall & U. Schmidt (Eds.), *Natural History of Vampire Bats* (pp. 111–131). Florida: CRC Press Inc.

Griffin, D. E. (2010). Measles virus-induced suppression of immune responses. *Immunology Reviews, 236*, 176–189.

Hahn, C. S., Lustig, S., Strauss, E. G., & Strauss, J. H. (1988). Western equine encephalitis virus is a recombinant. *PNAS, USA, 85*, 5997–6001.

Halpin, K., Hyatt, A. D., Plowright, R. K., Epstein, J. H., Daszak, P., Field, H. E., Wang, L., Daniels, P. W., & the Henipavirus Ecology Research Group (2007). Emerging viruses: Coming in on a wrinkled wing and a prayer. *Clin. Infect. Dis., 44*, 711–717.

Holmes, E. C., & Moya, A. (2002). Is the quasispecies concept relevant to RNA viruses? *J. Virol., 76*, 460–462.

Horimoto, T., & Kawaoka, Y. (1995). Molecular changes in virulent mutants arising from avirulent avian influenza viruses during replication in 14-day-old embryonated eggs. *Virology, 206*, 755–759.

Hubálek, Z. (2004). An annotated checklist of pathogenic microorganisms associated with migratory birds. *J. Wild. Dis., 40*, 639–659.

Johnson, N., Freuling, C., Marston, D. A., Tordo, N., Fooks, A. R., & Muller, T. (2007). Identification of European bat lyssavirus isolates with short genomic insertions. *Virus Research, 128*, 140–143.

Kapatai, G., & Murray, P. (2007). Contribution of the Epstein-Barr virus to the molecular pathogenesis of Hodgkin lymphoma. *J. Clin. Path., 60*, 1342–1349.

Kuhne, M., Brown, D. M., & Jin, L. (2006). Genetic variability of measles virus in acute and persistent infections. *Infect. Gen. Evol., 6*, 269–276.

Kuzmin, I. V., Bozick, B., Guagliardo, S. A., Kunkel, R., Shak, J. R., Tong, S., & Rupprecht, C. E. (2011). Bats, emerging infectious diseases, and the rabies paradigm revisited. *Emerging Health Threats, 4*, 7159.

Lauring, A. S., & Andino, R. (2010). Quasispecies theory and the behavior of RNA viruses. *PloS Pathogens, 6*, e1001005.

Leonard, J. A., Wayne, R. K., Wheeler, J., Valadex, R., Guillén, S., & Vilá, C. (2002). Ancient DNA evidence for Old World origins of New World dogs. *Science, 298*, 1613–1616.

Liao, J. B. (2006). Viruses and human cancer. *Yale Journal of Biology and Medicine, 79*, 115–122.

McKintosh, E., Tabrizi, S. J., & Collinge, J. (2003). Prion diseases. *J. Neurovirol, 9*, 183–193.

Mech, L. D., & Boitani, L. (2006). Wolf social ecology. In L. D. Mech & L. Boitani (Eds.), *Wolves: Behaviour, Ecology and Conservation* (pp. 1–34). Pub. Chicago, .

Meers, P. D. (1986). Yellow fever in Swansea, 1865. *J. Hyg., 97*, 185–191.

Mohd Nor, M. N., Gan, C. H., & Ong, B. L. (2000). Nipah virus infection of pigs in peninsular Malaysia. *Rev. sci. tech. Off. Int. Epiz, 19*, 160–165.

Morse, S. S. (1995). Factors in the emergence of infectious diseases. *Emerg. Infect. Dis., 1*, 7–15.

Moss, W. J., & Strebel, P. (2011). Biological feasibility of measles eradication. *J. Infect. Dis, 204*(Suppl. 1), S47–S53.

Ojosnegros, S., Perales, C., Mas, A., & Domingo, E. (2011). Quasispecies as a matter of fact: Viruses and beyond. *Virus Research, 162*, 203–215.

Randall, R. E., & Goodbourn, S. (2008). Interferons and viruses: an interplay between induction, signaling, antiviral responses and virus countermeasures. *J. Gen. Virol., 89*, 1–47.

Razonable, R. R. (2011). Rare, unusual and less common virus infections after organ transplantation. *Current Opinions in Organ Transplantation, 16*, 580–587.

Roeder, P. L. (2011). Rinderpest: the end of cattle plague. *Prevent. Vet. Med., 102*, 98–106.

Sanjuán, R., Nebot, M. R., Chirico, N., Mansky, L. M., & Belshaw, R. (2010). Viral mutation rates. *J. Virol., 84*, 9733–9748.

Savoilainen, P., Zhang, Y. -P., Luo, J., Lundeberg, J., & Leitner, T. (2002). Genetic evidence for an East Asian origin of domestic dogs. *Science*, *298*, 1610–1613.

Schaffner, F., Kauffman, C., Hegglin, D., & Mathis, A. (2009). The invasive mosquito *Aedes japonicus* in Central Europe. *Med. Vet. Entomol.*, *23*, 448–451.

Schimmer, B., Nygard, K., Eriksen, H. M., Lassen, J., Linstedt, B. A., Brandal, L. T., Kapperud, G., & Aavitsland, P. (2008). Outbreak of haemolytic uraemic syndrome in Norway caused by stx2-positive Escherichia coli O103:H25 traced to cured mutton sausages. *BMC Infect. Dis.*, *8*, 41.

Schneider, M. C., Santos-Burgoa, C., Aron, J., Munoz, B., Ruiz-Velazco, S., & Uieda, W. (1996). Potential force of infection of human rabies transmitted by vampire bats in the Amazonian region of Brazil. *Am. J. Trop. Med. Hyg.*, *55*, 680–684.

Simonsen, L., Kane, A., Lloyd, J., Zaffran, M., & Kane, M. (1999). Unsafe infections in the developing world and transmission of bloodborne pathogens: a review. *Bulletin of the World Health Organisation*, *77*, 789–800.

Smith, K. M., Anthony, S. J., Switzer, W. M., Epstein, J. H., Seimon, T., Jia, H., Sanchez, M. D., Huynh, T. T., Galland, G. G., Shapiro, S. E., Sleeman, J. M., McAloose, D., Stuchin, M., Amato, G., Kolokotronis, S. -O., Lipkin, W. I., Karesh, W. B., Daszak, P., & Marano, N. (2012). Zoonotic viruses associated with illegally imported wildlife products. *PLoS One*, e29505.

Tan, Y. -J. (2011). Hepatitis B virus infection and the risk of hepatocellular carcinoma. *World Journal of Gastroenterology*, *17*, 4853–4857.

Taylor, D. J., Dittmar, K., Balliner, M. J., & Bruenn, J. A. (2011). Evolutionary maintenance of filovirus-like genes in bat genomes. *BMC Evol. Biol.*, *11*, 336.

Taylor, L. H., Latham, S. M., & Woolhouse, M. E. (2001). Risk factors for human disease emergence. *Phil. Trans. R. Soc. Lond. B Biological Science*, *356*, 983–989.

Tesh, R. B. (1994). The emerging epidemiology of Venezuelan hemorrhagic fever and Oropouche fever in tropical South America. *Annals of the New York Academy of Science*, *740*, 129–137.

Thulke, H. H., Tischendorf, L., Stauback, C., Selhorst, T., Jeltsch, F., Müller, T., Schlüter, H., & Wissel, C. (1999). The spatio-temporal dynamics of a post-vaccination resurgence of rabies in foxes and emergency vaccination planning. *Prev. Vet. Med.*, *47*, 1–21.

Weaver, S. C., Kang, W., Shirako, Y., Runemapg, T., Strauss, E. G., & Strauss, J. H. (1997). Recombinational history and molecular evolution of Western equine encephalomyelitis complex alphaviruses. *J. Virol.*, *71*, 613–623.

Webster, R. G., & Rott, R. (1987). Influenza virus A pathogenicity: the pivotal role of hemagglutinin. *Cell*, *50*, 665–666.

Wolfe, N. D., Dunavan, C. P., & Diamond, J. (2007). Origins of major human infectious diseases. *Nature*, *447*, 279–283.

Chapter 2

Patterns of Foot-and-Mouth Disease Virus Distribution in Africa

The Role of Livestock and Wildlife in Virus Emergence

Miriam B. Casey[1,2], Tiziana Lembo[1], Nick J. Knowles[2], Robert Fyumagwa[3], Fredrick Kivaria[4], Honori Maliti[3], Christopher Kasanga[5], Raphael Sallu[6], Richard Reeve[1], Satya Parida[2], Donald P. King[2] and Sarah Cleaveland[1]

[1]Institute of Biodiversity, Animal Health and Comparative Medicine, College of Medical, Veterinary and Life Sciences, University of Glasgow, United Kingdom, [2]The Pirbright Institute, United Kingdom, [3]Tanzania Wildlife Research Institute, Tanzania, [4]Ministry of Livestock and Fisheries Development, Tanzania, [5]Sokoine University of Agriculture, Tanzania, [6]Tanzania Veterinary Laboratory Agency, Tanzania.

THE DISEASE

Foot-and-mouth disease (FMD) is a highly contagious disease of cloven-hooved animals caused by FMD virus (FMDV), a positive-sense, single-stranded RNA virus of the family *Picornaviridae* (genus *Aphthovirus*) that exists as seven serotypes (O, A, C, Asia 1, SAT 1, SAT 2 and SAT 3). Disease in susceptible animals is characterized by a high fever and development of blisters in the mouth and on hooves. Weight loss and reduction in milk production are commonly observed. While the disease has been reported since the 16th century (Francastorius, 1546), FMD poses an increasing challenge to the international community with circulation of highly divergent virus serotypes and strains that have great potential for transboundary spread. FMDV has many of the characteristics of a successful emergent pathogen: it has high genetic and antigenic variability (Carrillo, 2012; Vosloo et al., 2010); it has a spectrum of variants suited to very different epidemiological conditions; it can infect over 70 different species (Hedger 1981; Shimshony, 1988; Bengis & Erasmus, 1988; Pinto, 2004; Arzt et al., 2011a; Karesh, 2012) and it is highly contagious in the acute stages of disease, but can also survive subclinically

The Role of Animals in Emerging Viral Diseases. http://dx.doi.org/10.1016/B978-0-12-405191-1.00002-8

21

for years in persistently infected animals, so-called "carriers" (Burrows, 1966; Bengis et al., 1986; Alexandersen et al., 2003).

Carriers are defined by the World Organisation for Animal Health (Office International des Épizooties: OIE) as persistently infected animals which are recovered, vaccinated or exposed and from which FMDV can be isolated from the oropharynx for more than 28 days after acute stages of disease (OIE, 2009). About 50% of ruminants are thought to become persistent carriers (Arzt et al., 2011b). Cattle are capable of maintaining the virus for up to 3.5 years, sheep for at least 9 months, goats for 4 months and African buffalo (*Syncerus caffer*) for 5 years (Condy et al., 1985; Alexandersen et al., 2002; Arzt et al., 2011b). However, there is much uncertainty about whether these animals transmit FMDV to other animals, and, if they do, which particular factors cause a persistently infected animal to recommence virus shedding to the extent that it can infect another animal (Thomson, 1996).

GLOBAL DISTRIBUTION AND POTENTIAL FOR EMERGENCE

While FMD was eradicated in most of Western Europe by the late 1980s, five out of the seven known FMDV serotypes (O, A, SAT 1, SAT 2 and SAT 3) are present in Africa, whereas A, O and Asia 1 serotypes are found in Asia. Serotypes A and O have the widest global distribution. Conversely, serotype C has been very rarely reported over the past 15 years, the last confirmed outbreaks occurring in Brazil and Kenya in 2004 (Rweyemamu et al., 2008). In Asia, South America and Africa, FMDV can be further divided into seven major pools of infection (Paton et al., 2009). It is generally considered that FMDV originated in Africa due to the long-term subclinical infection of African buffalo (involving co-evolution with that species) and the greater genetic diversity of the SAT serotypes compared to the Eurasian types (Vosloo et al., 2002); however, the earliest reliable descriptions of FMD come from Europe, leading others to conclude that its origin lies on that continent (Tully & Fares, 2008). Additionally, it has been suggested that FMD was present in the 11th century in India since Lokopakara (1025 AD) compiled by Chavundaraya (Ayangarya, 2006) described "boils of gum and hoof" as a distinct disease in cattle (Nene, 2007).

The genome of FMDV is highly plastic and evolves rapidly as a consequence of errors that are introduced and inherited during replication. These characteristics allow nucleotide sequence data to be used to reliably reconstruct the relationship between viruses recovered from different locations, or at different times. At the broadest scale, analyses of sequences encoding a capsid protein (VP1/1D) are widely used to categorize field strains into discrete variants (or topotypes) that frequently show geographical clustering based on the historical distribution of FMD. The pattern of serotypes and variants around the world is not static and sequencing of these viruses allows us to precisely characterize new isolates of FMDV and trace their origin and movements across international boundaries (Samuel & Knowles, 2001; Knowles & Samuel, 2003).

The escape of FMDV strains from their endemic pools into other regions is a matter of great concern due to the potential for disease emergence in new areas previously naïve to those strains. These introductions can have considerable consequences in terms of disease spread and severity even if resident FMDV strains are already present, because of poor cross-protection against exotic strains (Vosloo et al., 2010). The recent outbreaks of SAT 2 in the Middle East and North Africa or the PanAsia strain of serotype O in the UK in 2001 are examples of this (Knowles et al., 2001; Di Nardo et al., 2011; Valdazo-González et al., 2012). Host vulnerability to new strains, for instance, was evident in a recent incursion of SAT 2 into Egypt, where mortality rates as high as 20% were reported in livestock (Ahmed et al., 2012).

HISTORICAL EMERGENCE OF FMD IN AFRICA

Human activity has had major impacts on the epidemiology of FMD. This is particularly evident in sub-Saharan Africa largely as a consequence of movements of animals and infectious diseases following European colonization. The rinderpest (cattle plague) pandemic, which swept across Africa in the late 19th century following the importation of livestock from India into Ethiopia, decimated more than 90% of cattle, buffalo and other susceptible species in eastern and southern Africa. The pandemic has played a central role in the social and political history of Africa, in the epidemiology of many livestock and wildlife diseases present on the continent today (including FMD), and in shaping African ecosystems (Sinclair, 1979; Reid et al., 2005; Sinclair et al., 2007). Its repercussions are still observable today (African Union, 2010).

Reports of animals with FMD in southern Africa are as old as 1795 (reviewed by Knowles, 1990). However, the rinderpest pandemic largely removed populations susceptible to FMD and, as a result, FMD occurrence declined around the turn of the century, with cases in southern Africa only being reported again in 1931 (Thomson, 1995). It is likely that currently circulating lineages of SAT serotypes re-emerged from small numbers of buffalo that survived the rinderpest pandemic once buffalo and livestock numbers had recovered.

Anthropogenic factors are also likely to have been critical in the introduction and spread of other serotypes in Africa, and phylogenetic analyses are consistent with the interpretation that Eurasian FMDV serotypes (O, A and C) were re-introduced through trade and restocking of livestock from Asia or Europe following the ravages of rinderpest. For instance, there is evidence for a relatively recent (within the past 100 years) common ancestral history between FMDV O topotypes that are currently present in Africa, Asia and South America (Figure 2.1), consistent with emergence of O strains into susceptible animal populations of Africa as a result of introduction with cattle. Furthermore, a more diverse serotype O sequence obtained from a Sudanese FMD virus in the 1960s may be a sole representative sequence of FMDVs present in Africa prior to the rinderpest pandemic (J.M. Stirling and N.J. Knowles, unpublished data).

Over the past century, and continuing to this day, the unfenced rangelands of eastern Africa have supported abundant wildlife populations, with frequent opportunities for close contact between wildlife and livestock. The control of rinderpest through cattle vaccination in the 1950s and 1960s may have played a major role in livestock–wildlife interactions in the region. In the Serengeti, for example, rinderpest vaccination was associated with dramatic increases in wildebeest and buffalo numbers (Sinclair, 1979), with the potential for increased interactions with neighboring pastoral livestock populations. There are arguably now more susceptible hosts, more contact between them, and more intra- and

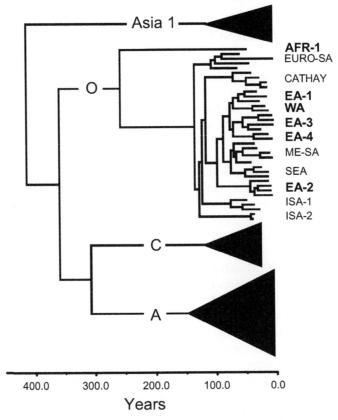

FIGURE 2.1   Impact of the rinderpest pandemic upon current FMDV distribution in Africa. Bayesian phylogenetic tree for representative VP1 (1D) sequences for serotype O FMD viruses. These results indicate that current FMDV topotypes present in Africa (highlighted in bold) have diverged from other global FMDV topotypes within the last 100 years. A single isolate representative of the putative FMDV strains that were present in sub-Saharan Africa prior to the rinderpest pandemic in 1889–1997 is also shown (AFR-1).

interregional livestock movements than at any other time in recent history. This creates an ideal environment for the emergence of novel FMDV strains and may explain the greater diversity of FMDV serotypes and topotypes than in any other regions. Genomic analyses may provide a useful approach to explore this hypothesis, and to gauge whether viral populations are diversifying more rapidly in areas with high levels of wildlife–livestock interactions, and/or high mobility of livestock. However, a true picture of diversity is difficult to obtain retrospectively, as there is much bias from the patchy sampling coverage and disease reports from many areas in the last century.

MAINTENANCE OF FMD IN DIFFERENT RESERVOIR POPULATIONS IN SUB-SAHARAN AFRICA

Although buffalo are considered the ancestral host of SAT serotypes and important maintenance host populations in southern Africa (Thomson et al., 1992; Vosloo et al., 2001, 2002, 2010), many features of the epidemiology of FMD in Africa remain unclear, particularly in relation to the role of livestock and wildlife in maintaining different FMDV serotypes in other parts of Africa (Figure 2.2). While SAT 1 and SAT 2 are known to be maintained in buffalo, these serotypes have also been able to "escape" from sub-Saharan Africa to cause extended livestock outbreaks in North Africa, the Middle East and Europe without involvement of buffalo or any other wildlife species (Ahmed et al., 2012; Bastos, 2003; Dimitriadis & Delimpaltas, 1992; Rweyemamu et al., 2008). This suggests that SAT 1 and SAT 2 can be maintained independently in both livestock and buffalo populations (Figure 2.2A). However, in the wildlife-rich rangelands of East Africa, the degree to which SAT 2 outbreaks are sustained by re-introduction from buffalo is still unclear. In contrast to SAT 1 and SAT 2, serotype SAT 3 appears to be mainly confined to buffalo with only a small number of outbreaks reported in domesticated species (Figure 2.2B) (Thomson, 1995; Bastos et al., 2003; Thomson et al., 2003). Conversely, although maintenance hosts for SAT serotypes, buffalo are not believed to be reservoirs of Eurasian FMDV serotypes (Anderson, 1979; Ayebazibwe et al., 2010) (Figure 2.2C).

The role of other wildlife hosts in the epidemiology of the disease is even less clear. In contrast to buffalo populations that show consistently high levels of exposure (Thomson et al., 1992, 2003; Bronsvoort et al., 2008; Ayebazibwe et al., 2010), seroprevalence in other wild ungulates, for example impala (*Aepyceros melampus*), giraffe (*Giraffa camelopardalis*), eland (*Taurotragus oryx*), tsessebe (*Damaliscus lunatus*), kudu (*Tragelaphus strepsiceros*), waterbuck (*Kobus ellipsiprymnus*), sable antelope (*Hippotragus niger*), bushbuck (*Tregelaphus sylvaticus*), nyala (*Nyala angasii*), warthog (*Phacochoerus africanus*), bushpig (*Potamochoerus larvatus*), redbuck (*Redunca spp.*) and wildebeest (*Connochaetes gnou*), is very low, suggesting that they are spill-over hosts rather than maintenance populations (Anderson et al., 1993) (Figure 2.2D). However, in some parts of southern Africa it is suggested that spill-over from

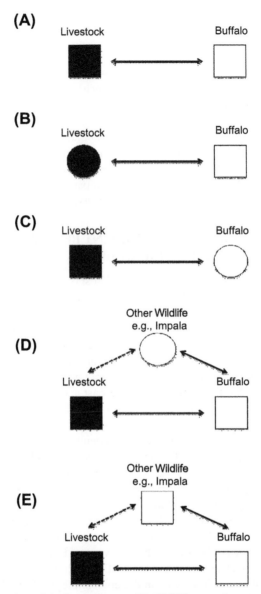

FIGURE 2.2 Simple models that outline possible FMDV reservoir systems in sub-Saharan Africa. Squares represent maintenance populations and circles show non-maintenance populations. Schematics show different scenarios where: (A) Livestock and buffalo can both maintain FMDV independently of one another, as is thought to be the case for SAT 2 in different parts of Africa; (B) Buffalo, but not livestock can maintain FMDV independently, for example in the case of SAT serotypes in South Africa where livestock control measures are in place; (C) Livestock, but not buffalo, can maintain FMDV independently, as is thought to be the case for serotypes A and O; (D) Livestock and buffalo can both maintain FMDV independently of one another. FMDV may spill over to other susceptible animals such as impala but cannot be maintained independently in this other wildlife population, as is the case in most non-buffalo wildlife in Africa; and (E) Livestock, buffalo and other wildlife can all maintain FMDV independently of one another but can also transmit it between each other, as is proposed for some high density impala populations in South Africa.

buffalo to impala may occur frequently and that denser impala populations may be capable of self-sustained circulation (Vosloo et al., 2009) (Figure 2.2E). Impala have also been implicated as intermediate hosts between buffalo and cattle (Bastos et al., 2000; Hargreaves et al., 2004; Vosloo et al., 2006).

OPPORTUNITIES FOR BUFFALO-TO-LIVESTOCK TRANSMISSION

African buffalo are of particular concern where they act as potential reservoirs of FMDV for livestock, and as a maintenance source of persistently infected animals (carriers) where antigenic diversity may be generated (Vosloo et al., 1996). Across southern Africa, where the disease is well controlled in livestock, buffalo are implicated as the likely source of many new livestock outbreaks (Bastos et al., 2000; Hargreaves et al., 2004; Thomson et al., 2003; Vosloo et al., 2001). However, much less is known about the role of buffalo elsewhere in Africa, and the importance of buffalo-to-livestock transmission in triggering new outbreaks and sustaining endemic cycles of infection.

Acutely infected buffalo develop FMD lesions that shed virus, albeit in quantities lower than cattle (Gainaru et al., 1986). Buffalo calves become infected with FMD between 3 and 6 months (Condy and Hedger, 1978), with the proportion of persistently infected animals peaking in the 1–3 year age group (Juleff et al., 2012a). It is speculated that acutely infected buffalo calves may be a source of virus for other animals (Thomson et al., 2003). However, clear experimental evidence for FMDV transmission from artificially infected buffalo to livestock has been elusive. In the two experiments where transmission was achieved, cattle only became infected 5 and 10 months after the acute stage of the disease in the buffalo (Dawe et al., 1994; Vosloo et al., 1996). A further four studies reported absence of infection in cattle despite protracted contact with persistently infected buffalo (Bengis et al., 1986; Gainaru et al., 1986; Condy & Hedger, 1974; Anderson et al., 1979). In the studies where transmission occurred, male buffalo were mixed with female cattle, and cattle became infected only after the buffalo reached sexual maturity. This led to the hypothesis that FMD can be transmitted by the sexual route. However, FMD virus was retrieved from semen and sheath wash from only one out of twenty FMDV seropositive male buffalo (Bastos et al., 1999), and therefore the importance of possible sexual transmission of FMD from buffalo to cattle remains inconclusive. Although there are few experimental reports of buffalo-to-cattle transmission, epidemiological field data and phylogenetic evidence in southern Africa demonstrates that transmission from buffalo to FMD-free cattle does occur (Bastos et al., 2000; Hargreaves et al., 2004; Thomson et al., 2003; Vosloo et al., 2009).

Tanzania, Zimbabwe, Zambia, Democratic Republic of Congo and South Africa represent the five countries with the highest estimated buffalo numbers, with Tanzania having at least six times more buffalo than any other country (Table 2.1, Figure 2.3). In most of these countries buffalo populations

TABLE 2.1 Estimated African Buffalo Population Sizes and Population Trends in the Ten African Countries with the Highest Buffalo Populations (East, 1999)

Country	Estimated total number of buffalo (in 1998)	Population trend
Tanzania	>342,450	Stable/decreasing
Zimbabwe	>50,330	Stable/decreasing
Zambia	>40,090	Stable/decreasing
Democratic Republic of Congo	>39,180	Decreasing
South Africa	>30,970	Increasing
Botswana	>26,890	Stable/decreasing
Uganda	>20,220	Stable/increasing
Kenya	>19,560	Decreasing
Gabon	>20,000	Stable/decreasing
Central African Republic	>19,000	Decreasing

are stable or decreasing (East, 1999). Livestock densities in these areas are also considerable. Ethiopia, Sudan and South Sudan and Tanzania, for example, have the highest populations of cattle in Africa and Nigeria, Sudan and South Sudan, and Ethiopia have the highest combined sheep and goat populations (FAO, 2013; Chilonda, 2005). Hence, together with maximal FMDV diversity (Rweyemamu et al., 2008), East Africa also contains the largest pool of susceptible hosts.

Achieving a better understanding of the relative importance of buffalo in the epidemiology of FMDV in East Africa is of particular relevance for disease control in livestock-keeping communities living at the wildlife–livestock interface, particularly given the ecological and economic importance of buffalo in these areas. Buffalo are bulk grazers, and open up habitats preferred by short grass grazers. They are one of the "big five," that are sought by tourists, both for game viewing and sport hunting. In 7 out of the 14 Southern African Development Community countries, the revenue from the game hunting industry is estimated to be worth $192 million, with wildlife-watching tourism revenue worth $3.2 billion for 10 of these countries where data are available (Booth, 2010). To develop effective FMDV control strategies that support both livestock-based livelihoods and wildlife conservation, much information is still needed on how wildlife species interact with livestock, how and where cross-species transmission occurs, and the possible role of wildlife, other than buffalo, as intermediaries in transmission.

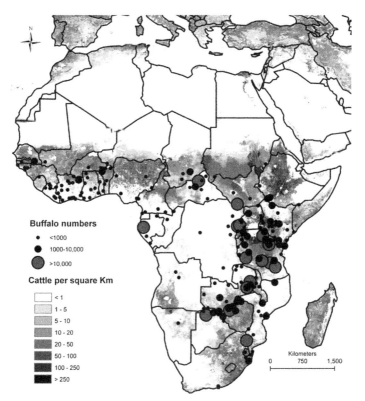

FIGURE 2.3   Estimated geographical distribution of buffalo and cattle in Africa. *Data from East, 1999; Robinson et al., 2007.*

IMPORTANCE OF FMD IN DEVELOPING COMMUNITIES IN AFRICA

Although the devastating consequences of FMD outbreaks in industrialized countries are well recognized, there is relatively little data available to quantify FMD impacts in endemic settings which are often developing countries. The clinical disease has often been regarded as of little significance to livestock health in traditional livestock-keeping systems in Africa, but it is clear that even in extensive, low-production systems, FMD has important consequences on livelihoods and food security, as a result of both direct and indirect effects of the disease. The impact on human poverty has been assessed on the basis of treatment costs, reduced productivity of animals, loss of draft power for tillage and transport, disruption of access to markets, the cost of risk management, limitation of land usage in areas with high disease risk, and risk adversity to embracing advances in animal management (Perry et al., 2002). Based on weighted analysis of socioeconomic criteria and national impacts that also affect the poor,

FMD was ranked third (after gastrointestinal helminths and neonatal mortality syndrome) among animal diseases having greatest impact on overall poverty (Perry et al., 2002).

Livestock owners in East Africa consistently rank FMD among the top five most important livestock diseases (Jost et al., 2010; Ohaga et al., 2007; Bedelian et al., 2007; Cleaveland et al., 2001) with anecdotal evidence for an increasing frequency of outbreaks in pastoral herds and flocks. Studies in Ethiopia, Cameroon, Sudan and Tanzania showed that endemic FMD is associated with calf deaths, reduced milk supply, poor reproductive performance and heat intolerance syndrome (Cleaveland et al., 2001; Catley et al., 2004; Barasa et al., 2008; Rufael et al., 2008). Milk yield may be reduced for the animal's entire lactation period after FMD infection, and lack of milk is likely to be a contributory factor to calf death in traditional livestock keeping regions (Barasa et al., 2008). FMD udder damage may also increase susceptibility to mastitis (Saini et al., 1992). Many people in Africa rely on unpasteurized milk from their animals as an important source of nutrition. Although this is the most common means whereby humans contract FMD (Bauer, 1997), and pastoral communities frequently report a self-limiting febrile disease in people at the time of FMD outbreaks in cattle (Shirma, 2005), no studies have investigated the prevalence of zoonotic FMD infections in African countries.

PROSPECTS FOR FMD CONTROL IN AFRICA

FMDV control presents multiple challenges across the African continent. In East Africa, the broad spectrum of FMDV diversity, large-scale animal movements, which are often unregulated, and an abundance of potential FMDV wildlife hosts makes the region a theater for FMDV emergence and one of the most challenging areas in the world to control the disease. However, new disease control issues are also emerging in southern Africa, with the development of trans-frontier conservation areas (TFCAs) that promote conservation and sustainable management of ecosystems with cross-border tourism. While TFCAs have a clear conservation and political rationale, they are at odds with conventional FMD control methods in southern Africa, such as veterinary fencing and movement controls. Physical segregation of buffalo and livestock, which has traditionally been used in southern Africa to prevent transmission, is not only incompatible with the TFCA vision, but would also be infeasible in many other parts of Africa (e.g., East Africa) due to concerns about negative consequences for wildlife migration and dispersal, which are central to the ecological integrity of many of these unfenced ecosystems (Ferguson et al., 2013).

The African countries that have established zonal FMD freedom have utilized a combination of animal movement control, separation of livestock and wildlife and vaccination of livestock (Brückner et al., 2004).

This geographically based approach may need to be adapted to balance the needs of people, livestock and wildlife, but could, for example, be introduced in areas far from wildlife-protected areas and exploit geographical features that could act as natural barriers to movement of animals and people. Sustainability of control measures also remains uncertain and challenging, particularly in the face of volatile dynamics in livestock markets and in areas with political instability (Thomson, 1995; Vosloo et al., 2002; Batho, 2003).

Another catalyst to FMD control in the East African region would be for more achievable targets to be allowed for entry into lucrative markets for livestock products. At present, African producers are locked out of many markets due to stringent, geographically based rules on FMD status and importation (OIE, 2011). Africa accounts for 7% of global beef consumption, but less than 2% of global trade (Morgan & Tallard, 2007), leaving much potential for growth of commodity-based trade in livestock products both within and outside of Africa. Incentives for positive steps in FMD control, such as commodity-based trade (Thomson et al., 2004; 2009), may provide a welcome injection of funds to further FMD control measures in a positive feedback loop.

Despite the potential for improving FMD control in Africa, the lack of effective FMDV vaccines remains a critical constraint. Current FMDV vaccines produce immunity lasting a maximum of 6 months, need a continuous cold-chain until inoculation, and give very little cross-protection between strains (Vosloo et al., 2002; Paton et al., 2009; Domenech et al., 2010), which limits their usefulness against the high diversity of circulating strains in East Africa. A key step towards effective control and potential elimination of FMD in Africa must be the development of stable vaccines that produce long-lived immunity to a broad spectrum of strains

CONCLUSIONS

East Africa, with its large populations of susceptible ungulates, vast movements of livestock and diversity of FMDV strains, presents the ideal cauldron from which novel strains of FMDV can emerge. Implementation of measures that are sympathetic to wildlife conservation and pastoralism presents a formidable challenge for the control of FMD and the emergence of new epidemic cycles. However, the incentives for control are great; it would indisputably contribute to an improved quality of life for humans and animals and support economic development, acting as an important tool to break cycles of poverty. Where FMD is well controlled in livestock, African buffalo appear to be a critical source of infection. However, in East African countries where FMDV is endemic, livestock and human-related factors are likely to contribute as additional important drivers of FMDV emergence.

Glossary of terms used in the text

Term	Definition	Reference	Possible example in context of FMD	Reference for FMD example
Carrier (epidemiology)	An animal that harbors a specific infectious agent without discernible clinical disease and serves as a potential source of infection	Martin et al., 1987	Examples of asymptomatic animals transmitting FMD are rare	Vosloo et al., 1996; Dawe et al., 1994
OIE FMDV carrier definition	Animals which are recovered, vaccinated or exposed and in which FMDV can be recovered from the oropharynx more than 28 days after acute stages of disease	OIE, 2009	Approximately 50% of domestic ruminants. High proportions of African buffalo (peaking at 1–3 years old)	Alexandersen et al., 2002; Condy et al., 1985; Arzt et al., 2011b; Juleff et al., 2012b
Critical community size	The minimum size of a closed population within which a pathogen can persist indefinitely	Bartlett, 1960		
Maintenance host	A population larger than the critical community size: disease will be maintained within the population even if transmission into the population from the outside is prevented. A combination of nonmaintenance communities can still combine to make a maintenance community	Haydon et al., 2002		

Reservoir	One or more epidemiologically connected populations or environments in which the pathogen can be permanently maintained and from which infection is transmitted to the defined population of interest (target population)	Haydon et al., 2002	Livestock and buffalo in FMD endemic countries. Buffalo in South Africa	Thomson et al., 2003
Spill-over transmission	Interspecies transmission from a maintenance host to a nonmaintenance host	Daszak, 2000; Power & Mitchell, 2004	Transmission from livestock and buffalo to impala and other susceptible wildlife. Amplification of FMDV may occur in spill-over hosts with secondary transmission to the target population. Independent maintenance proposed in high densities of impala in Kruger National Park	Anderson et al., 1993; Bastos et al., 2000; Vosloo et al., 2005
Commodity-based trade	A focus on the attributes of the product (quality, food safety) rather than the disease status of the place of origin. Transmission of diseases such as FMD	Thomson et al., 2004	Deboned beef poses little threat of FMD transmission. Proponents of commodity-based trade argue that countries that are taking all possible measures to reduce the risk of FMDV transmission from their products should be allowed to trade these products	Rich & Perry, 2011; Thomson et al., 2009

ACKNOWLEDGMENTS

The authors acknowledge the Biotechnology and Biological Sciences Research Council (BBSRC), the Department for International Development and the Scottish Government through their support for the Combating Infectious Diseases of Livestock for International Development initiative (projects BB/H009302/1 and BB/H009175/1), The UK Department of the Environment, Food and Rural Affairs (DEFRA) for providing support to the FMD Reference Laboratory and two research projects (SE2939 and SE2940) at the Pirbright Institute, and the Wellcome Trust who fund the Southern African Centre for Infectious Disease Surveillance (SACIDS). Doctoral training for MC is funded by a BBSRC Doctoral Training Grant.

REFERENCES

African Union. (2010). History of Rinderpest Eradication from Africa Impact, Lessons Learnt and Way Forward Position Paper. *Agriculture Ministers Conference. Entebbe/Uganda: InterAfrican Bureau for Animal Resources.* Retrieved from http://www.au-ibar.org/docs/20100514_Serecu_PositionpaperRinderpest.pdf.

Ahmed, H. A., Salem, S. A. H., Habashi, A. R., Arafa, A. A., Aggour, M. G. A., Salem, G. H., Gaber, A. S., Selem, O., Abdelkadar, S. H., Knowles, N. J., Madi, M., Valdez-Gonzalez, B., Wadsworth, J., Hutchings, G. H., Miolet, V., Hammond, J. M., & King, D. P. (2012). Emergence of foot-and-mouth disease virus SAT 2 in Egypt during 2012. *Transbound. Emerg. Dis., 59*, 476–481.

Alexandersen, S., Zhang, Z., & Donaldson, A. I. (2002). Aspects of the persistence of foot-and-mouth disease virus in animals—the carrier problem. *Microbes and Infection, 4*, 1099–1110.

Alexandersen, S., Zhang, Z., Donaldson, A., & Garland, A. J. (2003). The Pathogenesis and Diagnosis of Foot-and-Mouth Disease. *J. Comp. Pathol., 129*, 1–36.

Anderson, E. C., Doughty, W. J., Anderson, J., & Paling, R. (1979). The pathogenesis of foot-and-mouth disease in the African buffalo (Syncerus caffer) and the role of this species in the epidemiology of the disease in Kenya. *J. Comp. Pathol., 89*, 541–549.

Anderson, E. C., Foggin, C., Atkinson, M., Sorensen, K. J., Madekurozva, R. L., & Nqindi, J. (1993). The role of wild animals, other than buffalo, in the current epidemiology of foot-and-mouth disease in Zimbabwe. *Epidemiol. Infect., 111*, 559–563.

Arzt, J., Juleff, N., Zhang, Z., & Rodriguez, L. L. (2011a). The pathogenesis of foot-and-mouth disease I: viral pathways in cattle. *Transbound. Emerg. Dis., 58*, 291–304.

Arzt, J., Baxt, B., Grubman, M. J., Jackson, T., Juleff, N., Rhyan, J., Rieder, E., et al. (2011b). The pathogenesis of foot-and-mouth disease II: viral pathways in swine, small ruminants, and wildlife; myotropism, chronic syndromes, and molecular virus-host interactions. *Transbound. Emerg. Dis., 58*, 305–326.

Ayangarya, V. S. (Translation), (2006). Lokopakara (For the Benefit of People). Agri-History Bulletin No. 6. Asian Agri-History Foundation, Secunderabad 500 009, India. 134 pp.

Ayebazibwe, C., Tjørnehøj, K., Mwiine, F. N., Muwanika, V. B., Okurut, A. R. A., Siegismund, H. R., & Alexandersen, S. (2010). Patterns, risk factors and characteristics of reported and perceived foot-and-mouth disease (FMD) in Uganda. *Trop. An. Health Prod., 42*, 1547–1559.

Barasa, M., Catley, a, Machuchu, D., Laqua, H., Puot, E., Tap Kot, D., & Ikiror, D. (2008). Foot-and-mouth disease vaccination in South Sudan: benefit-cost analysis and livelihoods impact. *Transbound. Emerg. Dis., 55*, 339–351.

Bartlett, M. (1960). The Critical Community Size for Measles in the United States. *J. Royal Stat. Soc., 123*, 37–44.

Bastos, A. D., Bertschinger, H. J., Cordel, C., Van Vuuren, C. D., Keet, D., Bengis, R. G., Grobler, D. G., & Thomson, G. R. (1999). Possibility of sexual transmission of foot-and-mouth disease from African buffalo to cattle. *Vet. Rec.*, *145*, 77–79.

Bastos, A. D., Boshoff, C. I., Keet, D. F., Bengis, R. G., & Thomson, G. R. (2000). Natural transmission of foot-and-mouth disease virus between African buffalo (Syncerus caffer) and impala (Aepyceros melampus) in the Kruger National Park, South Africa. *Epidemiol. Infect.*, *124*, 591–598.

Bastos, A. D. (2003). The implications of virus diversity within the SAT 2 serotype for control of foot-and-mouth disease in sub-Saharan Africa. *J. Gen. Virol.*, *84*, 1595–1606.

Bastos, A. D., Anderson, E. C., Bengis, R. G., Keet, D. F., Winterbach, H. K., & Thomson, G. R. (2003). Molecular epidemiology of SAT3-type foot-and-mouth disease. *Virus Genes*, *27*, 283–290.

Bauer, K. (1997). Foot- and-mouth disease as zoonosis. *Arch. Virol.*, (Suppl. 13), 95–97.

Batho, H. (2003). Report to the OIE following a request by the Southern African Development Community (SADC) for an emergency audit on foot and mouth disease (FMD) in southern Africa. In OIE, (Vol. 33). Retrieved from http://www.oie.int/doc/ged/D11249.PDF.

Bedelian, C., Nkedianye, D., & Herrero, M. (2007). Maasai perception of the impact and incidence of malignant catarrhal fever (MCF) in southern Kenya. *Prev. Vet. Med.*, *78*, 296–316.

Bengis, R. G., & Erasmus, J. M. (1988). Wildlife diseases in South Africa : a review. *Rev. sci. tech. Off. Int. Epiz.*, *7*, 807–821.

Bengis, R. G., Thomson, G. R., Hedger, R. S., De Vos, V., & Pini, A. (1986). Foot-and-mouth disease and the African buffalo (Syncerus caffer). 1. Carriers as a source of infection for cattle. *Onderstep. J. Vet. Res.*, *53*, 69–73.

Booth V. Contribution of Wildlife to National Economies. pp. 1–72. CIC Technical Series Publication No. 8, In: Wollscheid K, Czudek R, editors. Budapest: Food and Agriculture Organization, Rome International Council for Game and Wildlife Conservation; 2010.

Bronsvoort, B. M., Parida, S., Handel, McFarland, S., Fleming, L., Hamblin, P., & Kock, R. (2008). Serological survey for foot-and-mouth disease virus in wildlife in eastern Africa and estimation of test parameters of a nonstructural protein enzyme-linked immunosorbent assay for buffalo. *Clin. Vaccine. Immunol.*, *15*, 1003–1011.

Brückner, G. K., Vosloo, W., Cloete, M., Dungu, B., & Du Plessis, B. J. A. (2004). Foot-and-mouth disease control using vaccination: South African experience. *Dev. Biol.*, *119*, 51–62.

Burrows, R. (1966). Studies on the carrier state of cattle exposed to foot-and-mouth disease virus. *J. Hyg.*, *64*, 81–90.

Carrillo, C. (2012). Foot and Mouth Disease Virus Genome. In M. L. Garcia & V. Romanowski (Eds.), *Viral Genomes - Molecular Structure, Diversity, Gene Expression Mechanisms and Host-Virus Interactions* (pp. 53–68). InTech.

Catley, A., Chibunda, R. T., Ranga, E., Makungu, S., Magayane, F. T., Magoma, G., Madege, M. J., & Vosloo, W. (2004). Participatory diagnosis of a heat-intolerance syndrome in cattle in Tanzania and association with foot-and-mouth disease. *Prev. Vet. Med.*, *65*, 17–30.

Chilonda, P. *Livestock sector brief: Sudan.* (2005). Retrieved from http://www.fao.org/ag/againfo/resources/en/publications/sector_briefs/lsb_SDN.pdf.

Cleaveland, S., Kusiluka, L., Kuwai, J., Bell, C., & Kazwala, R. *Assessing the impact of malignant catarrhal fever in Ngorongoro district, Tanzania: A study commissioned by the Animal Health Programme, Department for International Development.* (2001). Retrieved from http://www.participatoryepidemiology.info/MCF Report.pdf.

Condy, J., Hedger, R. (1974). The survival of foot-and-mouth disease virus in African buffalo with non-transference of infection to domestic cattle. *Res. Vet. Sci.*, *16*, 182–185.

Condy, J. B., & Hedger, R. S. (1978). Experiences in the establishment of a herd of foot-and-mouth-disease free African buffalo (*Syncerus caffer*). *South African Journal of Wildlife Research. 8*, 87–89.

Condy, J. B., Hedger, R. S., Hamblin, C., & Barnett, I. T. (1985). The duration of the foot-and-mouth disease virus carrier state in African buffalo (i) in the individual animal and (ii) in a free-living herd. *Comp. Immunol. Microbiol. Infect. Dis, 8*, 259–265.

Daszak, P. (2000). Emerging Infectious Diseases of Wildlife- Threats to Biodiversity and Human Health. *Science, 287*, 443–449.

Dawe, P. S., Sorensen, K., Ferris, N. P., Barnett, I. T., Armstrong, R. M., Knowles, N. J. (1994). Experimental transmission of foot-and-mouth disease virus from carrier African buffalo (Syncerus caffer) to cattle in Zimbabwe. *The Veterinary record 134*, 211–5.

Di Nardo, A., Knowles, N. J., & Paton, D. J. (2011). Combining livestock trade patterns with phylogenetics to help understand the spread of foot and mouth disease in sub-Saharan Africa, the Middle East and Southeast Asia. *Rev sci. tec. Int. off. Epiz., 30*, 63–85.

Dimitriadis, I. A., & Delimpaltas, P. (1992). Epizootiology of foot and mouth disease during a quarter century (1962–1988) in Greece. *Berliner und Münchener tierärztliche Wochenschrift, 105*, 90–95.

Domenech, J., Lubroth, J., & Sumption, K. (2010). Immune protection in animals: the examples of rinderpest and foot-and-mouth disease. *J. Comp. Pathol, 142*(Suppl.), S120–S124.

East, R. (1999). *African Antelope Database 1998 (pp. 106–115)*. Gland, Switzerland and Cambridge: UK: IUCN/SSC.

FAO. (2013). *FAOSTAT-Live animals in Africa 2011*.

Ferguson, K., Cleaveland, S., Haydon, D. T., Caron, A., Kock, R. A., Lemob, T., et al. Evaluating the potential for the environmentally sustainable control of food and mouth disease in sub-Saharan Africa. Ecohealth 2013. http://dx.doi.org/10.1007/s10393-013-0850-6.

Francastorius, H., 1546. De contagione et contagiosis morbis et curatione. *1* (p. 12).

Gainaru, M. D., Thomson, G. R., Bengis, R. G., Esterhuysen, J. J., Bruce, W., & Pini, A. (1986). Foot-and-mouth disease and the African buffalo (Syncerus caffer). II. Virus excretion and transmission during acute infection. *Onderstepoort. J. Vet. Res., 53*, 75–85.

Hargreaves, S. K., Foggin, C. M., Anderson, E. C., Bastos, a, D. S., Thomson, G. R., Ferris, N. P., & Knowles, N. J. (2004). An investigation into the source and spread of foot and mouth disease virus from a wildlife conservancy in Zimbabwe. *Rev. sci. tech. Int. Off. Epiz., 23*, 783–790.

Haydon, D. T., Cleaveland, S., Taylor, L. H., & Laurenson, M. K. (2002). Identifying reservoirs of infection: a conceptual and practical challenge. *Emerg. Infect. Dis, 8*, 1468–1473.

Hedger, R. S. (1981). Foot-and-mouth disease. In J. Davis, L. H. Karstad & D. Trainer (Eds.), *Infectious Diseases of Wild Mammals* (2nd ed., pp. 87–96). Ames, USA: Iowa State University Press.

Jost, C. C., Nzietchueng, S., Kihu, S., Bett, B., Njogu, G., Swai, E. S., & Mariner, J. C. (2010). Epidemiological assessment of the Rift Valley fever outbreak in Kenya and Tanzania in 2006 and 2007. *Am. J. Trop. Med. Hyg., 83*(Suppl. 2), 65–72.

Juleff, N. D., Maree, F. F., Waters, R., Bengis, R. G., & Charleston, B. (2012a). The importance of FMDV localisation in lymphoid tissue. *Vet. Immunol. Immunopath., 148*, 145–148.

Juleff, N., Maree, F., Bengis, R., De Klerk-Lorist, L., & Charleston, B. (2012b). Role of buffalo in the maintenance of foot-and-mouth disease virus. *Open sessions of the standing technical and research committees of the EuFMD commission, 172*.

Karesh, W. B. (2012). *Wildlife and Foot and Mouth Disease A look from the Wild Side. FAO/OIE Global Conference on Foot and Mouth Disease Control*. Bangkok, Thailand. Retrieved from http://www.oie.int/eng/A_FMD2012/FAO OIE Global Conference 2012 in Thailand/Day_1/1120–1140 B.Karesh.pdf.

Knowles, N. J. (1990). Molecular and antigenic variation of foot-and-mouth disease virus. Council for National Academic Awards. *MPhil Thesis*. Retrieved from http://www.picornaviridae.com/aphthovirus/fmdv/fmd_history.htm.

Knowles, N. J., Samuel, A. R., Davies, P. R., Kitching, R. P., & Donaldson, A. I. (2001). Outbreak of foot-and-mouth disease virus serotype O in the UK caused by a pandemic strain. *Vet. Rec.*, *148*, 258–259.

Knowles, N. J., & Samuel, A. R. (2003). Molecular epidemiology of foot-and-mouth disease virus. *Virus Res.*, *91*, 65–80.

Martin, S., Heck, A., & Willeberg, P. (1987). *Veterinary Epidemiology: principles and methods*. Ames, Iowa: Iowa State University Press.

Morgan, N., & Tallard, G. (2007). Cattle and Beef International Commodity Profile. In F. A. O. Food and Agriculture Association, Retrieved from http://siteresources.worldbank.org/INTAFRICA/Resources/257994-1215457178567/Cattle_and_beef_profile.pdf.

Nene, Y. L. (2007). A glimpse at viral diseases in the ancient period. *Asian Agri-History*, *11*(1), 33–46.

Ohaga, S. O., Kokwaro, E. D., Ndiege, I. O., Hassanali, A., & Saini, R. K. (2007). Livestock farmers' perception and epidemiology of bovine trypanosomosis in Kwale District, Kenya. *Prev. Vet. Med.*, *80*, 24–33.

OIE. (2009). *Foot-and-Mouth DIsease: Aetiology Epidemiology Diagnosis Prevention and Control References*, 1–5. Retrieved from http://www.oie.int/fileadmin/Home/eng/Animal_Health_in_the_World/docs/pdf/FOOT_AND_MOUTH_DISEASE_FINAL.pdf.

Paton, D. J., Sumption, K. J., & Charleston, B. (2009). Options for control of foot-and-mouth disease: knowledge, capability and policy. *Phil. Trans. Royal Soc. B: Biol. Sci.*, *364*, 2657–2667.

Perry, B. D., Randolph, T. F., McDermott, J. J., Sones, K. R., & Thornton, P. K. (2002). *Animal diseases and their impact on the poor. Chapter 5, pp 49–63 in Investing in animal health research to alleviate poverty*. Nairobi, Kenya: ILRI (International Livestock Research Institute). Retrieved from http://ilri.org/InfoServ/Webpub/fulldocs/InvestAnim/Book1/media/PDF_chapters/Book1_Contents.pdf.

Pinto, A. A. (2004). Foot-and-Mouth Disease in Tropical Wildlife. *Ann. NY Acad. Sci.*, *1026*, 65–72.

Power, A. G., & Mitchell, C. E. (2004). Pathogen spillover in disease epidemics. *The American naturalist*, *164*(Suppl.), S79–S89.

Reid, R. S., Serneels, S. M. N., & Hanson, J. *The changing face of pastoral systems in grass dominated ecosystems of eastern Africa*. (2005). Retrieved from http://www.fao.org/docrep/008/y8344e/y8344e06.htm.

Rich, K. M., & Perry, B. D. (2011). Whither Commodity-based Trade? *Development Policy Review*, *29*, 331–357.

Robinson, T. P., Franceschini, G., & Wint, W. (2007). The Food and Agriculture Organization's Gridded Livestock of the World. *Vet. Ital.*, *43*, 745–751.

Rufael, T., Catley, A., Bogale, A., Sahle, M., & Shiferaw, Y. (2008). Foot and mouth disease in the Borana pastoral system, southern Ethiopia and implications for livelihoods and international trade. *Trop. An. Health Prod.*, *40*, 29–38.

Rweyemamu, M., Roeder, P., Mackay, D., Sumption, K., Brownlie, J., Leforban, Y., Valarcher, J. -F., Knowles, N. J., & Saraiva, V. (2008). Epidemiological patterns of foot-and-mouth disease worldwide. *Transbound. Emerg. Dis.*, *55*, 57–72.

Saini, S. S., Sharma, J. K., & Kwatra, M. S. (1992). Actinomyces pyogenes mastitis among lactating cows following foot-and-mouth disease. *Vet. Rec.*, *131*, 152.

Samuel, A. R., & Knowles, N. J. (2001). Foot-and-mouth disease type O viruses exhibit genetically and geographically distinct evolutionary lineages (topotypes). *J. Gen. Vir.*, *82*, 609–621.

Shimshony, A. (1988). Foot and mouth disease in the mountain gazelle in Israel. *Rev. sci. tech. Int. off. Epiz.*, 7, 917–923.

Shirma, G. M. (2005). *The epidemiology of brucellosis in animals and humans in Arusha and Manyara regions, Tanzania. PhD.* Thesis: University of Glasgow.

Sinclair, A. R. E. (1979). *The eruption of the ruminants. Serengeti: Dynamics of an ecosystem* (pp. 82–103). Chicago: University of Chicago Press.

Sinclair, A. R. E., Mduma, S. A. R., Hopcraft, J. G. C., Fryxell, J. M., Hilborn, R., & Thirgood, S. (2007). Long-term ecosystem dynamics in the Serengeti: lessons for conservation. *Conservation Biology: The Journal of the Society for Conservation Biology*, 21, 580–590.

Thompson, D., Muriel, P., Russell, D., Osborne, P., Bromley, A., Rowland, M., Creigh-Tyte, S., & Brown, C. (2002). Economic costs of the foot and mouth disease outbreak in the United Kingdom in 2001. *Rev. Sci. Tech. Off. Int. Epiz.*, 21, 675–687.

Thomson, G. R., Vosloo, W., Esterhuysen, J. J., & Bengis, R. G. (1992). Maintenance of foot and mouth disease viruses in buffalo in Southern Africa. *Rev. Sci. Tech. Off. Int. Epiz.*, 11, 1097–1107.

Thomson, G. R. (1995). Overview of foot and mouth disease in southern Africa. *Rev. sci. tech. Int. off. Epiz.*, 14, 503–520.

Thomson, G. R. (1996). The role of carrier animals in the transmission of foot and mouth disease. *OIE Comprehensive Reports on Technical Items Presented to the International Committee or to Regional Commissions*, 87–103.

Thomson, G. R., Vosloo, W., & Bastos, A. D. S. (2003). Foot and mouth disease in wildlife. *Virus Res.*, 91, 145–161.

Thomson, G. R., Tambi, E. N., Hargreaves, S. K., Leyland, T. J., Catley, A. P., Van 't Klooster, G. G. M., & Penrith, M. L. (2004). International trade in livestock and livestock products: the need for a commodity-based approach. *Vet. Rec.*, 155, 429–433.

Thomson, G. R., Leyland, T. J., & Donaldson, A. I. (2009). De-Boned Beef—An Example of a Commodity for which Specific Standards could be Developed to Ensure an Appropriate Level of Protection for International Trade. *Transbound. Emerg. Dis.*, 56, 9–17.

Tully, D. C., & Fares, M. A. (2008). The tale of a modern animal plague: tracing the evolutionary history and determining the time-scale for foot and mouth disease virus. *Virology*, 382, 250–256.

Valdazo-González, B., Knowles, N. J., Hammond, J., & King, D. P. (2012). Genome sequences of SAT 2 foot-and-mouth disease viruses from Egypt and Palestinian Autonomous Territories (Gaza Strip). *J. Virol.*, 86, 8901–8902.

Vosloo, W., Bastos, A. D., Kirkbride, E., Esterhuysen, J. J., Van Rensburg, D. J., Bengis, R. G., Keet, D. W., & Thomson, G. R. (1996). Persistent infection of African buffalo *(Syncerus caffer)* with SAT-type foot-and-mouth disease viruses: rate of fixation of mutations, antigenic change and interspecies transmission. *J. Gen. Virol.*, 77, 1457–1467.

Vosloo, W., Boshoff, K., Dwarka, R., & Bastos, A. D. (2001). The possible role that buffalo played in the recent outbreaks of foot-and-mouth disease in South Africa. *Ann. NY Acad. Sci.*, 969, 187–190.

Vosloo, W., Bastos, A. D., Sangare, O., Hargreaves, S. K., & Thomson, G. R. (2002). Review of the status and control of foot and mouth disease in sub-Saharan Africa. *Rev. Sci. Tech. Int. off. Epiz.*, 21, 437–449.

Vosloo, W., Bastos, A. D. S., & Boshoff, C. I. (2006). Retrospective genetic analysis of SAT-1 type foot-and-mouth disease outbreaks in southern Africa. *Arch. Virol.*, 151, 285–298.

Vosloo, W., Thompson, P. N., Botha, B., Bengis, R. G., & Thomson, G. R. (2009). Longitudinal study to investigate the role of impala (Aepyceros melampus) in foot-and-mouth disease maintenance in the Kruger National Park, South Africa. *Transbound. Emerg. Dis.*, 56, 18–30.

Vosloo, W., Bastos, A. D. S., Sahle, M., Sangare, O., & Dwarka, R. (2010). Chapter 10 Virus Topotypes and the Role of Wildlife in Foot and Mouth Disease in Africa. *International Union for Conservation of Nature.*

Chapter 3

Parvoviruses of Carnivores

Their Transmission and the Variation of Viral Host Range

Andrew B. Allison and Colin R. Parrish

Baker Institute for Animal Health, Department of Microbiology and Immunology, College of Veterinary Medicine, Cornell University, USA

INTRODUCTION

Most viruses naturally infect and spread among the members of specific host species, and while those host ranges may be narrow or broad, the emergence of viral variants with significantly altered host ranges that initiate large-scale epidemics or pandemics is a rare event (Parrish et al., 2008). The small number of examples that are well documented include the transfer of influenza viruses from birds and swine to humans, as well as to some other mammals, HIV-1 which transferred from nonhuman primates to humans, as well as the canine parvovirus (CPV) which emerged as a pandemic pathogen of dogs in the late 1970s. The biological barriers that initially prevented the infection and spread of the viruses in the new hosts varied, and appear to be quite high for the influenza viruses and to require several genetic changes of the viruses to be overcome (Matrosovich et al., 2009; Parrish and Kawaoka, 2005), and relatively low in the case of HIV-1 spreading from chimpanzees to humans (Keele et al., 2006; Nomaguchi et al., 2012). In the case of CPV, some of the major obstacles that were overcome to allow the pandemic spread of the virus have been determined and not only involved changes in the ancestral virus needed to cross the species barrier to dogs, but also the differences between carnivore host receptors that provided the initial barrier.

THE HOSTS

The parvoviruses of interest are known to infect members of the order Carnivora which includes over 280 species comprising 13 terrestrial and 3 aquatic families, although the taxonomic structure of the order may differ among systematists (Eizirik et al., 2010; Flynn et al., 2005; Nowak, 2005; Nyakatura and

The Role of Animals in Emerging Viral Diseases. http://dx.doi.org/10.1016/B978-0-12-405191-1.00003-X

Bininda-Emonds, 2012; Wozencraft, 2005) (Figure 3.1). Carnivores are believed to have split from other mammals approximately 65 million years ago and now occupy virtually all habitats on earth and can be found from tropical jungles and arid grasslands to montane forests and polar regions (Nowak, 2005). The modern Carnivora can be divided into two suborders: the Feliformia, which contains the "cat-like" carnivores such as large and small cats, hyenas, mongooses, civets and related groups, and the Caniformia, which includes the "dog-like" carnivores such as dogs, wolves, foxes, skunks, mustelids, raccoons, bears and allies. Collectively referred to as pinnipeds, the marine carnivores, which include seals, sea lions, and the walrus, are also members of the suborder Caniformia (Figure 3.1).

THE VIRUS

Parvoviruses are small viruses with a 25nm capsid that packages a genome of ~5,100 nucleotides of linear ssDNA that contains terminal hairpin structures. The genome encodes two major genes, the nonstructural protein gene on the 3′ half (i.e., of the minus strand DNA) and structural protein gene on the 5′ half, each of which produces at least two different proteins. The nonstructural gene encodes the nonstructural 1 and 2 (NS1 and NS2) proteins, the latter derived from alternative splicing, that are involved in DNA replication, capsid assembly, and transport within the cell. The structural gene encodes the virus proteins 1 and 2 (VP1 and VP2) that assemble to make the infectious capsid. Although the VP1 and VP2 proteins overlap in most of their sequence, so that they form the same structural unit in the T=1 capsid, VP1 has an additional 145 residue N-terminal sequence not found in VP2. This sequence is essential for viral cell infection, as it includes an active phospholipase A2 enzyme thought to be used in escape from the endosome (Canaan et al., 2004; Zadori et al., 2001), as well as a potential nuclear transport sequence (Vihinen-Ranta et al., 2002). The capsid is quite robust and can persist in the environment (Bouillant, 1965a, b). Despite the relatively simple composition, the capsid has a number of morphological and functional structures and forms, which include the packaging of the ssDNA through one of the pores at the fivefold axis of symmetry (Cotmore and Tattersall, 2005), the exposure of a proportion of the N-terminal sequences of the VP2 molecule to the outside of the capsid (where it may be cleaved off by proteinases), and divalent ions (likely Ca^{2+}) at two or three positions within each capsid asymmetric unit (Simpson et al., 2000). Once the viral DNA is released from the capsid, by a mechanism which is not currently known, the viral DNA replicates through dsDNA intermediates in the nucleus of the cell, through a process that appears to involve the activities of DNA polymerases delta and alpha, in a process referred to as "rolling hairpin" replication (Cotmore, 1996). The virus is only able to complete its replication cycle in cells that are undergoing cell division, and since infection does not induce mitosis, this restricts

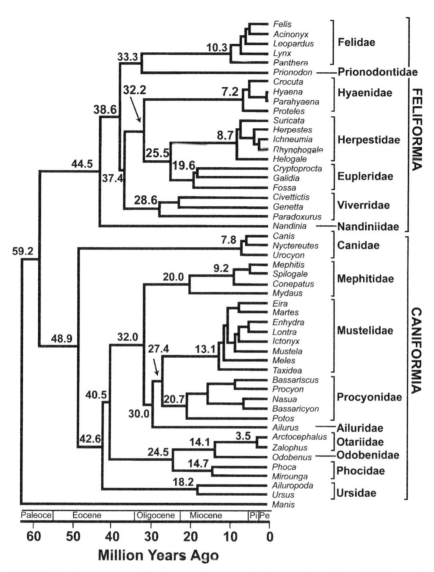

FIGURE 3.1   A phylogeny of the hosts in the order Carnivora, adapted from (Eizirik et al., 2010), indicating the diversification of the hosts over 60 million years. The hosts that are susceptible to parvoviruses in this group include members of both the suborders, and also most of the families. Until recently the dogs, coyotes, and wolves within the family Canidae resisted infection by these viruses, and that resistance was related to the presence of a unique glycosylation site in the apical domain of the TfR in those animals that prevented the previous viruses from binding.

replication to tissues that contain proliferating cells. As those cells will vary in number and location depending on the age of the animal, there is a strong age dependence to the virus tropism and the severity of disease that occurs (Parrish, 1995).

PARVOVIRUSES OF CARNIVORES—THE EARLY YEARS

The parvoviruses of interest are classified within the feline panleukopenia virus group (species) of the genus Parvovirus, family Parvoviridae, and those include viruses that have been named after the hosts which they were originally isolated from: feline panleukopenia virus (FPV) from cats, mink enteritis virus (MEV) from mink, canine parvovirus (CPV) from dogs, and raccoon parvovirus (RPV) from raccoons (Tijssen et al., 2012). However, the current nomenclature system may be misleading, as the host range of these viruses is not limited to the initial species of isolation and, in some cases, the current taxonomic listings for separate strains may be the same virus. For instance, FPV infects members of both the Feliformia and Caniformia, as does CPV, demonstrating that neither virus is a specialist in terms of host range (Figure 3.2). In addition, RPV, in which the name suggests it is a distinct raccoon strain of parvovirus, is essentially identical genetically to FPV isolates recovered from cats, demonstrating that FPV can infect many different carnivore species in multiple families. Although these viruses have likely been circulating in carnivores for millions of years, infection in cats and raccoons was first reported in the 1920s and 1940s, respectively, and the first isolations of the viruses in tissue culture were made in the 1960s. The disease in mink appeared to be a new syndrome when it was first observed in the late 1940s in Canada, and subsequently spread worldwide among mink in the next few years (Gorham and Hartsough, 1955). Similar to RPV, many of the viruses isolated from mink do not appear to be significantly different from FPV in cats or other hosts (Allison et al., 2013) (Figure 3.2), but as most were collected several years to decades after the original outbreak it is not clear what the original virus in mink was, or whether this was truly a new virus in mink at that time, or if this represented the recognition of a virus that had long been infecting mink.

EMERGENCE OF CPV—A NEW VIRUS IN DOGS

The emergence of CPV in 1978 was clearly associated with the emergence of a parvovirus in dogs that caused a previously unrecognized disease, and it was recognized very early that the disease and the virus were similar to FPV in cats. Testing of dog sera collected during different years between 1975 and 1980 showed that there were no anti-CPV antibodies in dogs in most parts of the world before the middle of 1978, although there were antibodies in dogs in Europe in 1976 and 1977, suggesting that the virus was circulating there prior

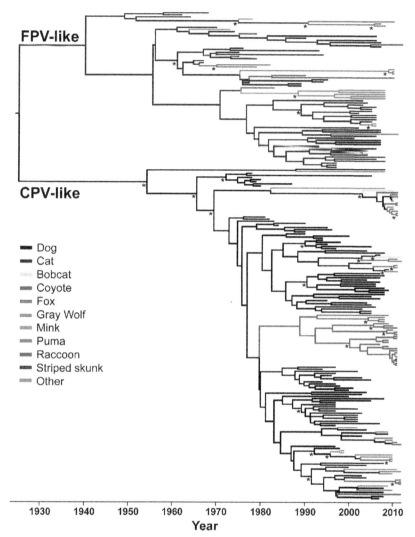

FIGURE 3.2   The relationships of the viruses or viral sequences isolated from different hosts, showing the broad distribution of the viruses among the animals in the order Carnivora—this is a conservative representation of the viruses, as only a small proportion of the known carnivores have been examined at this time. The phylogenetic history (Bayesian MCC tree) of carnivore parvoviruses was inferred from 234 complete VP2 sequences. Clusters of viruses are labeled and colored according to host species. Hosts in the "other" group which includes singleton viruses are lion, palm civet, monkey and tiger. Because the tree was inferred using a relaxed molecular clock, all tip heights are scaled to the year of sampling. Posterior probability values > 0.9 at major nodes or which connect multiple species are marked by a * symbol. A time-scale in years is given by the X-axis. *From Allison et al., 2013 with permission.*

to spreading worldwide during the first part of 1978 (reviewed in Parrish, 1990). The virus spread just as quickly into dogs in countries that have strict quarantine for live dogs (New Zealand, Australia and Japan) as it did into other countries, suggesting that the mode of transmission was likely by the transport of contaminated fomites (e.g., shoes) through air travel, rather than through the transport of infected live dogs. CPV also spread rapidly into populations of wild dogs around the world, as seen in the widespread infection of wild coyotes in the USA within a year of first emerging (Thomas et al., 1984), as well as infecting wild wolves (Zarnke and Ballard, 1987), indicating that the virus was not just being spread by direct human contact with the animals that became infected.

This suggested that the agent (CPV) was a new virus with the ability to spread efficiently among both domestic and wild dogs, and the pathogenesis of the enteric disease was very similar to that seen in cats infected with FPV, suggesting a parvovirus was responsible. This was soon confirmed by analysis of the viral isolates from dogs, which showed that they were closely related antigenically to FPV isolates from cats, and subsequently it was shown that CPV and FPV were >99% identical in DNA sequence (Parrish, 1991; Truyen et al., 1995). Much of the subsequent study of CPV and FPV has been focused on those viruses isolated from domestic dogs and cats, as these were the species that were initially recognized as important hosts in the maintenance and spread of the viruses. While these comparisons have allowed us to understand much of the biology of the host range shift of the viruses, it is now clear that different virus–host relationships likely differ in details, such that a complete understanding would need to take into account the number of diverse wild and domesticated hosts, other than domestic dogs and cats, that are naturally susceptible to these viruses, and the specific changes that occur in the viruses as they replicate and spread within and between different hosts (Allison et al., 2012, 2013).

Host range reflected at the cellular level

Testing for infection of the viruses in vitro in different hosts showed that the canine host range of the viruses was reflected in the tissue cultured cells, as the viruses isolated from cats or mink did not infect canine cells, but the isolates from dogs infected both feline and canine cells with similar efficiency. Using genetic analyses of the viruses, the difference in host range between the two viruses resulted from a small number of changes in the capsid protein, with the differences shown in residues Lys93Asn, Val103Ala, and Asp323Asn being most important for the control of the ability to infect canine cells, and for infection of dogs (Chang et al., 1992). Subsequently, the ability to infect canine cells was also shown to be associated with changes in another region of the capsid structure, where a mutation of VP2 residue 299 (Gly to Glu) completely blocked the ability to infect canine cells (Parker and Parrish, 1997). Those residues were all displayed on the surface of the capsids, but fell into three different regions that were separated by 20-30Å in the structure so there was no obvious direct connection between them (Agbandje et al., 1993). In addition, each changed residue caused only a small change in the protein structure, with the main effect being in the presence or absence of hydrogen bonds, but little effect on the overall folding of the protein structures (Agbandje et al., 1993; Govindasamy et al., 2003; Llamas-Saiz et al., 1996).

The role of the transferrin receptor type-1 (TfR) capsid binding in host tropism

The receptor that the viruses use to bind and infect cells was shown to be the transferrin receptor type-1 (TfR) (Parker et al., 2001), a surface-expressed membrane protein that binds and transports iron-loaded transferrin into the cells through clathrin-mediated endocytotic pathways. Once under the influence of the low pH within the cell, the iron is released and transported into the cytoplasm, while the iron-free transferrin remains bound to the receptor and is recycled to the cell surface and is released at the neutral pH of the extracellular medium (Luck and Mason, 2012; Steere et al., 2012). Transferrin, along with the human hemochromatosis (HFE) protein, binds to an overlapping site in the helical domain of the TfR, although iron-loaded transferrin also binds to the base of the protease-like domain (Eckenroth et al., 2011; Giannetti et al., 2003). However, mutational analysis of the TfR shows that positions in the apical domain, and likely on one side and the top of that domain, are used as binding sites for parvoviruses to gain entry into their hosts (Goodman et al., 2010; Kaelber et al., 2012; Palermo et al., 2003) (Figure 3.3). It was shown that there was a close relationship between the ability of a virus to bind the TfR and its ability to infect the cells, and comparing binding of FPV and CPV capsids with TfR showed that there was a clear difference in the ability of the viruses to bind the TfR expressed on canine cells (Hueffer et al., 2003). Examining the receptor sequence, preparing mutants of different positions in the receptor, and comparison with the structure of the human TfR showed that the region of the receptor that influenced virus binding was located on one side of the TfR apical domain (Figure 3.3). While several residues influenced the binding of the capsid, a single change in the apical domain that introduced a novel glycosylation site into the receptor blocked the binding of FPV capsids, while allowing the continued binding of CPV (Palermo et al., 2003). Removing that site from the canine TfR allowed binding of FPV, as did adding that mutation alone to the feline TfR or to the TfR of black-backed jackal, which was closely related to dogs but lacked the glycosylation site (Kaelber et al., 2012; Palermo et al., 2006) (Figure 3.4). The explanation for the multiple separate positions in the capsid surface appears to be related to the multiple and apparently broadly distributed contacts between the TfR and the capsid surface, and this was also supported by a low resolution structure of the TfR in complex with the capsid, where there appeared to be a broad footprint of contact between the TfR and

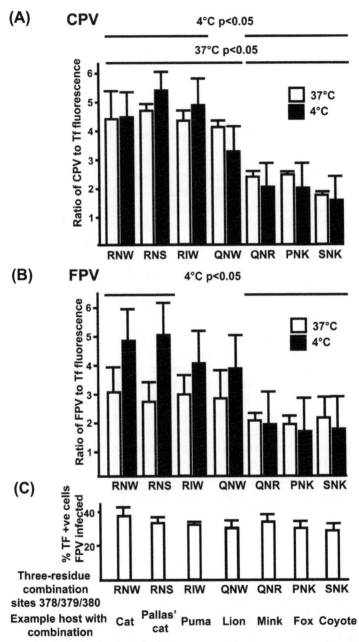

FIGURE 3.3 Determining the effects of varying residues that do not alter the glycosylation of the feline TfR apical domain on parvovirus binding. The mutant forms of the TfR contained up to three changes in the apical domain that were seen in the receptors from various animal hosts—the name of one host species which contains the combination of residues shown is given. CPV (A) or FPV (B) and canine transferrin (Tf) were incubated with TRVb cells

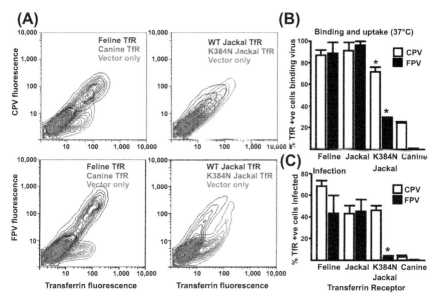

FIGURE 3.4   The presence of a specific glycosylation at position 384 has a significant effect on the binding and infection of viruses in the FPV-like clade. Here a Lys was introduced at that position in the TfR sequence from a black-backed jackal TfR, introducing the glycosylation site, and the effect on FPV and CPV binding and infection was determined, and compared to feline or canine TfRs under the same conditions. (A) Fluorescently labeled CPV or FPV and transferrin were bound to cells expressing empty vector, feline TfR, canine TfR, wild-type jackal TfR, or Lys384Asn mutant jackal TfR at 37°C. The binding of FPV to the 384Asn black-backed jackal TfR resembles the binding of CPV to the canine TfR. (B) Fluorescently labeled FPV or CPV capsids were incubated with cells expressing feline TfR, canine TfR, wild type or Asn384 mutant black-backed jackal TfR at 37°C. The binding was compared to that of fluorescently labeled transferrin. (C) Cells expressing these receptors were inoculated with FPV or CPV, and then infection measured by staining for the parvoviral NS1 expression, and the expression of TfR determined. Error bars = mean ± 1SD. * indicates statistically significant difference in frequency of binding or infection. *From Kaelber et al., 2012 with permission.*

expressing receptors with different combinations of the residues 378, 379, and 380 (feline TfR numbering). Ligands were incubated with the cells at 37°C (white) or 4°C (black). The level of binding is shown, with an adjustment for the level of the TfR expression estimated from the fluorescence of bound transferrin to account for differential receptor expression. Brackets connect groups of receptors that were statistically different. (C) Effects of variant residues in the feline TfR on FPV infection. Cells expressing exogenous TfR with different three-amino acid combinations at residues 378–380 were inoculated and the ratio of infected cells (expressing NS1) and those expressing TfR is shown. Those showed that the variant TfRs were generally used efficiently by the viruses for infection, and only one pairwise comparison was statistically significant; 34% of cells expressing TfR containing QNR (as seen in the mink TfR) were infected by FPV, while 26% of cells containing RNS (as Pallas' cat) were infected. *From Kaelber et al., 2012 with permission.*

capsid in the region where the host-range controlling mutations were found (Hafenstein et al., 2007).

EVOLUTION OF THE HOST RECEPTOR, AND THE ROLE OF SPECIFIC VIRUS BINDING

As mentioned, the order Carnivora contains over 280 species grouped into 16 families (Figure 3.1). However, it is not clear exactly how many of these species are susceptible to infection when exposed to various viruses, are routinely infected in the wild, or act as hosts for sustained parvovirus transmission. As many carnivore species are elusive nocturnal or crepuscular predators, are relatively nonabundant and live mostly solitary lives, little may be known about their life history or the viruses they may harbor (Nowak, 1999). Caniform ("dog-like") carnivores which have been verified to be infected with the feline panleuk-openia group parvoviruses, either through virus isolation or detection of viral DNA in tissues, include members of the families Canidae (e.g., dogs, wolves, coyotes, foxes, raccoon dogs), Mustelidae (e.g., mink, marten, honey badger), Mephitidae (skunks), Procyonidae (raccoons), and Ailuridae (red panda), while evidence of infection in the family Ursidae (bears) is limited to presumptive exposure through antibody positives from serum samples (Allison et al., 2012; Allison et al., 2013; Dunbar et al., 1998; Frolich et al., 2005; Madic et al., 1993; Marsilio et al., 1997; Mech et al., 2008; Neuvonen et al., 1982; Steinel et al., 2000; Veijalainen, 1988) (Figure 3.2). Although parvoviruses such as sea lion bocaviruses have been recently detected in pinnipeds (Li et al., 2011), infection with FPV or CPV has not been reported from marine carnivores. Feliform ("cat-like") species that have been demonstrated to be infected include members of the families Felidae (e.g., small and large cats) and Viverridae (Asian palm civet) (Allison et al., 2012; Demeter et al., 2009; Ikeda et al., 1999; Ikeda et al., 2000; Miyazawa et al., 1999; Steinel et al., 2000) (Figures 3.1 and 3.2). For other feliform families, such as the Hyaenidae (hyenas) or Herpestidae (mongooses), there is currently no compelling evidence, other than serological data, that they are naturally infected with these viruses (Harrison et al., 2004; Millan et al., 2009). Whether the absence of virus detection in these families represents a lack of thorough sampling or natural resistance to infection (see below) remains to be determined.

The open reading frame in the TFCR gene that encodes the TfR differs by up to 10% in DNA sequence for currently available carnivore species, and there are coding changes in various positions throughout the gene, including the apical domain region that is associated with parvovirus binding (Kaelber et al., 2012). When the sequences were examined for selected codons, those were shown to be distributed throughout the protein structure, including the apical domain, with the strongest evidence for selection found in the lineage leading to the dogs and wolves (Kaelber et al., 2012). Among those changes was the mutation of residue 384 (in the apical domain of the canine TfR), which was acquired by the ancestor of coyotes, wolves and dogs between 2 and 5 million years ago, which blocked the binding of capsids similar to the FPV sequence (Palermo et al., 2006; Palermo et al., 2003). In addition to the dominant host range determining change at position 384, other selected changes in the apical domain of various hosts were also seen that modified the interaction with the capsid, although those appeared to have the greatest effect in a synergistic fashion where there were multiple changes in the TfR which acted together to alter virus binding (Kaelber et al., 2012) (Figure 3.3). The nature of the selection that gave rise to the changes in the TfR is not clear. It is possible that this was due to genetic drift, to some other host selection, or to selection by ancient pathogens, including possible ancient parvoviruses. Additionally, as other viruses such as mouse mammary tumor virus and some of the New World arenaviruses are known to use the TfR as a receptor (Radoshitzky et al., 2008; Ross et al., 2002), it is possible that viruses other than parvoviruses may have given rise to the changes.

In this scenario of the evolution of virus susceptibility, many different carnivores including the ancestors of dogs were susceptible to the ancestor of FPV, and those became resistant to the virus when the glycosylation site mutation was acquired, likely between 2 and 4 million years ago. The canine lineage then remained resistant to the viruses until the mid-1970s, when CPV emerged as a new virus of dogs by gaining the ability to bind to the modified TfR found in dogs. A question that arises from this model is why the canine infecting virus emerged recently, and not during the previous millions of years. It is quite possible that viruses infecting the canine ancestor did emerge in former times, but that the populations of dogs or wolves were too small to sustain the virus in those animals alone, and that the domestication of dogs and the recent population expansion of dogs has allowed the virus to become established and spread widely.

EVOLUTION OF THE VIRUSES IN DEEP TIME AND OVER THE PAST DECADES

Ancient evolution and paleoviruses

It appears that the parvovirus family has been in existence for many millions of years. Although we do not have clear evidence of the FPV like viruses before the 1920s, all of the parvoviruses and adeno-associated viruses sequenced share a common ancestor, and it is assumed that this was present in animals tens of millions of years ago (Emerman and Malik, 2010; Katzourakis and Gifford, 2010). An ancient ancestry of the viruses is also supported by the finding of integrated parvovirus sequences or remnants in the genomes of various vertebrate (and invertebrate) hosts, some of which are suggested to have been integrated millions of years ago (Kapoor et al., 2010; Liu et al., 2011). Those paleovirus sequences may be difficult to characterize and none is closely related to the contemporary parvoviruses of carnivores. However, they do confirm a deep ancestry for the parvoviruses, which supports the idea that the viruses are highly host adapted, and that there has likely been co-evolution of viruses and the host sequences—so the idea that ancient parvoviruses selected the receptors of their hosts millions of years ago is reasonable.

Recent variation and evolution of the FPV and CPV

The recent evolution of the viruses is quite clearly defined, and samples of the FPV-like viruses exist from the 1950s, while many CPV isolates or sequences have been collected around the world since 1977, when the virus first emerged. The sequences include a great many VP2 gene sequences, as well as a smaller number of NS1 genes and almost complete genome sequences. Comparative analysis of these viruses demonstrated they were remarkably similar, and all are 98% or greater identical in DNA sequence. The nearest relative that is known is porcine parvovirus, which is 57% identical in VP2 sequence (Lukashov and Goudsmit, 2001; Simpson et al., 2002), indicating that the carnivore viruses have been separated for a long time from the other extant parvoviruses. Until recently most viral sequences have been obtained from isolates associated with clinical disease, while more recently it has become obvious that many infections are mild or subclinical, and that the viral DNA may persist for long periods (perhaps years) in the tissues of animals that have recovered from infections, from which it may be amplified by PCR (Allison et al., 2013). Many studies have examined the viral sequences using phylogenetic analysis, and while the results may differ in details depending on which samples are included, those all show that there are two distinct clades among the viruses. One clade contains the viruses from cats and other hosts going back to the first isolates that were collected in the early 1960s, while the other contained the CPV-like viruses that first emerged in dogs in the mid-1970s. The common ancestor of the CPV-like viruses existed during the period between 1969 and 1975 (Allison et al., 2013; Shackelton et al., 2005), and viruses in that lineage were initially termed CPV type-2 to distinguish them from other distantly related parvoviruses infecting dogs (Canine minute virus) (Hoelzer and Parrish, 2010; Parrish et al, 1982). Retrospective testing of stored serum samples showed clear parvovirus antibodies in dogs in Europe around 1976 and 1977, and then positive sera results and viruses were seen to spread worldwide in a pandemic during 1978 (reviewed in Parrish, 1990). However, the virus in domestic dogs underwent a significant shift in genotype during 1979 and 1980, when a genetic variant emerged that contained several differences from the original CPV-2, and that variant (designated CPV-2a) quickly spread worldwide and replaced the original virus (Parrish et al., 1985, 1988, 1991). The CPV-2a variant was clearly the ancestor of all of the subsequent viruses isolated from domestic dogs, and there have been a number of mutations in the sequences of the canine viruses that have spread worldwide (Hoelzer et al., 2008; Shackelton et al., 2005). Several questions that were open for many years were how the CPV-2a variant arose, whether the group of mutations in that virus were all functional, and whether there was a specific mutational pathway that was followed. When viruses with intermediate genotypes between CPV-2 and CPV-2a were prepared and tested for their viability and growth, it was seen that most showed some small reduction in the ability to replicate in cells in culture compared to the wildtype viruses (Stucker et al., 2012). However, the question of how the CPV-2a may have arisen was partially resolved when viruses and viral DNAs were isolated from raccoons and from other wild carnivores, and a number of those viruses proved to be intermediate genotypes between CPV-2 and CPV-2a, suggesting that the virus that became the dominant strain in dogs actually evolved in another host, and that this may have been because intermediate viruses had a higher fitness in those hosts than they did in dogs (Allison et al., 2012). Interestingly, the CPV-derived raccoon isolates lacked the ability to infect dog cells, and for those to transfer to dogs they would have had to acquire one or two mutations that allowed them to regain the canine host range (Allison et al., 2012).

Mutations that have become widespread

In the 35 years since CPV was first recognized in dogs, there have been a number of changes that have become widespread among the canine isolates in addition to those that define the CPV-2a variant (Hoelzer et al., 2008). Many of those changes alter residues that are exposed on the surface of the capsid, in positions that would interact with the TfR, affect the binding to sialic acid, or interact with host antibodies. Specific changes that have been seen include VP2 residue 426, which was an Asn in the FPV-like viruses that became mutated to an Asp in many CPV-like viruses around the world around 1984. Residue 426 then mutated at a second position in that codon to Glu, which also spread to other parts of the world after about 1995 (Buonavoglia et al., 2001; Gallo Calderon et al., 2011), although with various replacements at that position occurring in the different populations (Maya et al., 2013; Soma et al., 2013). Those mutated viruses have sometimes been referred to as CPV-2b (426Asp) and CPV-2c (426Glu), although those designations (which in the case of CPV-2b, was used to identify its antigenic difference from CPV-2a prior to the known sequence differences) have become less useful as other mutations have combined with the changes at VP2 residue 426. Additional mutations became distributed through multiple continents in dogs and some other hosts, including VP2 residues Ala300Gly/Val/Asp, Gly224Arg, Ser297Ala, and Thr440Ala (Allison et al., 2012; Qin et al., 2006). Some changes were closely positioned near the host range changes that are known to control canine cell infection including Ile324Thr, adjacent to the critical Asp323Asn mutation (Soma et al., 2013), as well as several different natural variants of the codon at VP2 residue 300 (Allison et al., 2012). The effects of those changes on host range and viral fitness have not been rigorously tested, but it is assumed that they influence binding to the receptors from different hosts (Goodman et al., 2010; Hafenstein et al., 2007).

Antigenic variation

The immunity that protects animals against infection appears to be primarily antibody-mediated, and a key protection is provided in maternal antibodies that are delivered to the kitten or puppy from the mother, primarily through transfer in colostrum shortly after birth. That antibody wanes at a steady rate due to growth of the animal and metabolism of the circulating IgG, so that in most animals, antibodies have dropped to low levels between 8 and 12 weeks after birth and they become susceptible to virus infection (Pollock and Carmichael, 1982). The pathogenesis of the infection involves systemic circulation, so it is assumed that the virus would be recognized by that antibody so that its infection would be aborted or greatly diminished. However, it appears that viruses may infect in the presence of low but detectable levels of circulating antibodies, hemagglutination inhibition titers of 1:20 to 1:40, a feature that has been taken advantage of with the application of vaccination even when there is some antibody in circulation (De Cramer et al., 2011; Hoare et al., 1997; Pratelli et al., 2000).

The capsids of the CPV and FPV-like viruses are highly immunogenic, and they stimulate the rapid production of antibodies. The antigenic structure of the capsid has been defined using a number of approaches, including the mapping of antigenic escape mutations, which fell into two discrete antigenic sites containing only small portions of the capsid surface, which were termed the A and B sites (Parrish and Carmichael, 1983; Strassheim et al., 1994). However, structural analysis of the Fab fragments of several different mouse monoclonal antibodies in complex with the capsids showed that the eight antibodies examined in that study covered about 70% of the capsid surface, suggesting that there are strong constraints on where effective escape mutations can fall on the capsid (Hafenstein et al., 2009). In addition, there was significant overlap of the sites that altered TfR binding and those that altered antibody binding, suggesting that many mutations affect both functions simultaneously. Many of the changes on the capsid surface within the major antibody binding regions that alter host range also have effects on antibody epitopes, so they are likely also subject to antigenic selection in nature (Hafenstein et al., 2009; Strassheim et al., 1994). However, it is also clear that there is a reasonable level of cross protection between the different virus strains in animals, such that vaccines made from any of these are able to protect animals against any of the circulating viruses with at least reasonable efficiency (Siedek et al., 2011). It is likely that the cross-protection efficiency and age of infection in young animals varies depending on the antigenic structure of the virus that infected or immunized the mother, and the comparative structure of the virus that challenges the kitten or puppy.

Host range adaptation, and capsid variation

The viruses from domestic cats and dogs have relatively well-defined forms, and the variation since 1977 is also now well defined. However, we know much less about the susceptibility of most other carnivore hosts to these parvoviruses, and also know less about the properties of the viruses that naturally infect them. This was likely in part due to the lack of distinction between the viruses in wild animal populations which may have larger or smaller populations, those that are being farmed where there may be large populations of animals that can sustain an ongoing outbreak, and viruses that are isolated from captive animals which are most likely exposed to viruses from domestic animals. It is now clear that most viruses that are transmitted within or between wild or farmed animal populations show some changes in their capsid protein genes, and also elsewhere in the genome (Allison et al., 2013). The most common changes are found in the region of the capsid surface around VP2 residue 300, which is within a raised region on the shoulder of the threefold spike of the capsid. This region has been shown to influence the binding of the virus to the canine TfR (Goodman et al., 2010; Palermo et al., 2003, 2006), suggesting that those will also influence the interactions with their various host receptors. When we sequenced the TfRs from various hosts, those showed variation of up to 10% in sequence, and variation in the apical domain in the position that interacts with the capsid (Goodman et al., 2010; Kaelber et al., 2012) (Figure 3.3). By phylogenetic analysis the viruses from different hosts were not found to define host related clusters, suggesting that they readily transfer from one host to another, but that during the growth or passage of the virus in the alternative host there is selection for a small number of changes in the capsids, in regions that interact specifically with the receptor.

Global spread and phylogeographic variation

The original CPV-2 spread worldwide during 1978, infecting dogs in countries where there were strict quarantines for live dogs, indicating that the virus may be readily distributed over long distances. Although the variation among the viral sequences is often widely distributed among the different continents over time, there is often some geographic signature to the changes that are seen. This was seen in the general phylogeny where there were some clades that appeared to be found primarily in certain continents and regions such as South America, Europe, and Asia (Hoelzer et al., 2008), although the clades were often defined by only one or a small number of signature mutations so that the bootstrap support for those was often quite weak. However, the temporal and geographic distribution of certain signature mutations could be tracked in some cases, and those appeared in certain regions of the world before they become widely distributed in other continents; for example, the VP2 Asp426Glu mutation was present in Europe for several years before it became widely distributed in other parts of the world (Decaro and Buonavoglia, 2011), and then that mutation has been seen in variable levels in other countries and regions (Hong et al., 2007; Kapil et al., 2007; Nandi et al., 2010; Perez et al., 2007), but in some cases that mutation was observed for a period and then replaced by other variants (Maya et al., 2013).

A question that remains to be resolved is the role of the extensive spread of particular mutant viruses which result in sweeps of the genomes that carry those mutations. Such sweeps are likely to result in the periodic reduction in the diversity of the viruses in those areas, with the recovery of diversity through gaining individual mutations, or potentially through the occurrence of recombination with other virus strains (Mochizuki et al., 2008; Ohshima and Mochizuki, 2009; Shackelton et al., 2007; Wang et al., 2012). An apparent example of such a sweep occurred with the emergence of the CPV-2a variant in dogs (Parrish et al., 1988, 1991; Stucker et al., 2012). Although it is likely that parvoviruses related to CPV and FPV have been infecting carnivores for possibly millions of years, all of the viruses isolated to date are less than 2% variant in DNA sequence, and the apparent common ancestor of the viruses has been dated to around the 1920s (Allison et al., 2013). The cause of the limited variation of the viruses may therefore be due to the spread of a virus through cats and other carnivore hosts in the early 20th century. Whether the most recent FPV ancestor was a new virus in cats at that time, or was a variant strain that spread during that period and replaced the existing diversity of viruses is not known, but the disease in cats was first reported in the 1920s (Verge and Cristoforoni, 1928).

EPIDEMIOLOGY, CONTROL, VACCINES AND PROTECTION

The epidemiology of the viruses involves fecal–oral transmission, with virus replicating to high titers in the intestine of infected animals, and being shed in very high titers (up to 10^9 infectious units per gram) (Meunier et al., 1985a, 1985b). The virus is resistant in the environment, and in cool moist conditions can likely persist for weeks or months. The infection of animals is through the oronasal route, and appears to involve infection of the tonsils and pharyngeal lymphoid tissues, after which it spreads systemically to the other tissues containing dividing cells (Parrish, 1995). For domestic cats and dogs, in the absence of vaccination, it appears that most or all animals will be infected with wildtype parvoviruses in the first several months of life, and that even indoor animals become infected—most likely through exposure to environmental virus. The host-specific epidemiology of the viruses is still being defined, but wild animals may be exposed to viruses from domestic animals, or to host-specific viruses that are adapted to their own species. The significance of each may depend on the degree of exposure to other hosts, the population density of the specific host animal, and the degree of host adaptation required to infect and spread between individuals of the different hosts.

Control can be achieved with varying degrees of success by reducing exposure of young animals to the virus, as well as through vaccination (Carmichael et al., 1983; De Cramer et al., 2011; Mouzin et al., 2004a, b; O'Brien, 1994). Virus may reach high levels in facilities where there are large numbers of susceptible young animals, and there may be an amplifying effect where inoculation with higher titers of virus results in more severe disease and increased shedding. Control of virus in the environment can be achieved through basic cleaning and decontamination with water and detergent, as well as through inactivation with hypochlorite or hot water (which needs to be above 70°C to inactivate the virus). Isolation of young animals during the period of susceptibility when maternal antibody has waned, but vaccination is not successful, can help to control the infection. Similar approaches can protect wild animals in captivity, but control of infections of animals in the wild is more difficult. Where they have been tested, the modified live vaccines are safe for use in wild animals, although that use is not currently licensed. In areas where there are small numbers of wild animals and many domestic animals, vaccination of the domestic animals may reduce the level of exposure to the wildtype virus, and delay the infection until the wild animals are older and less susceptible to the most severe disease.

Vaccines to parvoviruses are now mostly modified live attenuated viruses, which infect animals when they have low levels of or no maternal immunity, and when the vaccine viruses take, they establish complete protection against infection and disease. The different vaccines may contain different viral variants, including the antigenic variants of the CPV, CPV-2 and CPV-2a, as well as variants with mutations of VP2 residue 426 to the Asp or Glu variants. All of those vaccines provide protective immunity to the animal that is vaccinated against all of the circulating strains. Some cases of infection and disease in animals that have been extensively vaccinated with modified live vaccines have been reported, but those appear to be relatively rare, and the reasons for their apparent susceptibility are not yet understood.

ACKNOWLEDGMENTS

This research has been supported by NIH grants R01 GM080533, R01 AI092571, and R01 AI028385. ABA is supported on an NRSA fellowship (F32AI100545) from the NIH.

REFERENCES

Agbandje, M., McKenna, R., Rossmann, M. G., Strassheim, M. L., & Parrish, C. R. (1993). Structure determination of feline panleukopenia virus empty particles. *Proteins*, *16*, 155–171.

Allison, A. B., Harbison, C. E., Pagan, I., Stucker, K. M., Kaelber, J. T., Brown, J. D., Ruder, M. G., Keel, M. K., Dubovi, E. J., Holmes, E. C., & Parrish, C. R. (2012). Role of multiple hosts in the cross-species transmission and emergence of a pandemic parvovirus. *J. Virol.*, *86*, 865–872.

Allison, A. B., Kohler, D. J., Fox, K. A., Brown, J. D., Gerhold, R. W., Shearn-Bochsler, V. I., Dubovi, E. J., Parrish, C. R., & Holmes, E. C. (2013). Frequent cross-species transmission of parvoviruses among diverse carnivore hosts. *J. Virol.*, *87*, 2342–2347.

Bouillant, A., & Hanson, R. P. (1965a). Epizootiology of mink entertis: I. Stability of the virus in feces exposed to natural environmental factors. *Can. J. Comp. Med. Vet. Sci.*, *29*, 125–128.

Bouillant, A., Lee, V. H., & Hanson, R. P. (1965b). Epizootiology on mink enteritis: II. Musca domestica L as a possible vector of virus. *Can. J. Comp. Med. Vet. Sci.*, *29*, 148–152.

Buonavoglia, C., Martella, V., Pratelli, A., Tempesta, M., Cavalli, A., Buonavoglia, D., Bozzo, G., Elia, G., Decaro, N., & Carmichael, L. (2001). Evidence for evolution of canine parvovirus type 2 in Italy. *J. Gen. Virol.*, *82*, 3021–3025.

Canaan, S., Zadori, Z., Ghomashchi, F., Bollinger, J., Sadilek, M., Moreau, M. E., Tijssen, P., & Gelb, M. H. (2004). Interfacial enzymology of parvovirus phospholipases A2. *J. Biol. Chem.*, *279*, 14502–14508.

Carmichael, L. E., Joubert, J. C., & Pollock, R. V. H. (1983). A modified canine parvovirus vaccine 2. Immune response. *Cornell Vet.*, *73*, 13–29.

Chang, S. F., Sgro, J. Y., & Parrish, C. R. (1992). Multiple amino acids in the capsid structure of canine parvovirus coordinately determine the canine host range and specific antigenic and hem-agglutination properties. *J. Virol.*, *66*. 6858–6567.

Cotmore, S. F. (1996). Parvovirus DNA replication. In M. L. DePamphilis (Ed.), *DNA Replication in Eukaryotic Cells* (pp. 799). Cold Spring Harbor, NY: Cold Spring Harbor Laboratory Press; 1996.

Cotmore, S. F., & Tattersall, P. (2005). Encapsidation of minute virus of mice DNA: aspects of the translocation mechanism revealed by the structure of partially packaged genomes. *Virology*, *336*, 100–112.

De Cramer, K. G., Stylianides, E., & van Vuuren, M. (2011). Efficacy of vaccination at 4 and 6 weeks in the control of canine parvovirus. *Vet. Microbiol.*, *149*, 126–132.

Decaro, N., & Buonavoglia, C. (2011). Canine parvovirus—A review of epidemiological and diagnostic aspects, with emphasis on type 2c. *Vet Microbiol.*, *155*, 1–12.

Demeter, Z., Gal, J., Palade, E. A., & Rusvai, M. (2009). Feline parvovirus infection in an Asian palm civet (*Paradoxurus hermaphroditus*). *Vet. Rec.*, *164*, 213–216.

Dunbar, M. R., Cunningham, M. W., & Roof, J. C. (1998). Seroprevalence of selected disease agents from free-ranging black bears in Florida. *J. Wildl. Dis.*, *34*, 612–619.

Eckenroth, B. E., Steere, A. N., Chasteen, N. D., Everse, S. J., & Mason, A. B. (2011). How the binding of human transferrin primes the transferrin receptor potentiating iron release at endosomal pH. *Proc. Natl. Acad. Sci. USA*, *108*, 13089–13094.

Eizirik, E., Murphy, W. J., Koepfli, K. P., Johnson, W. E., Dragoo, J. W., Wayne, R. K., & O'Brien, S. J. (2010). Pattern and timing of diversification of the mammalian order Carnivora inferred from multiple nuclear gene sequences. *Mol. Phylogenet. Evol.*, *56*, 49–63.

Emerman, M., & Malik, H. S. (2010). Paleovirology—modern consequences of ancient viruses. *PLoS Biol*, *8*, e1000301.

Flynn, J. J., Finarelli, J. A., Zehr, S., Hsu, J., & Nedbal, M. A. (2005). Molecular phylogeny of the carnivora (mammalia): assessing the impact of increased sampling on resolving enigmatic relationships. *Syst. Biol.*, *54*, 317–337.

Frolich, K., Streich, W. J., Fickel, J., Jung, S., Truyen, U., Hentschke, J., Dedek, J., Prager, D., & Latz, N. (2005). Epizootiologic investigations of parvovirus infections in free-ranging carnivores from Germany. *J. Wildl. Dis.*, *41*, 231–235.

Gallo Calderon, M., Wilda, M., Boado, L., Keller, L., Malirat, V., Iglesias, M., Mattion, N., & La Torre, J. (2011). Study of canine parvovirus evolution: comparative analysis of full-length VP2 gene sequences from Argentina and international field strains. *Virus Genes*, *44*, 32–39.

Giannetti, A. M., Snow, P. M., Zak, O., & Bjorkman, P. J. (2003). Mechanism for multiple ligand recognition by the human transferrin receptor. *PLoS Biol*, *1*, E51.

Goodman, L. B., Lyi, S. M., Johnson, N. C., Cifuente, J. O., Hafenstein, S. L., & Parrish, C. R. (2010). Binding site on the transferrin receptor for the parvovirus capsid and effects of altered affinity on cell uptake and infection. *J. Virol.*, *84*, 4969–4978.

Gorham, J. R., & Hartsough, G. R. (1955). Infectious enteritis of mink. *J. Am. Vet. Med. Assoc.*, *126*, 467.

Govindasamy, L., Hueffer, K., Parrish, C. R., & Agbandje-McKenna, M. (2003). Structures of host range-controlling regions of the capsids of canine and feline parvoviruses and mutants. *J. Virol.*, *77*, 12211–12221.

Hafenstein, S., Palermo, L. M., Kostyuchenko, V. A., Xiao, C., Morais, M. C., Nelson, C. D., Bowman, V. D., Battisti, A. J., Chipman, P. R., Parrish, C. R, & Rossmann, M. G, (2007). Asymmetric binding of transferrin receptor to parvovirus capsids. *Proc. Natl. Acad. Sci. USA*, *104*, 6585–6589.

Hafenstein, S., Bowman, V. D., Sun, T., Nelson, C. D., Palermo, L. M., Chipman, P. R., Battisti, A. J., Parrish, C. R., & Rossmann, M. G. (2009). Structural comparison of different antibodies interacting with parvovirus capsids. *J. Virol.*, *83*, 5556–5566.

Harrison, T. M., Mazet, J. K., Holekamp, K. E., Dubovi, E., Engh, A. L., Nelson, K., Van Horn, R. C., & Munson, L. (2004). Antibodies to canine and feline viruses in spotted hyenas (*Crocuta crocuta*) in the Masai Mara National Reserve. *J. Wildl. Dis.*, *40*, 1–10.

Hoare, C. M., DeBouck, P., & Wiseman, A. (1997). Immunogenicity of a low-passage, high-titer modified live canine parvovirus vaccine in pups with maternally derived antibodies. *Vaccine*, *15*, 273–275.

Hoelzer, K., & Parrish, C. R. (2010). The emergence of parvoviruses of carnivores. *Vet. Res.*, *41*, 39.

Hoelzer, K., Shackelton, L. A., Parrish, C. R., & Holmes, E. C. (2008). Phylogenetic analysis reveals the emergence, evolution and dispersal of carnivore parvoviruses. *J. Gen. Virol.*, *89*, 2280–2289.

Hong, C., Decaro, N., Desario, C., Tanner, P., Pardo, M. C., Sanchez, S., Buonavoglia, C., & Saliki, J. T. (2007). Occurrence of canine parvovirus type 2c in the United States. *J. Vet. Diagn. Invest.*, *19*, 535–539.

Hueffer, K., Parker, J. S., Weichert, W. S., Geisel, R. E., Sgro, J. Y., & Parrish, C. R. (2003). The natural host range shift and subsequent evolution of canine parvovirus resulted from virus-specific binding to the canine transferrin receptor. *J. Virol.*, *77*, 1718–1726.

Ikeda, Y., Miyazawa, T., Nakamura, K., Naito, R., Inoshima, Y., Tung, K. C., Lee, W. M., Chen, M. C., Kuo, T. F., Lin, J. A., & Mikami, T. (1999). Serosurvey for selected virus infections of wild carnivores in Taiwan and Vietnam. *J. Wildl. Dis.*, *35*, 578–581.

Ikeda, Y., Mochizuki, M., Naito, R., Nakamura, K., Miyazawa, T., Mikami, T., & Takahashi, E. (2000). Predominance of canine parvovirus (CPV) in unvaccinated cat populations and emergence of new antigenic types of CPVs in cats. *Virology*, *278*, 13–19.

Kaelber, J. T., Demogines, A., Harbison, C. E., Allison, A. B, Goodman, L. B., Ortega, A. N., Sawyer, S. L., & Parrish, C. R. (2012). Evolutionary reconstructions of the transferrin receptor of Caniforms supports canine parvovirus being a re-emerged and not a novel pathogen in dogs. *PLoS Pathog*, *8*, e1002666.

Kapil, S., Cooper, E., Lamm, C., Murray, B., Rezabek, G., Johnston, L., 3rd, Campbell, G., & Johnson, B. (2007). Canine parvovirus types 2c and 2b circulating in North American dogs in 2006 and 2007. *J. Clin. Microbiol.*, *45*, 4044–4047.

Kapoor, A., Simmonds, P., & Lipkin, W. I. (2010). Discovery and characterization of Mammalian endogenous parvoviruses. *J. Virol.*, *84*, 12628–12635.

Katzourakis, A., & Gifford, R. J. (2010). Endogenous viral elements in animal genomes. *PLoS Genet*, *6*, e1001191.

Keele, B. F., Van Heuverswyn, F., Li, Y., Bailes, E., Takehisa, J., Santiago, M. L., Bibollet-Ruche, F., Chen, Y., Wain, L. V., Liegeois, F., Loul, S., Ngole, E. M., Bienvenue, Y., Delaporte, E., Brookfield, J. F., Sharp, P. M., Shaw, G. M., Peeters, M., & Hahn, B. H. (2006). Chimpanzee reservoirs of pandemic and nonpandemic HIV-1. *Science, 313*, 523–526.

Li, L., Shan, T., Wang, C., Cote, C., Kolman, J., Onions, D., Gulland, F. M., & Delwart, E. (2011). The fecal viral flora of California sea lions. *J. Virol., 85*, 9909–9917.

Liu, H., Fu, Y., Xie, J., Cheng, J., Ghabrial, S. A., Li, G., Peng, Y., Yi, X., & Jiang, D. (2011). Widespread endogenization of densoviruses and parvoviruses in animal and human genomes. *J. Virol., 85*, 9863–9876.

Llamas-Saiz, A. L., Agbandje-McKenna, M., Parker, J. S. L., Wahid, A. T. M., Parrish, C. R., & Rossmann, M. G. (1996). Structural analysis of a mutation in canine parvovirus which controls antigenicity and host range. *Virology, 225*, 65–71.

Luck, A. N., & Mason, A. B. (2012). Transferrin-mediated cellular iron delivery. *Curr. Top. Membr., 69*, 3–35.

Lukashov, V. V., & Goudsmit, J. (2001). Evolutionary relationships among parvoviruses: virus-host coevolution among autonomous primate parvoviruses and links between adeno-associated and avian parvoviruses. *J. Virol., 75*, 2729–2740.

Madic, J., Huber, D., & Lugovic, B. (1993). Serologic survey for selected viral and rickettsial agents of brown bears (Ursus arctos) in Croatia. *J. Wildl. Dis., 29*, 572–576.

Marsilio, F., Tiscar, P. G., Gentile, L., Roth, H. U., Boscagli, G., Tempesta, M., & Gatti, A. (1997). Serologic survey for selected viral pathogens in brown bears from Italy. *J. Wildl. Dis., 33*, 304–307.

Matrosovich, M., Stech, J., & Klenk, H. D. (2009). Influenza receptors, polymerase and host range. *Rev. Sci. Tech., 28*, 203–217.

Maya, L., Calleros, L., Francia, L., Hernandez, M., Iraola, G., Panzera, Y., Sosa, K., & Perez, R. (2013). Phylodynamics analysis of canine parvovirus in Uruguay: evidence of two successive invasions by different variants. *Arch. Virol.* in press.

Mech, L. D., Goyal, S. M., Paul, W. J., & Newton, W. E. (2008). Demographic effects of canine parvovirus on a free-ranging wolf population over 30 years. *J. Wildl. Dis., 44*, 824–836.

Meunier, P. C., Cooper, B. J., Appel, M. J., & Slauson, D. O. (1985a). Pathogenesis of canine parvovirus enteritis: the importance of viremia. *Vet. Pathol., 22*, 60–71.

Meunier, P. C., Cooper, B. J., Appel, M. J. G., Lanieu, M. E., & Slauson, D. O. (1985b). Pathogenesis of canine parvovirus enteritis: sequential virus distribution and passive immunization studies. *Vet. Pathol., 22*, 617–624.

Millan, J., Candela, M. G., Palomares, F., Cubero, M. J., Rodriguez, A., Barral, M., de la Fuente, J., Almeria, S., & Leon-Vizcaino, L. (2009). Disease threats to the endangered Iberian lynx (Lynx pardinus). *Vet. J., 182*, 114–124.

Miyazawa, T., Ikeda, Y., Nakamura, K., Naito, R., Mochizuki, M., Tohya, Y., Vu, D., Mikami, T., & Takahashi, E. (1999). Isolation of feline parvovirus from peripheral blood mononuclear cells of cats in northern Vietnam. *Microbiol. Immunol., 43*, 609–612.

Mochizuki, M., Ohshima, T., Une, Y., & Yachi, A. (2008). Recombination between vaccine and field strains of canine parvovirus is revealed by isolation of virus in canine and feline cell cultures. *J. Vet. Med. Sci., 70*, 1305–1314.

Mouzin, D. E., Lorenzen, M. J., Haworth, J. D., & King, V. L. (2004a). Duration of serologic response to five viral antigens in dogs. *J. Am. Vet. Med. Assoc., 224*, 55–60.

Mouzin, D. E., Lorenzen, M. J., Haworth, J. D., & King, V. L. (2004b). Duration of serologic response to three viral antigens in cats. *J. Am. Vet. Med. Assoc., 224*, 61–66.

Nandi, S., Chidri, S., Kumar, M., & Chauhan, R. S. (2010). Occurrence of canine parvovirus type 2c in the dogs with haemorrhagic enteritis in India. *Res. Vet Sci., 88*, 169–171.

Neuvonen, E., Veijalainen, P., & Kangas, J. (1982). Canine parvovirus infection in housed raccoon dogs and foxes in Finland. *Vet. Rec.*, *110*, 448–449.

Nomaguchi, M., Doi, N., Matsumoto, Y., Sakai, Y., Fujiwara, S., & Adachi, A. (2012). Species tropism of HIV-1 modulated by viral accessory proteins. *Front. Microbiol.*, *3*, 267.

Nowak, R. M. (1999). *Carnivora: Family Procyonidae, Walker's Mammals of the World* (6th ed.). Baltimore: Johns Hopkins University Press. 694–704.

Nowak, R. M. (2005). *Walker's Carnivores of the World*. Baltimore, MD: John Hopkins University Press.

Nyakatura, K., & Bininda-Emonds, O. R. (2012). Updating the evolutionary history of Carnivora (Mammalia): a new species-level supertree complete with divergence time estimates. *BMC Biol.*, *10*, 12.

O'Brien, S. E. (1994). Serologic response of pups to the low-passage, modified-live canine parvovirus-2 component in a combination vaccine. *J. Am. Vet. Med. Assoc.*, *204*, 1207–1209.

Ohshima, T., & Mochizuki, M. (2009). Evidence for recombination between feline panleukopenia virus and canine parvovirus type 2. *J. Vet. Med. Sci.*, *71*, 403–408.

Palermo, L. M., Hueffer, K., & Parrish, C. R. (2003). Residues in the apical domain of the feline and canine transferrin receptors control host-specific binding and cell infection of canine and feline parvoviruses. *J. Virol.*, *77*, 8915–8923.

Palermo, L. M., Hafenstein, S. L., & Parrish, C. R. (2006). Purified feline and canine transferrin receptors reveal complex interactions with the capsids of canine and feline parvoviruses that correspond to their host ranges. *J. Virol.*, *80*, 8482–8492.

Parker, J. S., Murphy, W. J., Wang, D., O'Brien, S. J., & Parrish, C. R. (2001). Canine and feline parvoviruses can use human or feline transferrin receptors to bind, enter, and infect cells. *J. Virol.*, *75*, 3896–3902.

Parker, J. S., & Parrish, C. R. (1997). Canine parvovirus host range is determined by the specific conformation of an additional region of the capsid. *J. Virol.*, *71*, 9214–9222.

Parrish, C. R. (1990). Emergence, natural history, and variation of canine, mink, and feline parvoviruses. *Adv. Virus. Res.*, *38*, 403–450.

Parrish, C. R. (1991). Mapping specific functions in the capsid structure of canine parvovirus and feline panleukopenia virus using infectious plasmid clones. *Virology*, *183*, 195–205.

Parrish, C. R. (1995). Pathogenesis of feline panleukopenia virus and canine parvovirus. Baillieres Clin. *Haematol.*, *8*, 57–71.

Parrish, C. R., & Carmichael, L. E. (1983). Antigenic structure and variation of canine parvovirus type-2, feline panleukopenia virus, and mink enteritis virus. *Virology*, *129*, 401–414.

Parrish, C. R., & Kawaoka, Y. (2005). The origins of new pandemic viruses: the acquisition of new host ranges by canine parvovirus and influenza A viruses. *Annu. Rev. Microbiol.*, *59*, 553–586.

Parrish, C. R., Carmichael, L. E., & Antczak, D. F. (1982). Antigenic relationships between canine parvovirus type-2, feline panleukopenia virus and mink enteritis virus using conventional antisera and monoclonal antibodies. *Arch.Virol.*, *72*, 267–278.

Parrish, C. R., O'Connell, P. H., Evermann, J. F., & Carmichael, L. E. (1985). Natural variation of canine parvovirus. *Science*, *230*, 1046–1048.

Parrish, C. R., Have, P., Foreyt, W. J., Evermann, J. F., Senda, M., & Carmichael, L. E. (1988). The global spread and replacement of canine parvovirus strains. *J. Gen. Virol.*, *69*, 1111–1116.

Parrish, C. R., Aquadro, C., Strassheim, M. L., Evermann, J. F., Sgro, J. -Y., & Mohammed, H. (1991). Rapid antigenic-type replacement and DNA sequence evolution of canine parvovirus. *J. Virol.*, *65*, 6544–6552.

Parrish, C. R., Holmes, E. C., Morens, D. M., Park, E. C., Burke, D. S., Calisher, C. H., Laughlin, C. A., Saif, L. J., & Daszak, P. (2008). Cross-species virus transmission and the emergence of new epidemic diseases. *Microbiol. Mol. Biol. Rev.*, *72*, 457–470.

Perez, R., Francia, L., Romero, V., Maya, L., Lopez, I., & Hernandez, M. (2007). First detection of canine parvovirus type 2c in South America. *Vet. Microbiol.*, *124*, 147–152.

Pollock, R. V. H., & Carmichael, L. E. (1982). Maternally derived immunity to canine parvovirus infection: transfer, decline and interference with vaccination. *J. Am. Vet. Med. Assoc.*, *180*, 37–42.

Pratelli, A., Cavalli, A., Normanno, G., De Palma, M. G., Pastorelli, G., Martella, V., & Buonavoglia, C. (2000). Immunization of pups with maternally derived antibodies to canine parvovirus (CPV) using a modified-live variant (CPV-2b). *J. Vet. Med. B Infect. Dis. Vet. Public Health*, *47*, 273–276.

Qin, Q., Loeffler, I. K., Li, M., Tian, K., & Wei, F. (2006). Sequence analysis of a canine parvovirus isolated from a red panda (Ailurus fulgens) in China. *Virus Genes*, *34*, 299–302.

Radoshitzky, S. R., Kuhn, J. H., Spiropoulou, C. F., Albarino, C. G., Nguyen, D. P., Salazar-Bravo, J., Dorfman, T., Lee, A. S., Wang, E., Ross, S. R., Choe, H., & Farzan, M. (2008). Receptor determinants of zoonotic transmission of New World hemorrhagic fever arenaviruses. *Proc. Natl. Acad. Sci. USA*, *105*, 2664–2669.

Ross, S. R., Schofield, J. J., Farr, C. J., & Bucan, M. (2002). Mouse transferrin receptor 1 is the cell entry receptor for mouse mammary tumor virus. *Proc. Natl. Acad. USA. Sci. USA*, *99*, 12386–12390.

Shackelton, L. A., Parrish, C. R., Truyen, U., & Holmes, E. C. (2005). High rate of viral evolution associated with the emergence of carnivore parvovirus. *Proc. Natl. Acad. USA. Sci.*, *102*, 379–384.

Shackelton, L. A., Hoelzer, K., Parrish, C. R., & Holmes, E. C. (2007). Comparative analysis reveals frequent recombination in the parvoviruses. *J. Gen. Virol.*, *88*, 3294–3301.

Siedek, E. M., Schmidt, H., Sture, G. H., & Raue, R. (2011). Vaccination with canine parvovirus type 2 (CPV-2) protects against challenge with virulent CPV-2b and CPV-2c. *Berl. Munch. Tierarztl. Wochenschr.*, *124*, 58–64.

Simpson, A. A., Chandrasekar, V., Hebert, B., Sullivan, G. M., Rossmann, M. G., & Parrish, C. R. (2000). Host range and variability of calcium binding by surface loops in the capsids of canine and feline parvoviruses. *J. Mol. Biol.*, *300*, 597–610.

Simpson, A. A., Hebert, B., Sullivan, G. M., Parrish, C. R., Zadori, Z., Tijssen, P., & Rossmann, M. G. (2002). The structure of porcine parvovirus: comparison with related viruses. *J. Mol. Biol.*, *315*, 1189–1198.

Soma, T., Taharaguchi, S., Ohinata, T., Ishii, H., & Hara, M. (2013). Analysis of the VP2 protein gene of canine parvovirus strains from affected dogs in Japan. *Res. Vet. Sci.*, *94*, 368–371.

Steere, A. N., Byrne, S. L., Chasteen, N. D., & Mason, A. B. (2012). Kinetics of iron release from transferrin bound to the transferrin receptor at endosomal pH. *Biochim. Biophys. Acta.*, *1820*, 326–333.

Steinel, A., Munson, L., van Vuuren, M., & Truyen, U. (2000). Genetic characterization of feline parvovirus sequences from various carnivores. *J. Gen. Virol.*, *81*, 345–350.

Strassheim, L. S., Gruenberg, A., Veijalainen, P., Sgro, J. -Y., & Parrish, C. R. (1994). Two dominant neutralizing antigenic determinants of canine parvovirus are found on the threefold spike of the virus capsid. *Virology*, *198*, 175–184.

Stucker, K. M., Pagan, I., Cifuente, J. O., Kaelber, J. T., Lillie, T. D., Hafenstein, S., Holmes, E. C., & Parrish, C. R. (2012). The role of evolutionary intermediates in the host adaptation of canine parvovirus. *J. Virol.*, *86*, 1514–1521.

Thomas, N. J., Foreyt, W. J., Evermann, J. F., Windberg, L. A., & Knowlton, F. F. (1984). Seroprevalence of canine parvovirus in wild coyotes from Texas, Utah, and Idaho (1972 to 1983). *J. Am. Vet. Med. Assoc.*, *185*, 1283–1287.

Tijssen, P., Agbandje-McKenna, M., Almendral, J. M., Bergoin, M., Flegal, T. W., Hedman, K., Kleinschmidt, J., Li, Y., Pintel, D. J., & Tattersall, P. (2012). Family Parvoviridae. In A. M. Q. King, M. J. Adams, E. B. Carstens & E. J. Lefkowitz (Eds.), *Virus Taxonomy: Ninth Report of the International Committee on Taxonomy of Viruses* (pp. 405–425). San Diego, CA: Elsevier Academic Press.

Truyen, U., Gruenberg, A., Chang, S. F., Obermaier, B., Veijalainen, P., & Parrish, C. R. (1995). Evolution of the feline-subgroup parvoviruses and the control of canine host range in vivo. *J. Virol, 69*, 4702–4710.

Veijalainen, P. (1988). Characterization of biological and antigenic properties of raccoon dog and blue fox parvoviruses: a monoclonal antibody study. *Vet. Micro., 16*, 219–230.

Verge, J., & Cristoforoni, N. (1928). La gastro enteritis infectieuse des chats; est-elle due à un virus filterable? *Compt. Rend. Soc. Biol., 99*, 312.

Vihinen-Ranta, M., Wang, D., Weichert, W. S., & Parrish, C. R. (2002). The VP1 N-terminal sequence of canine parvovirus affects nuclear transport of capsids and efficient cell infection. *J. Virol., 76*, 1884–1891.

Wang, J., Cheng, S., Yi, L., Cheng, Y., Yang, S., Xu, H., Zhao, H., Yan, X., & Wu, H. (2012). Evidence for natural recombination between mink enteritis virus and canine parvovirus. *Virol. J., 9*, 252.

Wozencraft, W. C. (2005). Order Carnivora. In D. E. Wilson & D. M. Reeder (Eds.), *Mammal species of the world: a taxonomic and geographic reference* (pp. 532–628). Baltimore, MD: The Johns Hopkins University Press.

Zadori, Z., Szelei, J., Lacoste, M. -C., Raymond, P., Allaire, M., Nabi, I. R., & Tijssen, P. (2001). A viral phospholipase A2 is required for parvovirus infectivity. *Developmental Cell, 1*, 291–302.

Zarnke, R. L., & Ballard, W. B. (1987). Serological survey for selected microbial pathogens of wolves in Alaska, 1975-1982. *J. Wildl. Dis., 23*, 77–85.

Chapter 4

Rabies

Animal Reservoirs of an Ancient Disease

Conrad Freuling[1], Ad Vos[2], Nicholas Johnson[3], Ralf-Udo Mühle[4] and Thomas Müller[1]

[1]*Institute of Molecular Biology, Friedrich-Loeffler-Institut, Federal Research Institute for Animal Health, Greifswald-Insel Riems, Germany,* [2]*IDT Biologika GmbH, Dessau-Rosslau, Germany,* [3]*Animal Health and Veterinary Laboratories Agency, Surrey, United Kingdom,* [4]*Institute for Biochemistry and Biology, University of Potsdam, Potsdam, Germany*

INTRODUCTION

Rabies is a zoonotic disease known to mankind since antiquity. It is caused by lyssaviruses of the *Rhabdoviridae* family in the order *Mononegavirales*. The *Lyssavirus* genus contains a small but growing number of viruses that have a similar genomic structure (Marston et al., 2007). Recently our understanding of lyssavirus epidemiology has fundamentally changed thanks to advancements in virus detection, characterization and phylogeny indicating that it is much more complex than originally believed (Banyard et al., 2011). Within the past few decades, in addition to classical Rabies Virus (RABV) the causative agent of the vast majority of human infections, 14 novel and one putative lyssavirus species have been detected around the globe with distinct geographic and host range distributions. These include Lagos bat virus (LBV), Mokola virus (MOKV), Duvenhage virus (DUVV), Shimoni bat virus (SHIV) and Ikoma virus (IKOV) in Africa, European bat lyssavirus type 1 and 2 (EBLV-1, -2), West Caucasian bat Virus (WCBV), Bokeloh bat Lyssavirus (BBLV), and Lleida bat lyssavirus (LBLV) in Europe, Irkut virus (IRKV), Aravan virus (ARAV) and Khujand virus (KHUV) in Asia, and Australian bat lyssavirus (ABLV) in Australia (Kuzmin et al., 2010; Freuling et al., 2011; ICTV, 2012; Aréchiga Ceballos et al., 2012; Marston et al., 2012). All with the exception of MOKV and IKOV are associated with bats and all are neurotropic, predominantly infecting neurons, with the brain and spinal cord being the most affected organs during infection.

As a member of the Mononegavirales, they have a negative sense single-stranded RNA genome, which codes for five genes in the order nucleoprotein (N), phosphoprotein (P), matrix (M), glycoprotein (G) and polymerase (L). Each gene

The Role of Animals in Emerging Viral Diseases. http://dx.doi.org/10.1016/B978-0-12-405191-1.00004-1

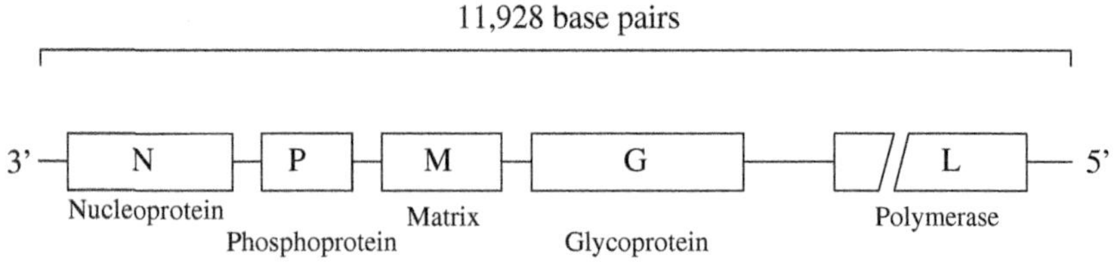

11,928 base pairs

FIGURE 4.1   Schematic view of the Lyssavirus genome.

is transcribed separately with signals encoded within the intergenic regions that initiate and stop RNA transcription (Figure 4.1). The virion particle is enveloped by a host derived membrane with only the virus glycoprotein protruding from it. This is responsible for binding to host cell receptors on target neurons. The glycoprotein is also the main target for neutralizing antibodies produced following vaccination. Following infection, the virus enters the peripheral nervous system and ascends to the nerve bodies in the dorsal root ganglia where replication takes place. The virus then tracks through axons of the central nervous system, eventually reaching the brain and causing widespread infection of neurons. At this point, disease signs develop into fatal encephalitis. The only available treatment is post-exposure vaccination and rabies immunoglobulin that must be given shortly after exposure to virus. In the absence of prompt post-exposure prophylaxis (PEP), patients that become infected invariably die shortly after the development of disease.

While bats represent the reservoirs for almost all lyssavirus species, human rabies is predominately associated with infection with classical rabies virus (RABV). Besides bats in the Americas, the reservoirs for RABV are in domestic and wild canids, most notably domestic dogs, and viverrid mammals worldwide. The exception to this is Antarctica that lacks representatives of these species. Dog-mediated rabies is also referred to as urban rabies, although rabies is also very common in rural areas in Africa and Asia. Sylvatic rabies, a term used synonymously for wildlife rabies, is found widely. Numerous carnivorous species act as independent reservoirs for RABV in different parts of the world (Table 4.1). Molecular phylogeny has revealed that all harbor distinct virus variants that are co-evolving with the reservoir host. Lyssavirus phylogeny implies at least two ancient spill-over events, both within the species RABV resulting in the known RABV variants and multispecies reservoirs throughout the world (Badrane and Tordo, 2001). Epidemiological data on present RABV epizootics indicate that 70–80% of all rabies cases are detected in their respective carnivore reservoir species. The remaining 20–30% of rabies cases are identified in nonreservoir species including humans suggesting a high spill-over rate from reservoir to nonreservoir mammals. In contrast, the spill-over rate of bat-associated lyssaviruses is considerably lower.

As a prime example of infectious zoonoses, rabies has shadowed mankind since antiquity, posing a severe public health risk. Transmission of lyssaviruses occurs through exposure to infectious saliva, from bites and scratches

TABLE 4.1 Examples of Sylvatic Reservoirs of Rabies Virus

Common name	Latin name	Continental Distribution
Red fox	Vulpes vulpes	North America, Europe, Asia
Arctic fox	Alopex lagopus	Arctic
Gray fox	Urocyon cinereoargenteus	North America
Raccoon	Procyon lotor	North America
Striped skunk	Mephitis mephitis	North America
Coyote	Canis latrans	North America
Raccoon dog	Nyctereutes procyonoides	Asia, Europe
Black-backed jackal	Canis mesomelas	Southern Africa
Small Indian mongoose	Herpestes edwardsii	Caribbean
Yellow mongoose	Cynictis penicillata	Southern Africa
Slender mongoose	Galerella sangoinea	Southern Africa

or contact with broken skin. The incubation period ranges between 2 and 3 months (2 weeks to 6 years have been reported) depending on the site where a bite was inflicted, the amount of virus and the virus species or variant. Dogs (Figure 4.2A) are the key animal reservoir responsible for transmission of rabies to humans. Currently, especially in the developing countries of Africa and Asia, virus transmissions via dog bites still cause thousands of human deaths per year, of which the majority are children. Critically, animals are the source of infection of humans and changes in the biology of reservoir species has a direct influence on the number of human rabies cases but may also alter virus evolution, spread and diversity. There is strong evidence that human sociocultural evolution and population growth, related activities and interventions, including the domestication of dogs in particular, have substantially contributed to the spread of RABV and its present-day global distribution. For several millennia the breeding and geographic distribution of dogs depended on man (Verginelli et al , 2005). The dog, with its origin in East Asia, is the only domesticated species that was distributed across both Eurasia and the Americas before transoceanic travel began during the 15th century (Leonard et al., 2002, Savolainen et al., 2002). It is not known how widespread rabies was in prehistoric times. Historical documents, however, provide evidence for the presence of rabies in the Mediterranean Basin and China for millennia (Neville, 2004, Wu et al., 2009). It is speculated that RABV had spread throughout the world during human colonization (Nel & Rupprecht, 2007). Modern molecular characterizations of isolates have provided evidence for the circulation of a so-called cosmopolitan RABV lineage that may

FIGURE 4.2   Photographs of a range of rabies reservoir species including the domestic dog (A), the striped skunk (B), the raccoon dog (C) and the red fox (D).

have its origin in the colonial dispersal of rabies infected dogs. Whether the occurrence of rabies in the New World was introduced from the Old World as a consequence of colonization or if it was already present in pre-Columbian times is still debated (Vos et al., 2011). However, the absence of ancient North and South American dog haplotypes from a large diversity of modern breeds illustrates the considerable impact that invading Europeans had on native cultures (Leonard et al., 2002). In contrast in Africa, particularly in sub-Saharan Africa, rabies appeared to have been absent prior to colonization (Brown, 2011). In fact, recent analyses from the continent provide further evidence that the establishment and intensification of travel and trade routes between African countries following colonization and during the first half of the 20th century have been accompanied by the spread of rabies in dogs across a large part of West/central Africa (Talbi et al., 2009). In the following sections we will attempt to explain how human sociocultural evolution as well as population growth and human related activities—e.g., the introduction of alien wildlife species, livestock, the translocation of domestic and wild animals, urbanization or climate change—have influenced or will affect the (re)emergence of rabies and its epidemiology and control. Also we consider how this has or will put human populations at risk of exposure to rabies.

HUMAN TRANSLOCATION OF DOMESTIC RABID ANIMALS

Human action has had a profound effect on the transport of rabies at a local and transcontinental level, particularly to those areas that were previously not affected by rabies in dogs. Rabies virus in carnivores is now found on virtually every continent although this was not always the case. The key event in the ascension of rabies in many parts of the world was the transport of dogs incubating rabies during the European colonial period, introducing the virus into previously disease-free regions. Human mediated movement of dogs has been implicated with the introduction of dog-rabies into the Americas resulting from the Spanish discovery of the New World (Vos et al., 2011) and into Africa during the colonial era (Smith et al., 1992). This makes the assumption that Europe was the source of the most recent common ancestor (MRCA) of the cosmopolitan strain. It is likely that the presence of rabies virus in many North American bat species is unrelated to the introduction of rabies in dogs. The explanation for this distribution of RABV in the New World remains to be fully explained if, as some researchers have proposed, the virus originated in bats (Badrane and Tordo, 2001). The association of almost all other lyssaviruses with bat species support this hypothesis.

The extended incubation period from exposure to disease development and virus shedding is measured in weeks to months for rabies so that in the absence of appropriate vaccination or border controls, translocation of asymptomatic but infected animals is a major concern. Precautionary measures such as quarantine and animal vaccination have been used to prevent such events. To emphasize the contribution of humans to the spread of rabies, road distance has been shown to be a strong predictor for the movement of rabies within North Africa (Talbi et al., 2010). This is combined with the ability of dogs to spread disease over surprisingly long distances, creating waves of infection (Hampson et al., 2007). The implication is that human-mediated spread of the vector is making a major contribution to disease dispersal. Currently the problem of human movement of infected animals continues in many countries that have strived to achieve "rabies-free status," often through extensive oral vaccination of wildlife. Therefore the World Organization for Animal Health (OIE) requires strict measures before movement of pet animals between countries, and additional measures are required for certain countries and regions. Six-month quarantine was historically used to detect infected animals, reflecting the long incubation period before the development of overt disease, prior to release into the receiving country. However, this has largely been replaced by antirabies vaccination and a check for virus-neutralizing antibodies. Both measures were greatly effective in avoiding rabies introduction.

Achieving rabies-free status inevitably leads to relaxation of disease control measures as countries reap the economic and health benefits of freedom from disease. For rabies, disease elimination leads to removal of the need to vaccinate indigenous pets, a reduction in wildlife vaccination campaigns, usually restricted to border areas, diminishing awareness of disease signs and a scaling down of diagnostic capacity for humans and animals. The development of road and air travel in combination with greater movement of people for recreational purposes has meant that apparently healthy animals can be rapidly imported into a rabies-free country with the potential to cause an epidemic in the large number of unvaccinated susceptible animals in that country. An ongoing example affecting the European Union (EU) is the human-mediated movement of rabies-infected dogs from Morocco into the EU (Johnson et al., 2011a). Rabies is a notifiable disease in the EU and control is supported by legislation, which includes extensive rules for excluding animals that do not comply with measures such as rabies vaccination. However, through a combination of increased tourism to Morocco, where rabies is endemic, and the ease with which cars are ferried across the Straits of Gibraltar, there are numerous examples of dogs being brought into Europe that subsequently develop rabies (Table 4.2). The most common route is by road although surprisingly, in an age of increased security in response to terrorism, there are even examples of dogs being brought into the EU illegally as hand luggage on commercial airline flights. Next to tourism, there is increasing evidence of an illegal trade in juvenile dogs with movement from Eastern Europe into the EU using falsified or manipulated documents. The criminal energy and professionalism behind this make border controls even more difficult, and it is most likely that the cases seen so far are the tip of the iceberg (Tietjen et al., 2011).

TABLE 4.2 Incidence of Transport of Rabid Dogs from Morocco to France, Belgium and Germany, 2001–2011

Year	Country	Age	Route	Reference
2001	France	Juvenile	Road (via Spain)	Bruyere-Masson et al., 2001
2002	France	Juvenile	?	Rabies Bulletin Europe
2004	France	Adult	Road (via Spain)	Servas et al., 2005
2004	France	Juvenile	?	Rabies Bulletin Europe
2004	France	Adult	?	Rabies Bulletin Europe
2004	Germany	?	Air (hand luggage)	Rabies Bulletin Europe
2007	Belgium	?	Air (hand luggage)	Le Roux et al., 2008
2008	France	Adult	Road (via Spain & Portugal)	Anon, 2008
2008	France	Juvenile	Road (via Spain)	Reported by OIE
2012	France		Road (via Spain)	Mailles et al., 2011

?: No data

There are both public health and economic consequences associated with the introduction of a rabid dog into a rabies-free country. Firstly, direct contact with a rabid animal could lead to human and animal infection that in the case of the latter could lead to disease spread. Contact tracing is required to ensure that those that have been in contact with a rabid animal receive vaccination and animals that may have been infected by the animal are treated appropriately. In a single incident of rabies introduction into France, 187 PEP treatments were required as a result of contact tracing (Servas et al., 2005). If the disease does spread most countries will instigate a control program, usually through dog control and vaccination. This in turn brings a financial cost to the affected country. In addition to direct measures to control disease, there is likely to be increased demand for diagnostic testing, provision of PEP and the costs associated with animal vaccination campaigns. In a further example, as part of an international animal rescue operation a number of dogs and two cats were transported to the United States from Iraq, without valid rabies vaccination certificates. One of the dogs became ill, was eventually destroyed and tested positive for rabies virus. All others were advised to be vaccinated and placed in quarantine (CDC, 2008). A similar incident occurred in the UK following transportation by an animal welfare charity of a juvenile dog from Sri Lanka. The dog died shortly after arrival in a quarantine facility without showing disease signs consistent with rabies, but was shown to be infected with the virus (Johnson et al., 2011b). By its nature, quarantine provides a controlled environment that restricts animal movement and should contain disease. However, as these incidents illustrate, robust measures need to be implemented to ensure that persons in contact with rabid animals are vaccinated and other animals within the quarantine facility need to be investigated for possible contact with the rabid animal that may have led to infection.

In other parts of the world the introduction of rabid dogs has sparked rabies epidemics with severe impact on public health including hundreds of human deaths. For instance, the previously rabies-free Indonesian island of Flores experienced a canine rabies outbreak following the importation of three dogs from rabies endemic Sulawesi. Even massive culling of the dog population, without an intensive vaccination campaign of the survivors, was not sufficient to end this epidemic (Windiyaningsih et al., 2004). As a consequence of this outbreak, a rabid dog was brought to another island, the tourist center of Bali, where a further rabies epidemic ensued causing deaths and the administration of tens of thousands of PEPs (Clifton, 2010). This "island hopping" of dogs in the Southeast Asian region forms a continuous risk for the re-emergence of dog rabies on rabies-free(d) islands.

Human-mediated movement of rabid animals has also been implicated in shaping the strain variation observed in the rabies viruses present in Ghana (Hayman et al., 2011). Three distinct variants were present within the dog population of the capital Accra when a cross-sectional survey was conducted on rabies viruses collected over a two-year period. Phylogenetic analysis linked these three lineages to viruses in Benin, Niger and the Ivory Coast, all countries in proximity to Ghana although separated by hundreds of miles. It is hypothesized that these viruses have been introduced into Ghana by humans crossing the borders from neighboring countries bringing with them infected dogs.

HUMAN TRANSLOCATION OF RABID WILDLIFE

In addition to the translocation of domestic animals, translocation of wild animals by humans can also lead to the long-distance spread of rabies. The translocation of wild animals, defined as the capture and transfer of a wild animal from one area to another, plays an important role in the management of wildlife. This is particularly true in the United States where hundreds of thousands of individual animals, usually representing many common species, are intentionally moved across the landscape each year for the purpose of recreational hunting and land management (reviewed in Chipman et al., 2008).

The most striking example of rabies translocation was the introduction of rabid raccoons (*Procyon lotor*) from the American state of Florida, where rabies was endemic within the native population, to a number of northern states that were free of disease. During the 1970s the commercial movement of wild-caught raccoons was common within the United States for the purposes of hunting. In 1979, two raccoons transported from Florida to North Carolina were shown to be rabid (Nettles et al., 1979). It is believed that an incident such as this led to the release of rabid animals in the state of West Virginia (Figure 4.3), which in turn triggered a spreading epizootic that was eventually detected on the American-Canadian border near the town of Prescott, Ontario in 1999 (Wandeler and Salsberg, 1999). An oral rabies vaccination zone was established mainly along the Appalachian Ridge to prevent raccoon rabies moving westward and northward. However, the expense of this intervention is still considered a cost-saving measure as the negative effect of raccoon rabies spread is estimated to be more costly than no action (Slate et al., 2005). Through a combination of human activity and natural spread, the disease is now endemic throughout the raccoon population of the eastern United States and requires continued resources to prevent further spread West through the establishment of this oral vaccine barrier, and the cost of administering PEP to those bitten by raccoons. Another example is the translocation of coyotes from South Texas to a fenced fox pen in Florida in 1994 (Anon, 1995). Five hounds used for hunting coyotes in the pen were eventually confirmed to be infected with canine variant of rabies virus which was circulating in coyotes in Texas. The pen was depopulated and no further cases were documented (Chipman et al, 2007).

A special form of wildlife translocation is the (un)intentional release of animal species beyond their original distribution area. Such nonindigenous animal species, sometimes also referred to as invasive or alien species, can have a profound effect on the local ecosystems including the transmission of infectious diseases. For example, the small Indian mongoose (*Herpestes auropunctatus*)

FIGURE 4.3   **Spread of raccoon (*Procyon lotor*) rabies through Eastern North America. Long-distance human mediated translocation (mid-1970s) is shown by the red arrow. Natural dispersal of rabies by the raccoon population is shown by the blue arrow.**

can be found on at least 29 Caribbean islands and forms a significant, if not primary, reservoir host for rabies on several of these islands: Grenada, Cuba, Dominican Republic (and Haiti by extension) and Puerto Rico (Slate, 2011). The first major outbreak among mongoose in this region was reported from Puerto Rico in 1950 (Tierkel et al., 1952). Molecular analysis indicates close evolutionary ties to the regional rabies strains and suggests that mongoose rabies emerged from dog strains on these Caribbean islands (Smith et al., 1992, Nadin-Davis et al., 2006). The small Indian mongoose, a native to southern Asia, was introduced to these Caribbean islands in the late 19th and early 20th century, and to many other islands in the Pacific Ocean, the Indian Ocean and Adriatic Sea. Why there is sustained transmission among the Indian mongoose only on these four islands in the Caribbean Sea is not fully understood. For example, the small Indian mongoose has spread southwards from the coastal zone of mainland Croatia along the entire coast in Montenegro as well as deep inland in Bosnia Herzegovina (Cirovic et al., 2011). This region is heavily affected by fox rabies but so far no rabies case has been reported in mongoose in this area (WHO Rabies Bulletin Europe). Presumably the high population density on the affected islands in the Caribbean favors successful perpetuation of the infection. Also, high numbers of seropositive animals, as found in Grenada, appear to be indicative for a high infection rate combined with high frequency of biting among mongoose (Everard et al., 1974). Rabies control on these four Caribbean islands using population reduction was practiced, but was without sustained success. Oral vaccination of mongoose seems to offer the best promise of an alternative approach to the elimination of rabies in these island environments (Slate, 2011). Although efficacious vaccine candidates have been tested in the small Indian mongoose (Blanton et al., 2006), mongoose-specific vaccine baits that would be attractive to this species are still in the developmental stage.

Another example of an invasive species with reservoir potential is the raccoon (*Procyon lotor*), the most frequently reported rabid animal species in the USA (Blanton et al., 2011). Since their first successful introduction in Germany in 1934, it can be found throughout Central Europe. In Germany, extremely high population densities, reaching approximately 100 animals/km², have been reported, especially in (semi-) urban areas. However, during the last European fox rabies epizootic, rabid raccoons were rarely reported in Germany, probably resulting from spill-over infections from foxes. Most likely the population density of raccoons was at that time still too low for the maintenance of an independent transmission cycle. Other factors such as low susceptibility to the fox rabies (RABV) virus variant and low genetic diversity among German raccoons could also have played a role (Vos et al., 2012).

In contrast to the raccoon, another invasive species in Europe, the raccoon dog (*Nyctereutes procyonoides*), has proven to be a more effective rabies reservoir host (Figure 4.2C). It is now the second most affected wildlife species in rabies endemic regions in Europe (WHO Rabies Bulletin Europe). Raccoon dogs can be divided into several subspecies with a natural range covering large parts of China, Korea, eastern Siberia, Mongolia and Japan (Kauhala and Kowalczyk, 2011). From the 1920s, thousands of raccoon dogs were introduced as a game species, and were particularly valued for their fur, to the European part of the former Soviet Union where it adapted to the local environment. As a result of its great plasticity in adaptation to various climatic and environmental conditions *N. procyonoides* has colonized large parts of eastern and central Europe by secondary expansion (Nowak 1984), and is still spreading (Kauhala and Kowalczyk, 2011). Although the red fox (Figure 4.2D) is the principal rabies vector in Europe, the raccoon dog plays a significant role in the epidemiology of rabies in the Baltic countries where numbers of infected raccoon dogs can exceed that of foxes (WHO Rabies Bulletin Europe). During a rabies epizootic in Finland in the late 1980s, 77% of the cases identified were in raccoon dogs (Westerling, 1991). Fortunately, rabies wildlife control through oral vaccination is successful in raccoon dogs and has been used in a number of countries (Müller et al., 2012).

With respect to invasive species in Europe, it will be interesting to see whether the golden jackal (*Canis aureus*) becomes a significant rabies host and whether oral rabies vaccines are effective in this species. Previously, only a few rabies cases in Europe were reported from this species (Johnson et al., 2006). However, this animal appears to be expanding its distribution area by colonizing new habitats from the Balkan Peninsula into Central Europe, with evidence of reproduction in Italy and Austria, and sightings in Germany, Slovakia, and the Czech Republic (Arnold et al., 2012).

TRANSLOCATION BY ANIMAL MIGRATION

Many animal species migrate seasonally and this phenomenon has also been identified as one of the mechanisms for rapid spread of infectious diseases. An important example is the long-distance translocation of viruses, including avian influenza, West Nile virus and Sindbis virus, by migrating birds (Hubálek, 2004). This has shaped the dissemination of these viruses, influencing their evolution and led to explosive outbreaks of disease (Rappole and Hubálek, 2003).

Long-distance seasonal migration has so far received little attention in the re-emergence of rabies. Several bat species are known to migrate annually over long distances. For example, the straw-colored fruit bat (*Eidolon helvum*) is broadly distributed across the lowland rainforest and savannah zones of Africa. The species is frugivorous and needs to undergo long-distance migration to follow fruit ripening across Central Africa. Seasonal congregations in huge roosts have been described from many different African countries (Thomas, 1983) although the exact migration routes have not been tracked. Annually, at the end of the year an estimated 5–10 million *E. helvum* arrive at Kasanka National Park in Zambia (Richter & Cumming, 2006). There is evidence that Lagos bat virus (LBV), a virus closely related to rabies virus, circulates in this bat species (Dzikwi et al., 2010; Hayman et al., 2012). Isolations of this virus have been made from a number of African frugivorous bats including *Roussettus aegyptiacus, Epomorphorus wahlbergi* and *E. helvum*. The virus has been responsible for the death of a cat in South Africa (Markotter et al., 2008), reflecting the possible predation on bats by cats and creating a potential bridge vector that could lead to human infection. This has occurred for a number of bat associated lyssaviruses (Johnson et al., 2010a). The enormous aggregations of these bats together with long-distance movements could offer ample opportunities for spread of this virus throughout the distribution range of this bat species across much of sub-Saharan Africa.

Long distance movements of solitary living mammals can be instrumental in the spread of rabies. An example of this has been the repeated outbreaks of rabies on the normally rabies-free Svalbard Islands in the High Arctic. During the winter months, sea ice enables wildlife to reach the islands from the European and American continental land masses. Rabies outbreaks within the indigenous mammals, particularly reindeer (*Rangifer tarandus*) and red fox, have been reported in 1980, 1987, 1990 (reviewed by Mørk and Prestrud, 2004) and most recently in 2011 (Ørpetveit et al., 2011). The most recent rabies outbreak on the island of Svalbard was likely due to migration of arctic foxes (*Canis [Alopex] lagopus*) from Russia over sea ice (Macdonald et al., 2011). It is believed that the arctic fox follows polar bears across the ice, scavenging from seal kills and feeding on polar bear excrement that has a high fat content. The incubation period of rabies in infected arctic foxes can be as long as 6 months (Konovalov et al., 1965; Rausch, 1972) and this fox species is among those mammals with the longest migration and foraging distance (Eberhardt & Hanson, 1978, Tarroux et al., 2010). Molecular phylogeny of rabies isolates obtained from these outbreaks indicates that the most likely origin is the Russian mainland (Johnson et al., 2007) and that there is strong evidence for transcontinental movement of rabies isolates between northern Europe and Greenland (Mansfield et al., 2006), mediated by the migrational movement of arctic foxes. Interestingly, projections of climate change in the Arctic region could result in limited pack ice and restrict these long distance movements, thereby preventing the re-emergence of arctic fox rabies on isolated islands like Svalbard.

Although the Gray wolf (*Canis [Lupus] lupus*) is capable of covering enormous distances during dispersal (Wabakken et al., 2007), wolves have not been documented to be involved in long distance transmission of rabies.

WILDLIFE CONSERVATION

While all mammals are susceptible to rabies, the disease can also be transmitted from the natural hosts to other mammalian species, including highly endangered ones. Normally, such spill-over events are restricted to one or few cases but depending on the behavioral ecology of the affected animal species, such spill-over incidences can lead to considerable population losses and, without intervention, to local extinction. The Ethiopian wolf, also named Simien fox (*Canis [Simenia] simensis*), is endemic to Ethiopian Simien and Bale Mountains National Parks, where they mainly prey on grass rats (*Arvicanthis abyssinicus*). The species is considered the world's rarest canid species with fewer than 500 animals worldwide. Major rabies outbreaks as a result of dog-transmitted rabies have occurred in 1992 and 2003 with mortality rates of 75% in the affected population (Sillero-Zubiri et al., 1996; Randall et al., 2004). A recent report has recorded a further detection of rabies virus in Ethiopian wolves submitted for rabies diagnosis (Johnson et al., 2010b). The animals live in packs (2–18 adults) and communally share and defend a territory against conspecifics. Hence, once a member of a pack is infected, the virus can spread rapidly within the pack and lead to local extinction.

Another example of rabies causing extinction in small relict populations of a highly endangered group of living canid species throughout Africa is the African wild dog (*Lycaon pictus*) (Gascoyne et al., 1993a; Hofmeyr et al., 2000). In an approach to interrupt the chain of transmission and prevent further spread of the disease among these endangered species Ethiopian wolves and in some areas African wild dogs were captured and vaccinated against rabies (Randall et al., 2004; Gascoyne et al., 1993b). However, such direct interventions are controversial (Woodroofe, 2001) and other less invasive approaches such as oral vaccination seem more appropriate (Knobel et al., 2002).

LIVESTOCK

Livestock are not considered a reservoir hosts for rabies, hence rabies cases in livestock are self-limiting spill-over infections from another reservoir host, such as dogs or foxes. However, in rabies endemic areas livestock rabies can cause considerable socioeconomic losses and it is still possible for humans to be infected with rabies from contact with rabid livestock. Unfortunately, the lack of reliable data is a major constraint for estimating the economic impact of livestock rabies on local economies. Generally, livestock are not vaccinated against rabies because the potential risk of infection is usually low, making the annual costs of vaccination prohibitive to many livestock keepers. However, during re-emergence of rabies, mass vaccination campaigns of cattle should be considered not only to reduce disease burden but also to prevent further spread. This policy is followed extensively in Mexico in response to cases of rabies in cattle of vampire bat origin. The average incubation period of rabies in cattle is 15 days (Hudson et al., 1996) and therefore transportation of infected livestock incubating the disease can enhance the spatial spread of disease. Sometimes a sudden increase in the number of rabies cases in livestock in a particular area can be used as an indication of (re)emergence of rabies in its true host. During the emergence of fox rabies in the Aegean region of Turkey, a total of 605 cases of rabies were reported in cattle, whereas only 165 cases in foxes were reported between 1998 and 2007 (Vos et al., 2009).

The economic impact of rabies on livestock is highest in many parts of Latin America as a result of bovine paralytic rabies transmitted by the common vampire bat (*Desmodus rotundus*). This species is only found in Latin America, from northern Mexico to Argentina and preferentially feeds on livestock because these are generally larger and tend to remain stationary at night in contrast to indigenous mammals. Intensive observational studies from Argentina have shown that vampire bats prey on the most abundant livestock species in a particular area and are more numerous in cattle-raising than natural ecosystems (Delpietro et al., 1992).

The presence of rabies in the New World prior to the Spanish colonization is uncertain, leading to the possibility that it was introduced from the Old World (Vos et al., 2011). However, the introduction of livestock had a profound effect on the vampire bat population that in turn influenced the (re)emergence of vampire bat transmitted rabies. It has been postulated that following the arrival of the Spanish conquistadors, the introduction of Old World domestic livestock, such as horses, cattle and pigs, enabled the vampire population to grow exponentially. Livestock provided vampire bats with a new, sustained and almost limitless food supply disrupting the ecological balance and causing a substantial growth in numbers of vampire bats (Belwood & Morton, 1991). This profoundly changed the rabies transmission dynamic between vampires and their bite victims. A further consequence of livestock introduction was the dramatic modification of the landscape to accommodate livestock, including destruction of forest, the natural environment for vampire bat roosts, which brought vampire bats into closer proximity to human and livestock habitation.

Repeated attacks by vampire bats, particularly on young animals, can result in the animal becoming weak and anemic, increasing its susceptibility to other diseases. Wounds can become infected and attacked by screw-worm flies (*Cochliomyia hominivorax*). However, rabies is the most severe disease transmitted to livestock by vampire bats. The vast burden of bovine rabies due to vampire bats is found in Brazil with 1321 cases reported to the Pan American Health Organisation in 2002 (Belotto et al., 2005). The general trend for bovine rabies varies across Latin America. In Brazil the trend is currently downwards from a high in 1989 of 7959 cases, perhaps reflecting the success of vampire control methods such as treatment of cattle with anticoagulants and rabies vaccination. However, in countries such as Mexico and Peru the trend is increasing.

Humans are also victims of rabies resulting from vampire bat bites and in contrast to the trend observed in cattle, numbers of human cases of rabies of bat origin are increasing (Schneider et al., 2009). A number of risk factors have been identified that increase the likelihood of a rabies outbreak. Such outbreaks tend to occur in rural settings, remote from public health services, and often when livestock are suddenly removed, for example following control for another livestock disease (McCarthy, 1989). A further factor observed is related to human associated change when, for example, humans enter the rainforest for activities such as logging or mining (Nehaul, 1954). Temporary populations tend to live in poor housing accommodation that does not prevent access by vampire bats. Reports of such outbreaks are becoming increasingly common (Schneider et al., 2001; Badilla, et al., 2003; Gilbert et al., 2012).

The exception that proves the rule that only bats and certain terrestrial carnivores can act as reservoir hosts for rabies virus is provided by the Greater kudu (*Tragelaphus strepsiceros*). This antelope is found throughout eastern and southern Africa and, especially in Namibia, this nondomesticated wildlife species is kept and bred on (fenced) ranches for trophy-hunting, game meat and ecotourism. Here, its economic value exceeds the value of sheep and goats combined (Barnes et al., 2004). As a result of several favorable conditions the kudu density was extremely high in Namibia during the early 1970s. The first case of rabies in a kudu was observed in 1975 near Windhoek, the capital of Namibia (Barnard and Hassel, 1981). In the following years thousands of dead kudus were reported and this first kudu epizootic peaked in 1980 and subsided by 1985 with an estimated loss of 30,000–50,000 animals. Ever since, rabies cycles in kudus have occurred in Namibia (Scott et al., 2012). Although the original spill-over infection was most likely from jackals, it is suggested that the rabies in kudu population is spread through intraspecific non-bite transmission.

The complexity of interrelated factors that eventually lead to the emergence of virus infections has been demonstrated by the declines in the populations of vultures across India. Vulture decline was caused by the widespread use of the nonsteroidal anti-inflammatory drug (NSAID) diclofenac to treat livestock.

Since livestock carcasses provided the main food supply for vultures, but are also eaten by dogs, the dog populations have increased substantially in contrast to the decline in the vulture population. Dogs are the main source of rabies in humans in India and therefore an increase in the dog population results in a larger potential reservoir for rabies virus in the absence of any anthropogenic attempts to control disease. This in turn causes an increase in risk to human health (Markandya et al., 2008). In addition to the increase in rabies within the dog population, there are further consequences for mammalian disease transmission expected as a result of vulture declines (Ogada et al., 2012). Hence a human intervention in livestock has indirectly caused an increase in a zoonotic infection in a domestic species.

CHALLENGES OF URBAN WILDLIFE

For a number of decades urban wildlife has been increasing. The ongoing development at the periphery of cities and rural properties within commuting distance of cities has been causing dramatic changes to the landscape and the alteration of ecology of such areas, and a reduction in biodiversity. With the spread of suburbia, however, come opportunities for some species to exploit new resources (DeStefano and DeGraaf, 2003) including rabies reservoir species, e.g., red foxes and raccoons.

Reports of red foxes colonizing cities in Britain date as early as the 1930s. What was initially thought to be an isolated British phenomenon (Harris, 1977) has recently been experienced by many other countries in continental Europe, North America and even Australia as well (Marks and Bloomfield, 1991; Adkins and Stott, 1998; Gloor et al., 2001). Also, urban and suburban raccoon populations are many times greater than raccoon populations reported from other habitats. The raccoon density in an urban national park in Washington, USA, for example, was shown to range from 333.3 to 66.7 / km², with an overall park estimate of 125 / km² (Riley et al., 1998). Although an alien species in Germany (see above), the North American raccoon is now firmly established in the country, reaching similar population densities as in its ancestral habitat, in particular in the city of Kassel in the federal state of Hess (Vos et al., 2012). The fact that Kassel is not the only city with a raccoon problem in Germany highlights this unstoppable trend. Hence, red foxes, like raccoons, have adapted well to synanthropic life in larger cities (Deplazes et al., 2004). This recent invasion of foxes and raccoons into urban areas represents a considerable health risk with particular regard to rabies (Marks and Bloomfield, 1991; Smith and Harris, 1991; Hegglin et al., 2004; Vos et al., 2012).

What problems would be posed should rabies be introduced to the high-density fox and raccoon populations found in many cities is illustrated by a few examples. In Germany, with the highest density of urban settlements in Europe, the persistence of red fox rabies in such areas was one of the major obstacles to rabies elimination in the final phase of the control program and required enhanced efforts in ORV campaigns to resolve the problem (Müller et al., 2005; Müller et al. 2012). During an outbreak of raccoon rabies in Central Park in New York City, New York, USA, in 2009 a total of 133 rabies positive raccoons were reported, while five persons and two dogs were exposed but did not become infected. Only an elaborate trap-vaccinate-release program similar to that used in metropolitan Toronto (Rosatte et al., 2007) during which about 500 raccoons were vaccinated with an inactivated rabies vaccine contributed to the end of the epizootic (Slavinski et al., 2012).

An introduction of rabies into German urban raccoon populations would confront veterinary authorities with a significant challenge (Vos et al., 2012). Hunting alone would not resolve such an emergency as it is prohibited in urban areas. Although sizeable ORV projects in the USA were somewhat successful in containing raccoon rabies, preventing further spread, reducing incidence of rabies cases, and even nearly eliminating the disease in certain regions (Slate et al., 2005, 2009), they ultimately have not eliminated wildlife rabies in North America. This suggests that ORV developed for the red fox cannot simply be copied for raccoons in Germany for various reasons. Among the factors that need to be considered are: the distribution of baits containing live replication competent virus vaccines in urban areas; a lack of local experience in distributing baits targeted for animals living at extremely high densities; and most importantly, the licensed oral rabies vaccine baits in Germany are unlikely to be suitable or attractive for raccoons (Vos et al., 2012). As urban red foxes and raccoons appear increasingly to use anthropogenic food sources within villages and small towns, any re-introduction of rabies might have additional negative implications to bait uptake in the course of an ORV program in case of emergency (Hegglin et al., 2004). The bait density used in ORV programs in the USA to vaccinate raccoons is already high and is a multiple of what is applied for red foxes in Europe (Slate et al., 2009). Any attempt to combat an incursion of rabies into an urban raccoon population using ORV would require a high bait density. Because of the fragmented landscape, aerial distribution of baits is highly limited. As most urban areas are nonflying zones, baits would have to be distributed by hand. In certain places helicopters could be used but at great cost. This would result in enormous costs for ORV. Hence, additional oral vaccines, improved baits to reach target species, and optimized ORV strategies need to be developed (Slate et al., 2005, 2009; Vos et al., 2012). Furthermore, urban fox and raccoon density data should be updated in the near future in preparation for implementation of control programs (Wilkinson and Smith, 2001).

CLIMATE AND ENVIRONMENTAL CHANGE

While human activities—among others land use, change in land-cover, soil and water pollution and degradation, air pollution, habitat fragmentation, urbanization and introduction of alien species—have already caused and will continue to cause a loss in biodiversity, changes in climate as observed over the 20th century exert additional pressure up to potential extinction of species (Gitay et al., 2002; Thomas et al., 2004). In a public and animal health context, climate exerts both direct and indirect effects on the appearance and spread of many emerging and re-emerging human and animal infectious diseases, including zoonoses. Therefore, the impact of climate change and global warming on the transmission and geographical distribution of diseases has been associated with changes in the replication rate and dissemination of pathogen, vector and animal host populations (OIE, 2010; EASAC, 2010). It is well acknowledged that due to human-induced climate change the composition of most current ecosystems is likely to change in the near future by differently affecting the habitat of many species. It is assumed that this will likely result in different rates of species migration polewards or upwards through fragmented landscape (Gitay et al., 2002; Hickling et al., 2006). Although it is still unclear whether the effects of the suggested climate changes on the distribution of species including reservoir species can be predicted, this process is beginning to be discerned across the world (Araújo and Rahbeck, 2006). Only recently has this development been the focus of attention with regard to rabies.

The North Polar Region in North America and Eurasia are currently considered free of red fox-mediated rabies. Recently, arctic variant rabies in red foxes has been successfully eliminated from Ontario, Canada, using oral vaccination (MacInnes et al., 2001; Rosatte et al., 2007). However, there is increasing evidence that warming temperatures are allowing red foxes to migrate further north into tundra regions around the North Polar Region, where they set about dominating and ousting the smaller Arctic fox (Hersteinsson and MacDonald, 1992). As a result, numbers of Arctic fox are declining in parts of the Arctic. Displacement of Arctic foxes by invading red foxes have been reported from Finnmark, Norway, and St. Matthew Island, Alaska, USA, where they have recently established a permanent presence (Post et al., 2009). This could result in a dramatic spread of red fox-mediated rabies over vast as yet unaffected regions, e.g., in Russia's far North, on the one hand. On the other hand, the risk of sustained spill-overs of arctic variant RABV from Arctic into red foxes and re-incursion into already freed areas or incursion into unaffected areas in North Polar regions could increase. Extreme direct interactions such as direct observations of red fox intrusion on an Arctic fox breeding den and interspecific killing of an Arctic fox by a red fox, as recently reported from Alaska and Russia, respectively (Pamperin et al., 2006; Rodnikova et al., 2011) strongly support the latter hypothesis. And as warming extends to higher elevations, the red fox also will likely occupy higher mountain ranges where it previously did not venture. There is anecdotal evidence that, following a resurgence of fox rabies in northeastern Italy in 2008–2009, fox rabies hotspots in the Alpine mountain ranges were identified above 2000 meters above sea level, probably representing local residual rabies foci which may have contributed to maintaining the infectious cycle in areas not covered by ORV at higher altitudes (Mulatti et al., 2012).

Similar scenarios are imaginable for gray foxes, raccoons, coyotes and skunks (Figure 4.2B) in North America and golden jackals (*Canis aureus*) in Europe. Any further expansion in range of canine wildlife reservoir species would not only have serious public health implications but would also dramatically increase the costs for oral rabies vaccination programs in the future and challenge elimination of reservoir-associated RABV variants and protection of wildlife rabies free areas.

Next to nonvolant rabies reservoir species, climate change will also likely affect the true historical reservoirs for lyssaviruses—bats. The response of vampire bats (*Desmodus rotundus*) to climate change is a special focus of attention. Being an opportunistic species especially sensitive to low temperatures, the common vampire (limited to Mexico, Central and South America) will probably see its range expand dramatically north and southward. Initial results suggest that over the next few decades, the size of this bat's range is expected to increase by at least 260,000 square kilometers to areas in the USA (Mistry and Moreno-Valdez, 2008) and even in the Brazilian highlands, northern Argentina, and east of the Andes in Peru (Lee et al., 2012). This expansion of common vampire bats would present challenges on many fronts: ecological, commercial, medical and educational including probable impacts on other bat species, the livestock industry and public health concerns (Mistry and Moreno-Valdez, 2008). As people and cattle in those regions are likely to have a greater risk of contracting vampire-transmitted rabies (Lee et al., 2012; Schneider et al., 2009) costs for rabies prevention on the human and veterinary side would dramatically increase. In particular, implementing rabies prevention measures in the Amazon region must confront a series of challenges ranging from the remoteness of human populations due to the size of the Amazon basin and the absence of transport infrastructure, to the absence of modern public health care in many regions.

CONCLUDING REMARKS

Given the diversity of lyssavirus reservoirs, a true eradication of rabies is not foreseeable in the future (Rupprecht et al., 2008). In fact, as bats represent the reservoirs for almost all lyssavirus species, and targeted searches for viral diseases including lyssaviruses among bats have only recently been initiated, it can be expected that many other lyssaviruses will be identified in the near future. For some of the recently discovered lyssaviruses no efficacious vaccines are available. However, dogs represent the predominant reservoir and effort must be focused on control of rabies in this species.

While man-made environmental changes have contributed to the spread of this disease, swift action globally is needed to understand the consequences of these developments and modify existing successful strategies for application to rabies control, particularly for emerging vectors in new environments such as raccoons. Great efforts are being made worldwide to reduce the human burden by controlling dog rabies, which has proven successful in the Americas and

Europe. Also fox rabies has been controlled in large parts of western and central Europe and North America. With the existing anti-rabies human biologicals and the progress in animal rabies control, a reduction of human rabies cases to near zero is a matter of commitment, resources and political will. Sustainability of those activities can only be achieved through community engagement at a local level and governmental support at a national and international level. In all instances, control programs will need to take the human sociocultural evolution and population growth, together with other related activities and changing environmental circumstances that also alter virus dispersal, into account.

ACKNOWLEDGMENTS

This chapter was written in the frame of a national research network entitled "Lyssaviruses—a potential public health threat" funded by the German Ministry for Education and Research (BMBF, grant 01KI1016A).

REFERENCES

Anonymous. (1995). Translocation of coyote rabies—Florida, 1994. *Morbidity and Mortality Weekly Report*, *44*, 580–581.

Anonymous. (2008). Identification of a rabid dog in France illegally introduced from Morocco. *Eurosurveillance*, *13*, pi=8066.

Adkins, C. A., & Stott, P. (1998). Home ranges, movements and habitat associations of red foxes *Vulpes vulpes* in suburban Toronto, Ontario, Canada. *Can. J. Zool*, *244*, 335–346.

Araújo, M. B., & Rahbek, C. (2006). Ecology. How does climate change affect biodiversity? *Science*, *313*, 1396–1397.

Aréchiga Ceballos, N., Vázquez Morón, S., Berciano, J. M., Nicolás, O., Aznar, C., Juste, J., & Rodríguez (2012). *Novel lyssavirus in Miniopterus Schreibersii in Spain*. 15th Annual Meeting of the European Society for Clinical Virology, Spain: Madrid, 4 September 2012 – 7 September 2012, http://escv.ivdnews.net/public/show_abstract/1417.

Arnold, J., Humer, A., Heltai, M., Murariu, D., Spassov, N., & Hacklander, K. (2012). Current status and distribution of golden jackals *Canis aureus* in Europe. *Mamm. Rev.*, *42*, 1–11.

Badilla, X., Pérez-Herra, V., Quirós, L., Morice, A., Jiménez, E., Sáenz, E., Salazar, F., Fernández, R., Orciari, L., Yager, P., Whitfield, S., & Rupprecht, C. E. (2003). Human rabies: reemerging disease in Costa Rica? *Emerg. Infect. Dis.*, *9*, 721–723.

Badrane, H., & Tordo, N. (2001). Host switching in Lyssavirus history from the Chiroptera to the Carnivora order. *J. Virol*, *75*, 8096–8104.

Banyard, A. C., Hayman, D., Johnson, N., McElhinney, L., & Fooks, A. R. (2011). Bats and lyssaviruses. *Adv Virus Res.*, *79*, 239–289.

Barnard, B., & Hassel, R. (1981). Rabies in kudus (*Tragelaphus strepsiceros*) in South West Africa/Namibia. *J. S. Afr. Vet. Assoc.*, *52*, 309–314.

Barnes, J. I., Lange, G. M., Nhuleipo, O., Muteyauli, P., Katoma, T., Amapolo, H., Lindeque, P., & Erb, P. (2004). *Preliminary valuation of the wildlife stocks in Namibia: wildlife asset accounts*. Nambia: Directorate of Scientific Services, Ministry of Environment and Tourism. pp. 1–9.

Belotto, A., Leanes, L. F., Schneider, M. C., Tamoyo, H., & Correa, E. (2005). Overview of rabies in the Americas. *Virus Res.*, *111*, 5–12.

Belwood, J. J., & Morton, P. A. (1991). Vampires: The real story. *Bats Magazine*, *9*, 11–16.

Blanton, J. D., Meadows, A., Murphy, S. M., Manangan, J., Hanlon, C. A., Faber, M. L., Dietzschold, B., & Rupprecht, C. E. (2006). Vaccination of small Asian mongoose (*Herpestes javanicus*) against rabies. J Wildl. *Dis.*, *42*, 663–666.

Blanton, J. D., Palmer, D., Dyer, J., & Rupprecht, C. E. (2011). Rabies surveillance in the United States during 2010. *J. Am. Vet. Med. Assoc.*, *239*, 773–783.

Brown, K. (2011). *Mad Dogs and Meerkats: A History of Resurgent Rabies in Southern Africa.* Athens, Ohio: Ohio University Press.

Bruyere-Masson, V., Barrat, J., Cliquet, F., Rotivel, Y., Bourhy, H., Brie, Ph., Melik, N., Gibon, C., & Alvado-Brette, B. (2001). A puppy illegally imported from Morocco brings rabies to France. *Rabies Bulletin Europe, 25,* 12–13.

CDC. (2008). Rabies in a Dog Imported from Iraq—New Jersey, June 2008. *MMWR, 57,* 1076–1078. Available at http://www.cdc.gov/mmwr/preview/mmwrhtml/mm5739a3.htm.

Chipman, R., Slate, D., Rupprecht, C., & Mendoza, M. (2008). Downside risk of wildlife translocation. *Dev. Biol. (Basel), 131,* 223–232.

Cirovic, D., Rakovic, M., Milenkovic, M., & Paunovic, M. (2011). Small Indian mongoose *Herpestes auropunctatus* (Herpestidae, Carnivora): an invasive species in Montenegro. *Biological Invasions, 13,* 393–399.

Clifton, M. (2010). How not to fight a rabies epidemic: a history in Bali. *Asian Biomed, 4,* 663–670.

Delpietro, H. A., Marchevsky, N., & Simonetti, E. (1992). Relative population densities and predation of the common vampire bat (*Desmodus rotundus*) in natural and cattle-raising areas in north-east Argentina. *Prev. Vet. Med., 14,* 13–20.

DeStefano, S., & DeGraaf, R. M. (2003). Exploring the ecology of suburban wildlife. Front Ecol. *Environ., 1,* 95–101.

Deplazes, P., Hegglin, D., Gloor, S., & Romig, T. (2004). Wilderness in the city: the urbanization of *Echinococcus multilocularis. Trends. Parasitol., 20,* 77–84.

Dzikwi, A. A., Kuzmin, I. I., Umoh, J. U., Kwaga, J. K., Ahmad, A. A., & Rupprecht, C. E. (2010). Evidence of Lagos bat virus circulation among Nigerian fruit bats. *J. Wildl. Dis., 46,* 267–271.

EASAC. (2010). *Climate change and infectious diseases in Europe.* London: European Academies Science Advisory Council. 15p.

Eberhardt, L. E., & Hanson, W. C. (1978). Long-distance movements of arctic foxes tagged in northern Alaska. *Can. Field-Nat., 92,* 386–389.

Everard, C. O. R., Baer, G. M., & James, A. (1974). Epidemiology of mongoose rabies in Grenada. *J. Wildl. Dis., 10,* 190–196.

Freuling, C. M., Beer, M., Conraths, F. J., Finke, S., Hoffmann, B., Keller, B., Kliemt, J., Mettenleiter, T. C., Mühlbach, E., Teifke, J. P., Wohlsein, P., & Müller, T. (2011). Novel lyssavirus in Natterer's bat. *Germany. Emerg. Infect. Dis., 17,* 1519–1522.

Gascoyne, S. C., King, A. A., Laurenson, M. K., Borner, M., & Schildger, B. (1993a). Aspects of rabies infection and control in the conservation of the African wild dogs (*Lycaon pictus*) in the Serengeti region, Tanzania. *Onderstepoort J. Vet. Res., 60,* 415–420.

Gascoyne, S. C., Laurenson, M. K., Lelo, S., & Borner, M. (1993b). Rabies in African wild dogs (*Lycaon pictus*) in the Serengeti region, Tanzania. *J. Wildl. Dis., 29,* 396–402.

Gilbert, A. T., Petersen, B. W., Recuenco, S., Niezgoda, M., Gómez, J., Laguna-Torres, V. A., & Rupprecht, C. E. (2012). Evidence of rabies virus exposure among humans in the Peruvian Amazon. *Am. J. Trop. Med. Hyg., 87,* 206–215.

Gitay, H., Suárez, A., Watson, R. T., & Dokken, D. J. (2002). *Climate change and biodiversity.* Intergovernmental Panel on Climate Change, United Nations, IPCC technical paper. 77 p, http://www.ipcc.ch/pub/tpbiodiv.pdf.

Gloor, S., Bontadina, F., Hegglin, D., Deplazes, P., & Breitenmoser, U. (2001). The rise of urban fox populations in Switzerland. *Mamm. Biol.*, *66*, 155–164.

Hampson, K., Dushoff, J., Bingham, J., Brückner, G., Ali, Y. H., & Dobson, A. (2007). Synchronus cycles of domestic dog rabies in sub-Saharan Africa and the impact of control efforts. *Proc. Natl. Acad. Sci. USA.*, *104*, 7717–7722.

Harris, S. (1977). Distribution, habitat utilization and age structure of a suburban fox (*Vulpes vulpes*) population. *Mamm. Rev.*, *7*, 25–39.

Hayman, D. T. S., Johnson, N., Horton, D. L., Hedge, J., Wakeley, P. R., Banyard, A. C., Zhang, S., Alhassan, A., & Fooks, A. R. (2011). Evolutionary history of rabies in Ghana. *PLoS Neg. Trop. Dis.*, *5*, e1001.

Hayman, D. T., Fooks, A. R., Rowcliffe, J. M., McRea, R., Restif, O., Baker, K. S., Horton, D. L., Suu-Ire, R., Cunningham, A. A., & Wood, J. L. (2012). Endemic Lagos Bat virus infection in *Eidolon helvum*. *Epidemiol. Infect.*, *28*, 1–9.

Hegglin, D., Bontadin, F., Gloor, S., Romer, J., Müller, U., Breitenmoser, U., & Deplazes, P. (2004). Baiting red foxes in an urban area: A camera trap study. *J. Wild. Manag.*, *68*, 1010–1017.

Hersteinsson, P., & MacDonald, D. W. (1992). Interspecific competition and the geographical distribution of red and arctic foxes *Vulpes vulpes* and *Alopex lagopus*. *OIKOS*, *64*, 505–515.

Hickling, R., Roy, D. B., Hill, J. K., Fox, R., & Thomas, C. D. (2006). The distributions of a wide range of taxonomic groups are expanding polewards. *Global Change Biology*, *12*, 450–455. http://dx.doi.org/10.1111/j.1365-2486.2006.01116.x.

Hofmeyr, M., Bingham, J., Lane, E. P., Ide, A., & Nel, L. (2000). Rabies in African wild dogs (*Lycaon pictus*) in the Madikwe Game Reserve, South Africa. *Vet. Rec.*, *146*, 50–52.

Hubálek, Z. (2004). An annotated checklist of pathogenic microorganisms associated with migratory birds. *J. Wildl. Dis.*, *40*, 639–659.

Hudson, L. C., Weinstock, D., Jordan, T., & Bold-Fletcher, N. O. (1996). Clinical features of experimentally induced rabies in cattle and sheep. *Zentrbl. Veterinarmed. B.*, *43*, 85–95.

ICTV. (2012). *Taxonomy: Updates Since the 8th Report*. http://talk.ictvonline.org/files/ictv_official_taxonomy_updates_since_the_8th_report/m/vertebrate-official/4188.aspx.

Johnson, N., Fooks, A. R., Valtchovski, R., & Müller, T. (2006). Evidence for trans-border movement of rabies by wildlife reservoirs between countries in the Balkan Peninsular. *Vet. Microbiol.*, *120*, 71–76.

Johnson, N., Dicker, A., Mork, T., Marston, D. A., Fooks, A. R., Tryland, M., Fuglei, E., & Müller, T. (2007). Phylogenetic comparison of rabies virus from disease outbreaks on the Svalbard Islands. *Vector-Borne and Zoonotic Diseases*, *7*, 457–460.

Johnson, N., Vos, A., Freuling, C., Tordo, N., & Fooks, A. R. (2010a). Human rabies due to lyssavirus infection of bat origin. *Vet. Microbiol.*, *142*, 151–159.

Johnson, N., Mansfield, K. L., Marston, D. A., Wilson, C., Goddard, T., Selden, D., Hemson, G., Edea, L., van Kesteren, F., Shiferaw, F., Stewart, A. E., Sillero Zubiri, C., & Fooks, A. R. (2010b). A new outbreak of rabies in rare Ethiopian wolves (*Canis simensis*). *Arch. Virol*, *155*, 1175–1177.

Johnson, N., Freuling, C., Horton, D., Müller, T., & Fooks, A. R. (2011a). Imported rabies, European Union and Switzerland, 2001–2010. *Emerg. Infect. Dis.*, *17*, 753–754.

Johnson, N., Nunez, A., Marston, D. A., Harkess, G., Voller, K., Goddard, T., Hicks, D., McElhinney, L. M., & Fooks, A. R. (2011b). Investigation of an imported case of rabies in a juvenile dog with atypical presentation. *Animals*, *1*, 402–413.

Kauhala, K., & Kowalczyk, R. (2011). Invasion of the raccoon dog *Nyctereutes procyonoides* in Europe: History of colonization, features behind its success, and threats to native fauna. *Curr. Zool.*, *57*, 584–598.

Knobel, D. L., du Toit, J. T., & Bingham, J. (2002). Development of a bait and baiting system for delivery of oral rabies vaccine to free-ranging African wild dogs (*Lycaon pictus*). *J. Wildl. Dis.*, *38*, 352–362.

Konovalov, G. V., Kantorovich, R. A., Buzinov, I. A., & Riutova, V. P. (1965). Experimental investigations into rage and rabies in polar foxes, natural hosts of the infection. II An experimental morphological study of rabies in polar foxes. *Acta. Virol*, *9*, 235–239.

Kuzmin, I. V., Mayer, A. E., Niezgoda, M., Markotter, W., Agwanda, B., Breiman, R. F., & Rupprecht, C. E. (2010). Shimoni bat virus, a new representative of the *Lyssavirus* genus. *Virus Res.*, *149*, 197–210.

Lee, D. N., Papes, M., & Van Den Bussche, R. A. (2012). Present and potential future distribution of common vampire bats in the Americas and the associated risk to cattle. *PLoS ONE*, *7*, e42466.

Leonard, J. A., Wayne, R. K., Wheeler, J., Valadez, R., Guillen, S., & Vila, C. (2002). Ancient DNA evidence for Old World origin of New World dogs. *Science*, *298*, 1613–1616.

Le Roux, I., & Van Gucht, S. (2008). Two cases of imported canine rabies in the Brussels area within six months time. *Rabies Bulletin Europe*, *32*, 5–6.

Macdonald, E., Handeland, K., Blystad, H., Bergsaker, M., Fladberg, M., Gjerset, B., Nilsen, O., Os, H., Sandbu, S., Stokke, E., Vold, L., Ørpetveit, I., Gaup Amot, H., & Tveiten, O. (2011). Public health implications of an outbreak of rabies in arctic foxes and reindeer in the Svalbard archipelago, Norway, September 2011. *Eurosurveillance*, *16*, pii=19985.

MacInnes, C. D., Smith, S. M., Tinline, R. R., Ayers, N. R., Bachmann, P., Ball, D. G., Calder, L. A., Crosgrey, S. J., Fielding, C., Hauschildt, P., Honig, J. M., Johnston, D. H., Lawson, K. F., Nunan, C. P., Pedde, M. A., Pond, B., Stewart, R. B., & Voigt, D. R. (2001). Elimination of rabies from red foxes in eastern Ontario. *J. Wildl. Dis*, *37*, 119–132.

Mailles, A., Boisseleau, D., Dacheux, L., Michalewiscz, C., Gloaguen, C., Ponçon, N., Bourhy, H., Callon, H., Vaillant, V., Dabosville, I., & Morineau-Le Houssine, P. (2011). Rabid dog illegally imported to France from Morocco, August 2011. *Eurosurveillance*, *16*, pii=19946.

Mansfield, K. L., Racloz, V., McElhinney, L. M., Marston, D. A., Johnson, N., Rønsholt, L., Christensen, L. A., Neuvonen, E., Botvinkin, A. D., Rupprecht, C. E., & Fooks, A. R. (2006). Molecular epidemiological study of Arctic rabies virus isolates from Greenland and comparison with isolates from throughout the Arctic and Baltic regions. *Virus Res.*, *116*, 1–10.

Markandya, A., Taylor, T., Longo, A., Murty, M. N., Murty, S., & Dhavala, K. (2008). Counting the cost of vulture decline—An appraisal of the human health and other benefits of vultures in India. *Ecol. Econ.*, *67*, 194–204.

Marks, C. A., & Bloomfield, T. E. (1991). Distribution and density estimates for urban foxes (*Vulpes vulpes*) in Melbourne: implications for rabies control. *Wildlife Research*, *26*, 763–775.

Markotter, W., Kuzmin, I., Rupprecht, C. E., & Nel, L. (2008). Phylogeny of Lagos Bat virus: challenges for lyssavirus taxonomy. *Virus Res.*, *135*, 10–21.

Marston, D. A., McElhinney, L. M., Johnson, N., Müller, T., Conzelmann, K. K., Tordo, N., & Fooks, A. R. (2007). Comparative analysis of the full genome sequence of European bat lyssavirus type 1 and type 2 with other lyssaviruses and evidence for a conserved transcription termination and polyadenylation motif in the G-L 3' non-translated region. *J. Gen. Virol*, *88*, 1302–1314.

Marston, D. A., Ellis, R. J., Horton, D. L., Kuzmin, I. V., Wise, E. L., McElhinney, L. M., Banyard, A. C., Ngeleja, C., Keyyu, J., Cleaveland, S., Lembo, T., Rupprecht, C. E., & Fooks, A. R. (2012). Complete genome sequence of Ikoma lyssavirus. *J. Virol*, *86*, 10242–10243.

McCarthy, T. J. (1989). Human depredation by vampire bats (*Desmodus rotundus*) following a hog cholera campaign. *Am. J. Trop. Hyg. Med*, *40*, 320–322.

Mistry, S., & Moreno-Valdez, A. (2008). Climate change and bats. Vampire bats offer clues to the future. Bat Conservation International. *Bats*, *26*, 8–12. www.batcon.org.

Mørk, T., & Prestrud, P. (2004). Arctic rabies—a review. *Acta Veterinaria Scandinavia*, *45*, 1–9.

Mulatti, P., Müller, T., Bonfanti, L., & Marangon, S. (2012). Emergency oral rabies vaccination of foxes in Italy in 2009–2010: identification of residual rabies foci at higher altitudes in the Alps. *Epidemiol. Infect.*, *140*, 591–598.

Müller, T., Selhorst, T., & Pötzsch, C. (2005). Rabies in Germany—an update. *Eurosurveillance*, *10*, 229–231.

Müller, T., Bätza, H. J., Freuling, C., Kliemt, A., Kliemt, J., Heuser, R., Schlüter, H., Selhorst, T., Vos, A., & Mettenleiter, T. C. (2012). Elimination of terrestrial rabies in Germany using oral vaccination of foxes. *Berl. Muench. Tierärztl Wschr.*, *125*, 178–190.

Nadin-Davis, S. A., Torres, G., Ribas Mde, L., Guzman, M., De La Paz, R. C., Morales, M., & Wandeler, A. I. (2006). A molecular epidemiological study of rabies in Cuba. *Epidemiol. Infect.*, *134*, 1313–1324.

Nehaul, B. B. (1954). Rabies transmitted by bats in British Guiana. *Am. J. Trop. Dis. Hyg.*, *4*, 550–553.

Nel, L. H., & Rupprecht, C. E. (2007). Emergence of lyssaviruses in the Old World: the case of Africa. *Curr. Top. Microbiol. Immunol.*, *315*, 161–193.

Nettles, V. F., Shaddock, J. H., Sikes, K., & Reves, C. R. (1979). Rabies in translocated raccoons. *American Journal of Public Health*, *69*, 601–602.

Neville, J. (2004). Rabies in the ancient world. In A. A. King, A. R. Fooks, A. Aubert & A. I. Wandeler (Eds.), *Historical perspective of rabies in Europe and the Mediterranean Basin* (pp. 1–13). Paris: OIE.

Nowak, E. (1984). Development of populations and geographical distribution of the raccoon dog *Nyctereutes procyonoides* (Gray, 1834) in Europe. *Z. Jagdwissenschaft*, *30*, 137–154. (In German).

OIE. (2010). *Recommendation No. 1. Climate change and its link with animal diseases and animal production*. 20th Conference of the OIE Regional Commission for the Americas. Montevideo, Uruguay, 16–19 November 2010.

Ogada, D. L., Torchin, M. E., Kinnaird, M. F., & Ezenwa, V. O. (2012). Effects of vulture declines on facultative scavengers and potential implications for mammalian disease transmission. *Conserv. Biol.*, *26*, 453–460.

Ørpetveit, I., Ytrehus, B., Vikøren, T., Handeland, K., Mjøs, A., Nissen, S., Blysrad, H., & Lund, A. (2011). Rabies in an Arctic fox on the Svalbard archipelago, Norway, January 2011. *Eurosurveillance*, *16*, pii=19797.

Pamperin, N. J., Follmann, E. H., & Petersen, B. (2006). Interspecific Killing of an Arctic Fox by a Red Fox at Prudhoe Bay. *Alaska. Arctic.*, *59*, 361–364.

Post, E., Forchhammer, M. C., Bret-Harte, M. S., Callaghan, T. V., Christensen, T. R., Elberling, B., Fox, A. D., Gilg, O., Hik, D. S., Høye, T. T., Ims, R. A., Jeppesen, E., Klein, D. R., Madsen, J., McGuire, A. D., Rysgaard, S., Schindler, D. E., Stirling, I., Tamstorf, M. P., Tyler, N. J., van der Wal, R., Welker, J., Wookey, P. A., Schmidt, N. M., & Aastrup, P. (2009). Ecological dynamics across the Arctic associated with recent climate change. *Science*, *325*, 1355–1358.

Randall, D. A., Williams, S. D., Kuzmin, I. V., Rupprecht, C. E., Tallents, L. A., Tefera, Z., Argaw, K., Shiferaw, F., Knobel, D. L., Sillero-Zubiri, C., & Laurenson, M. K. (2004). Rabies in endangered Ethiopian wolves. *Emerg. Infect. Dis.*, *10*, 2214–2217.

Rappole, J. H., & Hubálek, Z. (2003). Migratory birds and West Nile virus. *J. App. Microbiol.*, *94*, 47–58.

Rausch, R. L. (1972). Observation on some natural-focal zoonoses in Alaska. *Arch. Environ. Health*, *25*, 246–252.

Richter, H. V., & Cumming, G. S. (2006). Food availability and annual migration of the straw-coloured fruit bat (*Eidolon helvum*). *J Zool.*, *268*, 35–44.

Riley, S. P. D., Hadidian, J., & Manski, D. A. (1998). Population density, survival, and rabies in raccoons in an urban national park. *Can. J. Zool.*, *76*, 1153–1164.

Rodnikova, A., Ims, R. A., Sokolov, A., Skogstad, G., Sokolov, V., Shtro, V., & Fuglei, E. (2011). Red fox takeover of arctic fox breeding den: an observation from Yamal Peninsula, Russia. *Polar Biol.*, *34*, 1609–1614.

Rosatte, R. C., Power, M. J., Donovan, D., Davies, J. C., Allan, M., Bachmann, P., Stevenson, B., Wandeler, A., & Muldoon, F. (2007). Elimination of arctic variant rabies in red foxes, metropolitan Toronto. *Emerg. Infect. Dis.*, *13*, 25–27.

Rupprecht, C. E., Barret, J., Briggs, D., Cliquet, F., Fooks, A. R., Lumlertdacha, B., Meslin, F. X., Müller, T., Nel, L., Schneider, C., Tordo, N., & Wandeler, A. I. (2008). Can rabies be eradicated? *Dev. Biol. (Basel)*, *131*, 95–121.

Savolainen, P., Zhang, Y. P., Luo, J., Lundeberg, J., & Leitner, T. (2002). Genetic evidence for an East Asian origin of domestic dogs. *Science*, *298*, 1610–1613.

Schneider, M. C., Aron, J., Santon-Burgoa, C., Uieda, W., & Ruiz-Velazco, S. (2001). Common vampire bat attacks on humans in a village of the Amazon region of Brazil. *Cad. Sáude Publica. Rio de Janeiro*, *17*, 1531–1536.

Schneider, M. C., Romijn, P. C., Uieda, W., Tamayo, H., da Silva, D. F., Belotto, A., da Silva, J. B., & Leanes, L. F. (2009). Rabies transmitted by vampire bats to humans: An emerging zoonotic disease in Latin America? *Rev. Panam. Salud Publica*, *25*, 260–269.

Scott, T., Hassel, R., & Nel, L. (2012). Rabies in kudu (*Tragelaphus strepsiceros*). *Berl. Münch. Tierärztl. Wochenschr*, *125*, 236–241.

Servas, V., Mailles, A., Neau, D., Castor, C., Manetti, A., Fouquet, E., Ragnaud, J. -M., Bourhy, H., Paty, M. -C., Melik, N., Astoul, J., Cliquet, F., Moiton, M. -P., Francois, C., Coustillas, M., Minet, J. -C., Parriaud, P., Capek, I., & Filleul, L. (2005). An imported case of canine rabies in Aquitaine: Investigation and management of the contacts at risk, August 2004 – March 2005. *Eurosurveillance*, *10*, pii=578.

Sillero-Zubiri, C., King, A. A., & Macdonald, D. W. (1996). Rabies and mortality in Ethiopian wolves (*Canis simensis*). *J. Wildl. Dis.*, *32*, 80–86.

Slate, D., Rupprecht, C. E., Rooney, J. A., Donovan, D., Lein, D. H., & Chipman, R. B. (2005). Status of oral rabies vaccination in wild carnivores in the United States. *Virus Res.*, *111*, 68–76.

Slate, D., Algeo, T. P., Nelson, K. M., Chipman, R. B., Donovan, D., Blanton, J. D., Niezgoda, M., & Rupprecht, C. E. (2009). Oral rabies vaccination in North America: opportunities, complexities, and challenges. *PLoS Negl. Trop. Dis.*, *3*, e549.

Slate, D. (2011). The exotic small Asian mongoose in the Caribbean: Rabies management opportunities and challenges. *RITA*, *2011*, 39.

Slavinski, S., Humberg, L., Lowney, M., Simon, R., Calvanese, N., Bregman, B., Kass, D., & Oleszko, W. (2012). Trap-vaccinate-release program to control raccoon rabies, New York, USA. *Emerg. Infect. Dis.*, *18*, 1170–1172.

Smith, G. C., & Harris, S. (1991). Rabies in urban foxes (Vulpes vulpes) in Britain: the use of a spatial stochastic simulation model to examine the pattern of spread and evaluate the efficacy of different control régimes. *Philos. Trans. R. Soc. Lond. B. Biol. Sci.*, *334*, 459–479.

Smith, J. S., Orciari, L. A., Yager, P. A., Seidel, H. D., & Warner, C. K. (1992). Epidemiologic and historical relationships among 87 rabies virus isolates as determined by limited sequence analysis. *J. Infect. Dis.*, *166*, 296–307.

Talbi, C., Holmes, E. C., Benedicts, P., de Faye, O., Nakoune, E., Gamatie, D., Diarra, A., Elmamy, B. O., Sow, A., Adjogoua, E. V., Sangare, O., Dundon, W. G., Capua, I., Sall, A. A., & Bourhy, H. (2009). Evolutionary history and dynamics of dog rabies virus in western and central Africa. *J. Gen. Virol*, *90*, 783–791.

Talbi, C., Lemey, P., Suchard, M. A., Abdelatif, E., Elharrak, M., Jalal, N., Faouzi, A., Echevar-ria, J. E., Moron, S. V., Rambaut, A., Campiz, N., Tatem, A. J., Holmes, E. C., & Bourhy, H. (2010). Phylodynamics and human-mediated dispersal of a zoonotic virus. *Plos Pathogens, 6*, e1001166.

Tarroux, A., Berteaux, D., & Bety, J. (2010). Northern nomads: Ability for extensive movements in adult arctic foxes. *Polar Biology, 33*, 1021–1026.

Thomas, D. W. (1983). The annual migrations of three species of West African fruit bats (Chirop-tera: Pteropodidae). *Can. J. Zool., 61*, 2266–2272.

Thomas, C. D., Cameron, A., Green, R. E., Bakkenes, M., Beaumont, L. J., Collingham, Y. C., Eras-mus, B. F. N., de Siqueira, M. F., Grainger, A., Hannah, L., Hughes, L., Huntley, B., van Jaars-veld, A. S., Midgley, G. F., Miles, L., Ortega-Huerta, M. A., Townsend Peterson, A., Phillips, O. L., & Williams, S. E. (2004). Extinction risk from climate change. *Nature, 427*, 145–148.

Tierkel, E. S., Arbona, G., Rivera, A., & De Juan, A. (1952). Mongoose rabies in Puerto Rico. *Public Health Reports, 67*, 274–278.

Tietjen, S., Kaufhold, B., Müller, T., & Freuling, C. M. (2011). Case report: illegal puppy trade and transport, Germany. *Rabies Bulletin Europe, 35*, 8–10.

Verginelli, F., Capelli, C., Coia, V., Musiani, M., Falchetti, M., Ottini, L., Palmirotta, R., Taglia-cozzo, A., De Grossi Mazzorin, I., & Mariani-Costantini, R. (2005). Mitochondrial DNA from prehistoric canids highlights relationships between dogs and South-East European wolves. *Mol. Biol. Evol., 22*, 2541–2551.

Vos, A., Freuling, C., Eskiizmirliler, C., Ün, H., Aylan, O., Johnson, N., Gürbüz, S., Müller, W., Akkoca, N., Müller, T., Fooks, A. R., & Askaroglu, H. (2009). Rabies in foxes, Aegean region, Turkey. *Emerg. Infect. Dis., 15*, 1620–1622.

Vos, A., Nunan, C., Bolles, D., Müller, T., Fooks, A. R., Tordo, N., & Baer, G. M. (2011). The occur-rence of rabies in pre-Columbian Central America: an historical search. *Epidemiol. Infect., 139*, 1445–1452.

Vos, A., Ortmann, S., Kretzschmar, A. S., Köhnemann, B., & Michler, F. (2012). The raccoon (Pro-cyon lotor) as potential rabies reservoir species in Germany: a risk assessment. *Berl. Muench. Tieraerztl. Wochenschr., 125*, 228–235.

Wabakken, P., Sand, H., Kojola, I., Zimmermann, B., Arnemo, J. M., Pedersen, H. C., & Liberg, O. (2007). Multistage, long-range natal dispersal by a global positioning system-collared Scandi-navian wolf. *J. Wild. Manag., 71*, 1631–1634.

Wandeler, A. I., & Salsberg, E. B. (1999). Raccoon rabies in eastern Ontario. *Can. Vet. J., 40*, 731.

Westerling, B. (1991). Rabies in Finland and its control 1988-90. *Suom. Riista., 37*, 93–100.

Wilkinson, D., & Smith, G. C. (2001). A preliminary survey for changes in urban Fox (Vulpes vulpes) densities in England and Wales, and implications for rabies control. *Mammal Review, 31*, 107–110.

Windiyaningsih, C., Wilde, H., Meslin, F. X., Saroso, T., & Hemachudha, T. (2004). The rabies epi-demic on Flores Island, Indonesia (1998–2003) (review). *J. Med. Assoc. Thai, 87*, 1530–1538.

Woodroofe, R. (2001). Assessing the risks of intervention: immobilization, radio-collaring and vac-cination of African wild dogs. *Oryx, 35*, 234–244.

Rabies Information System of the WHO Collaborating Centre for Rabies Surveillance and Research, Friedrich-Loeffler-Institut, Germany. http://www.who-rabies-bulletin.org/Queries/Distribution.aspx. Accessed 03.01.13.

Wu, X., Hu, R., Zhang, Y., Dong, G., & Rupprecht, C. E. (2009). Reemerging rabies and lack of systemic surveillance in People's Republic of China. *Emerg. Infect. Dis., 15*, 1159–1164.

Chapter 5

Lassa Fever

A rodent-human interaction

Elisabeth Fichet-Calvet

Bernhard-Nocht Institute of Tropical Medicine, Hamburg, Germany

INTRODUCTION

The first recorded human case of Lassa fever was in an American nurse, Laura Wine, who had lived in Lassa, Nigeria, for 4 years. On January 12, 1969, she complained of back pain. On January 19, the pain worsened and she was unable to leave her bed. On the Monday, January 20, she awoke with a severe sore throat but a physical examination revealed no peculiarity. On the morning of Tuesday, January 21, her general condition began to deteriorate, and her physician, Dr. Hamer, discovered groups of deep ulcerations in the throat and on oral mucosa. By Wednesday, January 22, lesions of the pharynx had increased in size and her temperature was recorded at 38.5°C. The following day, ulcerations of the throat were discolored, and her neck was swollen. Fever increased and dehydration worsened. The patient became drowsy and very lethargic. On Friday, January 24, her face was very congested, and examination of the lower limbs showed edema and petechiae. However, dyspnea settled and the patient was stable. Dr. Hamer decided to transfer the patient to the Evangel Hospital in Jos, because it was better equipped and with more experienced personnel. On Saturday, January 25, the patient developed several seizures that appeared to be signs of heart failure before boarding the plane. On admission in Jos, she was placed on a drip and under an oxygen tent. On Sunday, January 26, she developed convulsions and died at 21:30, 10 days after onset. This case was rapidly followed by a second case, Charlotte Shaw, who worked as a nurse in the hospital during the preceding night. On the night of Saturday, January 25, she pricked her finger while cutting roses in her garden and immediately disinfected the wound. She was on duty that night, and tried to relieve Laura Wine, through blotting ulcerations of the throat with her finger wrapped in gauze. This was followed by disinfection of herself following contact with the patient. On Monday, February 3, nine days later, she experienced severe back pain, and on the following day she was admitted to hospital as a patient. The same symptoms that had been observed in the first patient developed, and she died on February

13, 11 days after the onset of the disease. Following her death, all staff received an injection of gamma globulin on February 14. A third case, Lili Pinneo, the head nurse who took care of Charlotte Shaw throughout her illness, felt a vague unease, on February 20. She retired to bed for 2 days due to headache and nausea. She was finally admitted to hospital and began to develop the same symptoms as the second patient, but without edema. On February 26, her physicians, Drs. Troup, Christensen and White, decided to repatriate her to the USA. She arrived on March 3, in New York, where Dr. Frame from the Yale Arbovirus Research Unit (YARU), bled her immediately and took receipt of biopsies from the first two patients. Lili Pinneo was isolated in the Presbyterian Hospital of Columbia, where she remained for nine weeks. She recovered slowly after 2 months. In early June and late November 1969, two other persons working in YARU, Drs. J. Casals and J. Roman, were also infected, with the second individual dying after 2 weeks of disease. Finally, among the five Americans contracting the disease during this first outbreak, three died and two survived after a long period of recovery (Buckley et al., 1970; Frame et al., 1970). This tragedy led to the cessation of any research on what would become Lassa fever virus at YARU, and to its transfer to the new biosafety level 4 laboratory in the Center for Disease Control (CDC) in Atlanta. This was also the starting point for the research on Lassa fever, and the beginning of records of many subsequent outbreaks. It also lead to the introduction of surveillance for the virus in Nigeria, Sierra Leone, Liberia and Guinea (see reviews in Tables 5.1 and 5.2). Nevertheless, this disease was not new, as proven by retrospective diagnosis and the presence of antibodies in blood collected from before 1969 in missionaries living in West Africa (Frame, 1975; Henderson et al., 1972; Saltzmann, 1978). Now, only six laboratories in the world are currently working on Lassa virus: the CDC in Atlanta, USA; the Army Medical Research Institute of Infectious Diseases in Frederick, USA; the Bernhard-Nocht Institute in Hamburg, Germany; the Philips University in Marburg, Germany; the Merieux Institute in Lyon, France; and the Health Protection Agency in London and Porton Down, UK. Shortly after the first nosocomial outbreak, the American CDC undertook some serological studies in the localities of Panguma and Tongo, Sierra Leone, in the areas surrounding the affected hospitals (Fraser et al., 1974). They discovered that incidence of the disease was 2.2 cases per 1000 inhabitants. Thereafter, these investigations were extended throughout the country (McCormick et al., 1987b). However, even where seroprevalence reached levels of 52% in some villages, it appeared that the disease was often a benign, short flu-like illness. Monath estimated that of 60 persons contracting the disease, only 1 person died at the hospital. Further investigations mainly in Guinea, Liberia and Nigeria revealed locally high seroprevalence in the human population living in rural areas (Table 5.2, Figure 5.1). Taking into account the incidence rate, the lethality rate, and the human population living in West Africa, McCormick predicted that Lassa fever would affect 200,000–300,000 persons with 5,000–10,000 fatalities per annum (McCormick & Fisher-Hoch, 2002).

TABLE 5.1 Chronology of Lassa Fever Outbreaks and Hospital Surveillance in West Africa

Country	Year	Town/region	Hospital	Fatality rate	Reference
Benin	1977	North	Bambereke	25% (1/4)	Saltzmann, 1978
Guinea	1996–1999	Kindia/Faranah/Kissidougou/N'Zerekore	4 hospitals	18% (4/22)	Bausch et al., 2001
Liberia	1972	Lofa	Zorzor	36% (4/11)	Monath et al., 1973
	1980–1982	Lofa	Zorzor	14% (6/44)	Monson et al., 1984
Nigeria	1970	Benoue Plateau	Jos	54% (13/24)	Carey et al., 1972
	1974	Benoue Plateau	Onitsha	33% (1/3)	Bowen et al., 1975
	1975–1977	Benoue Plateau	Vom	33% (6/18)	Saltzmann, 1978
	2003–2004	Edo state	Irrua	0% (0/3)	Omilabu et al., 2005
	2005–2008	Edo state	Irrua	29% (7/25)	Ehichioya et al., 2012
	2009–2010	Edo state	Irrua	36% (61/170)	Asogun et al., 2012
Sierra Leone	1970–1972	Eastern province	Panguma/Tongo	38% (24/63)	Fraser et al., 1974
	1973–1976	Eastern province	Panguma	18% (28/156)	Keane & Gilles, 1977
	1975	Eastern province	Segbwema	23% (25/108)	Keane & Gilles, 1977
	1977–1979	Eastern province	Panguma/Segbwema	17% (76/441)	McCormick, et al., 1987a
	2004	Eastern province	Kenema	25% (80/321)	Khan et al., 2008
	2005	Eastern province	Kenema	26% (35/136)	Khan et al., 2008
	2006	Eastern province	Kenema	34% (17/51)	Khan et al., 2008
	2007	Eastern province	Kenema	58% (11/19)	Khan et al., 2008

TABLE 5.2 Lassa Seroprevalences Observed in Rural Localities in West Africa

Country	Year	Town/region	Village	IgG	Reference
Cote d'Ivoire	2000	Duekoue	Foret Classee	21% (8/39)	Akoua-Koffi et al., 2006
	2000	Guiglo	Guiglo	27% (33/124)	
Guinea	1982–1983	Madina Oula	Ouassou	14% (3/22)	Boiro et al., 1987
	1990–1992	Bofa	Walia	4% (6/160)	Lukashevich et al., 1993
	1990–1992	Boké	Hamdalaye	5% (5/102)	
	1990–1992	Pita	Bantignel	6% (10/165)	
	1990–1992	Labe	Sala	7% (8/11)	
	1990–1992	Mali	Mali	5% (8/147)	
	1990–1992	Mali	Deppal	4% (1/29)	
	1990–1992	Siguiri	Kintinian	8% (13/168)	
	1990–1992	Siguiri	Torekoro	5% (5/100)	
	1990–1992	Madina Oula	Madina Oula camp	34% (59/171)	
	1990–1992	Faranah	Gbetaya	43% (20/47)	
	1990–1992	Faranah	Kamaraya	25% (33/130)	
	1990–1992	Faranah	Sangoyah	42% (76/182)	
	1990–1992	Faranah	Tindo	33% (20/61)	

1990–1992	Siguiri	Balato	18% (27/150)	ter Meulen et al., 1996
1990–1992	Gueckedou	Bawa	55% (28/51)	
1990–1992	Gueckedou	Dawa Bombolo	48% (31/65)	
1990–1992	Gueckedou	Fangamandou	27% (32/112)	
1990–1992	Gueckedou	Nongoa Mbalia	40% (47/116)	
1990–1992	Gueckedou	Owet Djiba	37% (31/83)	
1990–1992	Gueckedou	Sassani Toli	34% (33/97)	
1990–1992	Gueckedou	Telekolo	30% (24/80)	
1990–1992	Yomou	Waita	26% (31/120)	
1990–1992	Yomou	Komore	25% (46/183)	
1990–1992	Yomou	Bamakama	30% (42/138)	
1990–1992	Lola	Thuo	33% (63/190)	
1990–1992	Lola	Gbenemou	19% (11/59)	
1990–1992	Lola	Gbah	26% (28/109)	
1993	Gueckedou	19 villages	14% (n = 751)	
1993	Pita	6 villages	3% (n = 232)	
2002	Dubreka	Ouassou/Tanene	8% (14/186)	Sylla, 2004
2000	Guéckédou	Koumoni	5% (2/40)	Kerneis et al., 2009
2000	Guéckédou	Kelema	4% (1/23)	

Continued

TABLE 5.2 Lassa Seroprevalences Observed in Rural Localities in West Africa — cont'd

Country	Year	Town/region	Village	IgG	Reference
Guinea – continued	2000	Guéckédou	Mangala	4% (1/26)	
	2000	Guéckédou	Sandia	4% (2/51)	
	2000	Guéckédou	Sokoro	6% (2/32)	
	2000	Guéckédou	Bandadou	2% (1/41)	
	2000	Guéckédou	Yelindou	8% (4/50)	
	2000	Guéckédou	Yaradou	14% (7/51)	
	2000	Guéckédou	Fandou Bendou	15% (5/33)	
	2000	Lola	Theasso	8% (2/25)	
	2000	Lola	Morigbedou	10% (4/40)	
	2000	Lola	Pine	42% (15/36)	
	2000	Lola	Gueasso centre	9% (1/11)	
	2000	Lola	Kokota	31% (16/51)	
	2000	Lola	Laine centre	5% (2/37)	
	2000	Lola	Gogota	0% (0/6)	
	2000	Lola	Zougueta	15% (2/13)	
	2000	Lola	Gote Koly	33% (7/21)	
	2000	Lola	Manga Bo	0% (0/19)	

Continued

	Year	Region	Location	Prevalence	Reference
	2000	Lola	N'zoo centre	6% (2/35)	
	2000	Lola	Tounkarata centre	2% (1/47)	
	2000	Yomou	Yekeni	25% (2/8)	
	2000	Yomou	Bemeye	11% (1/9)	
	2000	Yomou	Manangoya	0% (0/12)	
	2000	Yomou	Nawei	19% (10/53)	
	2000	Yomou	Bamakama	15% (4/26)	
	2000	Yomou	Bowe centre	27% (8/30)	
	2000	Yomou	Kowi	14% (4/29)	
	2000	Yomou	Diecke centre	4% (3/73)	
	2000	Yomou	Yomou centre	6% (3/49)	
	2004	Faranah	Bantou	42% (55/132)	Klempa et al., 2013
	2004	Faranah	Tanganya	39% (47/121)	
Liberia	1980	Lofa	Gbanway	7% (23/312)	Yalley-Ogunro et al., 1984
	1980	Lofa	Yapoa	1% (2/212)	
	1980	Lofa	Zuwulo	4% (12/293)	
	1980	Lofa	Balagualazu	3% (12/392)	
	1980	Lofa	Borlelo	5% (7/138)	
	1980	Lofa	Foya Kamara	14% (12/85)	

TABLE 5.2 Lassa Seroprevalences Observed in Rural Localities in West Africa—cont'd

Country	Year	Town/region	Village	IgG	Reference
Nigeria	1987?	Benue state	Gboko	37% (n = ?)	Tomori, et al., 1988
	1987?	Taraba state	Gongola	13% (n = ?)	
Sierra Leone	1977–1983	South	Yangema	8% (n = 109)	McCormick et al., 1987b
	1977–1983	East	Bomie	10% (n = 617)	
	1977–1983	East	Konia	37% (n = 733)	
	1977–1983	East	Kpandebu	15% (n = 578)	
	1977–1983	East	Landoma	11% (n = 157)	
	1977–1983	East	Lowoma	13% (n = 285)	
	1977–1983	East	Neama	15% (n = 105)	
	1977–1983	East	Niahun	19% (n = 1075)	
	1977–1983	East	Njakundoma	20% (n = 242)	
	1977–1983	East	Palima	52% (n = 217)	
	1977–1983	East	Semewabu	38% (n = 372)	
	1977–1983	East	Tongola	31% (n = 402)	
	1977–1983	North	Kamethe	15% (n = 115)	
	1977–1983	North	Kamabunyele	14% (n = 118)	
	1977–1983	North	Kathumpe	10% (n = 88)	

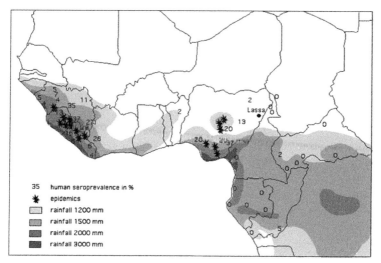

FIGURE 5.1   **West and Central Africa mean annual rainfall (1951–1989), Lassa Fever noso-comial outbreaks (stars) and human seroprevalence (numbers in %) (adapted from Fichet-Calvet & Rogers, 2009). The data are detailed in Tables 5.1 and 5.2, and null seroprevalences in Central Africa are compiled from (Gonzalez et al., 1983; Georges et al., 1985; Gonzalez et al., 1989; Talani et al., 1999; Nakounne et al., 2000).**

THE SYMPTOMS OF LASSA FEVER

Lassa fever has an atypical, nonspecific presentation. Incubation periods range from 7 to 10 days, followed by a flu-like illness that lasts 2 to 3 days, with a progressive fever, chills, malaise, weakness, headache and myalgia in the back or the limbs. At this stage, the symptoms are not specific and can be confused with an attack of malaria. Fever can then climb to 40°C, with gastrointestinal discomfort accompanied by nausea, vomiting, diarrhea and abdominal pain. Angina with significant dysphagia, coughing and chest pain is often observed. Finally, a swelling of the neck or face appears with a conjunctivitis and facial edema before the patient dies of endotoxin shock. However, the frequency and severity of these symptoms vary among patients. This is also dependent on the stage at which the patient presents at hospital. Later during the course of disease, dizziness and unilateral or bilateral deafness can occur. When deafness is permanent, this is the most long-lasting sequelae of disease, and can represent up to 18% of the survivors (Cummins et al., 1990).

If given early in the course of disease, intravenous ribavirin can improve survival in Lassa patients (McCormick et al., 1986); this drug is now routinely available in hospitals such as Kenema hospital in Sierra Leone and in Irrua Teaching Hospital in Nigeria.

THE GENOME

In 1969, Lassa virus was isolated from Vero cells infected with different tissues, serum, pleural fluid, urine and throat swabs, from Laura Wine, Lili Pinneo and Jordi Casals. After observation of cytopathic effects on Vero cells, the images from electron microscope revealed viral particles surrounded by a membrane containing opaque grains, which were ribosomes from the host cell (Buckley et al., 1970). Shortly before, studies on Lymphocytic Choriomeningitis virus (LCMV), Machupo virus and Tacaribe virus showed similar morphology (Murphy et al., 1969) suggesting a relationship with this diverse group of viruses. Meanwhile, inoculation experiments of Lassa virus into adult mice showed similar disease signs such as seizures, to those observed in mice inoculated with LCMV. A significant viruria was noted in those mice that survived infection. These results allowed these four viruses, LCMV, Machupo, Tacaribe and Lassa, to be grouped within the same family, the *Arenaviridae*. The prefix "arena" means "sand" in Latin, which reflects the grainy morphology common to all members of the family (Rowe et al., 1970). LCMV became the prototype of this family, because it was the first to be discovered in 1933 by Armstrong and Lillie (rev. in Oldstone & Campbell, 2011).

The genome is composed of two ambisense segments: the smaller segment (S) codes for the glycoprotein precursor and the nucleoprotein, and the larger segment (L) codes for the polymerase and the matrix protein. The coding regions are separated by an intergenic region with a strong secondary structure, and are flanked by noncoding regions (Figure 5.2), which are also suspected to have a role in the virulence of the virus (Albariño et al., 2011). The high affinity between the glycoprotein and the cellular receptor, dystroglycan, mainly present on dendritic cells, could be responsible for the high permeability of Lassa virus for cells, leading to increased viral replication and loss of immune system efficiency (Oldstone & Campbell, 2011).

This genomic structure has been corroborated through numerous studies on the genomes of other arenaviruses such as LCMV or Pichinde virus (Vezza et al., 1980; Auperin et al., 1982). They led Auperin and McCormick (1989) to derive the first S segment sequence of the strain Josiah from Sierra Leone. Nigerian strains were sequenced shortly after by Clegg and co-workers (Clegg et al., 1991), and in 2000, Bowen et al. (2000) proposed a classification of the different strains with a phylogeny based on the GP, NP and polymerase. They suggested the existence of four lineages: three in Nigeria and one in the Sierra Leone-Liberia-Guinea (Bowen et al., 2000). Lineage I consisted of a single virus that was isolated from Lili Pinneo. The remaining lineages contained multiple virus isolate sequences. These authors have proposed a maximum difference of 12% in the amino acids of the NP in order to group viruses into different Lassa virus clades. Until recently, this rule had been respected. However, the discovery of a new virus "Gbagroube" in the murine species *Mus setulosus* from Ivory Coast, suggests that this definition may require revision (Coulibaly-N'Golo et al., 2011). Only studies on the pathogenicity of Gbagroube will revise the threshold of 12%, because only Lassa in West Africa and Lujo in South Africa

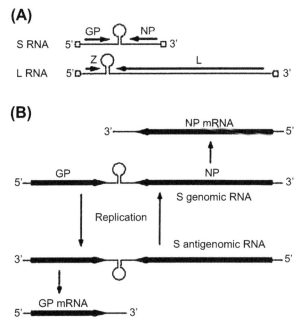

FIGURE 5.2   **(A) Ambisense gene organization of arenaviruses. Boxes represent the highly conserved RNA termini. The stable secondary structure in the intergenic region is shown schematically. (B) Replication and transcription of the arenavirus genome.** *Reproduced from Günther & Lenz 2004.*

are pathogenic for humans (Briese et al., 2009; Paweska et al., 2009). Currently, 11 strains from Nigeria including representatives from lineages I, II and III, 5 strains from Sierra Leone, Liberia, Guinea from lineage IV, and AV from Burkina Faso-Ghana-Ivory Coast are completely sequenced (Figure 5.3; Ehichioya et al., 2011).

Phylogenetic studies reveal that Lassa virus has migrated from East to West, the source being in Nigeria with lineage I, and the most divergent being in the Mano River region with lineage IV. All analyses to date, including a molecular clock study, conclude that Lassa virus has evolved over a relatively short period of around 700–900 years to the last common ancestor (Bowen et al., 2000; Günther & Lenz, 2004; Coulibaly-N'Golo et al., 2011; Lalis et al., 2012).

THE RESERVOIR

After the first outbreaks reported in Nigeria and Sierra Leone in the seventies, Monath et al. (1974a) identified the reservoir in Sierra Leone, by isolating the virus in the multimammate rat, *Mastomys natalensis*, a species indigenous to Africa. This species was living in the settlements and the surroundings of the villages of Panguma and Tongo, located in the mining region in the Eastern province. One year later, Wulff et al. (1975) described the same reservoir in the human

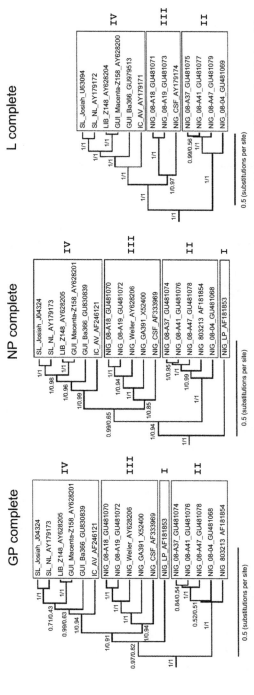

FIGURE 5.3   Phylogenetic analysis of Lassa virus, using complete nucleotide sequences of GP (1,473 nucleotides), NP (1,707 nucleotides) and L genes (6,654 nucleotides). *Adapted from Ehichioya et al. 2011.*

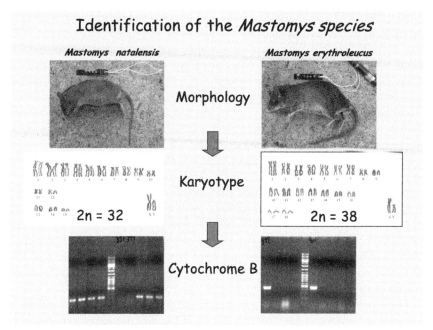

FIGURE 5.4   **Schematic for identification of** *Mastomys natalensis* **and** *Mastomys erythro-* *leucus* **using karyotyping or specific PCR on cytochrome b.** *Photos morphology:* ©*E. Fichet-Calvet, karyotypes:* © *V. Aniskin, gel:* © *E. Lecompte.*

dwellings in Benoue Plateau, Nigeria. In the eighties, McCormick evoked the possibility that several species of *Mastomys* could be the reservoirs, since new taxonomic revisions revealed the sympatric occurrence of *M. natalensis* (2n = 32), *M. erythroleucus* (2n = 38) and perhaps *M. huberti* (2n = 32) in the region (Robbins et al., 1983; Hubert et al., 1983; McCormick et al., 1987b). Complementary studies on the taxonomy of *Mastomys* were largely initiated to define the species, the original *Mastomys* from Natal province, *M. coucha* (2n = 36), being restricted in South Africa (Duplantier; 1990; Britton-Davidian; 1995; Granjon et al., 1996; Granjon et al., 1997). Because these species are difficult to recognize morphologically, the term *Mastomys* spp. or *Mastomys* complex was used to define the reservoir. This situation continued until the molecular era, allowing the differentiation of the four species by PCR (Figure 5.4; Lecompte et al., 2005).

With this new tool, the identification was definitive, and a large spatial survey of Lassa virus in Guinea confirmed *M. natalensis* as the unique reservoir (17 villages, 1591 small mammals, 847 *M. natalensis*, 202 *M. erythroleucus*; Lecompte et al., 2006). Nevertheless, Wulff in 1975 noted that other species such as the pygmy mice, *Mus minutoides*, and the black rat, *Rattus rattus*, can be infected by the virus, since 3 specimens, 2 *R. rattus* and 1 *M. minutoides*, have been found positive. Similar observations had not been reported until the recent discovery of a Lassa-like virus in *Mus baoulei*, another pygmy mouse tested in Ghana. The two species of *Mus* are impossible to distinguish morpho-logically, so the original species identification by Wulff must to be considered with caution. But it is very surprising to note that the GP and NP sequences found in *M. baoulei* are close to the initial virus sequence derived from one of the earliest victims, Lili Pinneo, who was infected from Lassa in the north-ern state, exactly the same area where Wulff found the positive pygmy mice. Consequently, it is possible that Lassa virus has several reservoirs, one in wild species such as the Boule's mouse, and another one in the more commensal species such as *M. natalensis*. Because these animals are numerous, are living close to humans, and are highly reproductive with 20 to 24 nipples and a mean litter size 9.2 (Fichet-Calvet et al., 2008), they could amplify the transmission of the virus towards humans, rather than from a more discrete source coming from savannah. The Boule's mouse appears to be very rare in comparison to other mice species: 2% (3/168) in Guinea, 3% (3/104) in Côte d'Ivoire and 4% (5/133) in Ghana. It is also possible that cross-species transmission occurs between different murine species although there is little evidence to demon-strate that one species is the principal reservoir and others become infected as a result of spill-over.

Until the present time, the pygmy mice were good candidates to spread other arenaviruses in Africa, since *M. minutoides* in Guinea, Tanzania, and Zambia, *M. setulosus* in Côte d'Ivoire and *M. mattheyi* in Ghana have been found posi-tive (Coulibaly-N'Golo et al., 2011; Gouy de Bellocq et al., 2010; Ishii et al., 2012; Lecompte et al., 2007; Kronmann et al. in press). Most of these new arenaviruses are close to LCMV, the prototype of the Arenaviridae family. Only Gbagroube virus in *M. setulosus* and Jirandogo virus in *M. baoulei* are basal to the Lassa clade, and *M. setulosus* is the closest species to *M. baoulei*.

Mastomys natalensis is widely distributed in Africa, from southeastern Senegal to South Africa (Figure 5.5). In sub-Saharan Africa, this species is found in all habitats except the mountains in East Africa, and lives commensally with man across its whole range. In Morogoro, Tanzania, the longitudinal studies of *M. natalensis* populations conducted by Telford, then Leirs, over a period of 10 years have revealed some large population fluctuations, depending on preced-ing rainfall and population densities (Leirs et al., 1994; Leirs et al., 1996; Leirs et al., 1997). During the outbreaks (demographic explosions) in the agricultural plain of Morogoro, trapping success could reach 80% some years, when it was only 8% in the maize fields surrounding the villages in Guinea. The compari-son is possible because the type of trap (Scherman), and method of capture in line with spacing every 5 m were similar (Fichet-Calvet et al., 2007; Leirs, 1994). This suggests that the commensal populations in West Africa are living at lower densities than the wild population in East Africa, with few variations between years. This could be due to restricted availability of optimal habitats in poorly developed areas. In this sense, deforestation and clearing vegetation for the purpose of increasing agricultural land would increase the densities of *M. natalensis*. This could be a cause for the emergence of Lassa fever cases in

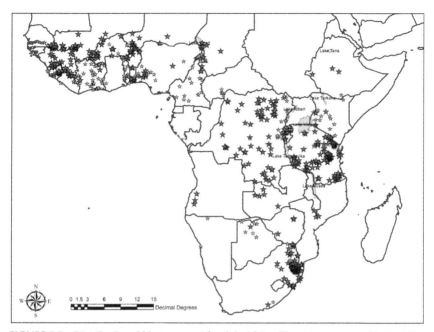

FIGURE 5.5   Distribution of *Mastomys natalensis* in Africa. The stars correspond to localities where the specimens have been collected: in red, a high confidence in the identification and in lilac a poor identification. This map is issued from a compilation of 42 references in the literature, the Belgian Biodiversity Information Facility database (GBIF portal) and the collection at the Tervuren Museum, Belgium. It was presented at the International Conference of Rodent Pest Management in Blomfountein, South Africa in 2010 (Fichet-Calvet, Leirs and Rogers, unpublished).

Sierra Leone and Liberia, where a large-scale deforestation was performed for mining activities and timber trade.

Reproduction is seasonal in Tanzania, starting shortly after the beginning of the rainy season (November to June) and ending in the dry season (July to October) (Leirs et al., 1994). Conversely in commensal populations, reproduction occurs throughout the year, although a highest fecundity was observed in the rainy season. Conscquently the average age of the rodent population was older at the beginning of the year and younger at the end of the rainy season (Fichet-Calvet et al., 2008). Studies on *M. natalensis* dynamics in the high Lassa endemic zone in Upper Guinea (see the details of the security procedure in Box 5.1) showed that the optimum habitats for the species were the houses and the proximal cultivations, with animals never found in forest (Figure 5.6). However, it is clear that the rodents are using the habitats differently according to season in Guinea: they were abundant inside and rare outside during the dry season in January. Conversely in the rainy season, from May to September, they were as numerous inside and outside in the proximal cultivations (Figure 5.7), (Fichet-Calvet et al., 2007).

Box 5.1 Security procedure for trapping rodents under tropical field conditions

This procedure was initiated in 2002 in Guinea, to be able to be run under African tropical conditions, at 30°C under shadow and without power supply. It follows most of the guidelines in the protocol for virologic testing in small mammals published by Mills et al. (1995). Up to now, it was successfully exported in Côte d'Ivoire (Coulibaly N'Golo et al. 2011), Ghana (Kronmann et al., in press) and Nigeria (Olayemi & Fichet-Calvet, unpublished).

Protocol visit: Visit the villagers and explain the purpose of the expedition, and introduce the team. Go around the village by walking and identify the different habitats where the traps will be placed. Ask the chief to delegate some people to accompany the field team. Find a shady spot near the village and ask for one or two maneuvers to clean a circle of 30 m² and to dig a well deep of 80 cm at 10 meters from the circle.

| Village of Bantou, Upper Guinea | Village of Tanganya, Upper Guinea | Morning activities in the village Bantou, Upper Guinea |

Setting the traps: The next day or several days later, during the afternoon, set 200 to 400 traps in houses, the surrounding fields, the savannah and woods. The traps are baited with peanut paste and dried fish. To allow comparison of the trapping success as a surrogate of abundance, the traps are distributed in line of 100 m with 20 traps each 5 m outside. Inside, 2 traps by room are set, and the houses are sampled in line.

| Preparation of the bait | Setting the traps | Setting the traps |

Checking the traps: In the morning, wear a gown, gloves and a P3 mask and check the traps. Put the full traps in plastic and close the others so that no catches happen during the day. Go back to the necropsy study camp, and lay the full traps on the ground.

Box 5.1 Security procedure for trapping rodents under tropical field conditions—cont'd

| Checking the traps | Checking the traps | Gathering the traps on the study camp |

Preparation of the necropsy session: Set up the autopsy table, latex gloves, dissection tools, the box containing the disinfectant, paper towels, spray with alcohol 70%, the squeeze bottle with alcohol 70%, cryo and normal tubes, racks, syringes, data sheets, pencils, permanent markers, a cork board with pins, a balance at the nearest g, and two bins, one for all comers, and one for the needles.

| Daily preparation of the study camp | Preparing the data sheets |

Necropsies: Four people are required for this phase: one performs euthanasia on trapped animals, one takes notes, one does the autopsy and one ensures all intermediate tasks are done, such as cleaning dissection instruments during the session, supplying the autopsy table, etc. All such persons should wear a gown, a P3 mask, gloves, and goggles. Only the person who undertakes the autopsy wears a face shield. The two persons at the table wear a double pair of gloves, the first one long and thick, and the second one normal.

The animal is humanely killed with an overdose of fluoran soaked in cotton. Introduce cotton into the trap and roll it in the plastic. Wait 30 seconds to 5 minutes, depending on the species. Set the dead animal on the table. Weigh, measure the head and body length, the tail length, the feet and ears, and give these values aloud. Provide a preliminary morphological identification. Extend the animal on its back, disinfect the belly with alcohol 70% and cut the chest

Continued

Box 5.1 Security procedure for trapping rodents under tropical field conditions—cont'd

at the sternum level. Perform a cardiac puncture and dispose of the blood in a cryotube or on a filter paper. Discard the syringe in the special waste, and continue the autopsy by fully opening the abdomen. Measure the width of uterus, the number of fetuses in females, the width and length of the seminal vesicles in males. Take the spleen, liver, kidneys and other organs if necessary and put them in cryotubes. Keep organ samples in alcohol (70%) for further molecular identification. Remove all organs, attach a tag to a hind leg and keep the carcass in a tank filled with alcohol at 70% or formalin at 3.7%. Put the set of tools, small and large scissors, and pincers in the disinfectant. Roll the paper towel with the intestine and discard. Remove and discard also the first pair of gloves. Disinfect the table and his/her hands. Put on a new pair of gloves, prepare a new set of tools and wait for the next animal.

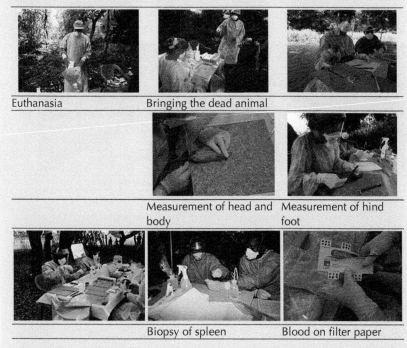

Euthanasia   Bringing the dead animal

Measurement of head and   Measurement of hind
body   foot

Biopsy of spleen   Blood on filter paper

Disinfection: During the daily necropsy session, all the traps which contained an animal are bleached for 5 minutes, and then rinsed and left to dry at the border of the study camp. After disinfection, the cryotubes are stored in the cool box during the day. At the end of the session, all the items set on the table must be disinfected. Soft items are disinfected with alcohol (70%), hard items with disinfectant (incidine, sekusept). The table and the chairs are also disinfected. The wastes,

Box 5.1 Security procedure for trapping rodents under tropical field conditions—cont'd

gowns, mask are burnt in situ in the well, and everybody is allowed to remove masks when no infectious material is remaining.

Disinfection of traps with bleach	Disinfection of personnel with alcohol 70%	Disinfection of the tubes with incidine 1%

Disinfection of the tools with incidine 1%	Disinfection of the tables with incidine 1%	Daily burning of the lab-coats and waste

Rebaiting: Empty and clean traps are set again on the same place as before and all the traps are set again for another night. The trapping session lasts 3 nights. At the end of the 3 days, all the traps are removed and bleached.

Liquid nitrogen: Each evening, dry and seal each cryotube in a piece of cryo-flex, and store them in a nitrogen tank.

Filling the tank before expedition	Sealing the tube before preservation in liquid nitrogen	Putting the tube inside the nitrogen tank

Spatial risk at local scale
spatial distribution of *M. natalensis*
according to a decreasing anthropic gradient

Species	houses	proximal cultivations	distal cultivations	forest
M. natalensis	447 (75%)	140 (23.5%)	9 (1.5%)	0
LASV positive	51	28	1	

M. natalensis confined to houses and gardens

FIGURE 5.6   Representation of the spatial risk according to a decreasing anthropic gradient.

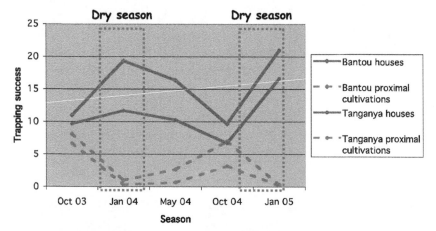

FIGURE 5.7   Dynamics in *Mastomys natalensis* living inside and outside in two villages located in Upper Guinea. January represented the middle of the dry season. *Adapted from Fichet-Calvet et al. 2007.*

At first, this finding appeared to be the key to understanding a higher risk of transmission during the dry season, as often claimed by the doctors working on Lassa fever. However, virus prevalence in the rodent population was two to three times higher during the rainy season, apparently contradicting a higher risk in dry season. Some other parameters such as the location of the village, the habitat, the age, sex and reproductive status of the rodents had an influence on virus prevalence (Fichet-Calvet et al., 2008). Lassa virus was observed in all age classes, including individuals of 1 month old, which is compatible with the vertical transmission within the species. Nevertheless, the Lassa prevalence was increased in young adults, suggesting that horizontal transmission also occurs. An increasing seroprevalence by age also supports this hypothesis (Demby et al., 2001).

FIGURE 5.8   Examples of rural dwellings in West Africa.

TRANSMISSION: RODENT TO HUMAN

Humans become infected in several ways and there is no clear evidence to suggest whether one route is more significant than the others. Individuals can be exposed through ingestion of food or touching objects that are contaminated with infected rodent excreta. The rodents shed the virus through urine, droplets and saliva for a period of a few days or few months, depending of the age at which they have been infected (Walker et al., 1975). During the night, rodents explore human dwellings, dispersing the virus on the ground, on surfaces, walls and roofs. This is particularly intense during the dry season when the houses are full of preserved food, and when many rodents converge towards this source of food (Figure 5.8).

Humans can be infected by ingestion of virus during the day immediately following distribution of virus by rodents, but they can also be infected several days after, since the virus can survive for several days outside the host. A recent experiment demonstrated that Lassa virus survived and remained virulent after 10 days on a solid surface (Sagripanti et al., 2010). Inside dwellings, the virus is also protected from the UV light, and can survive for a longer time period (Sagripanti & Lytle, 2011).

Another form of transmission can be the inhalation of virus-laden particles, which has been demonstrated by Stephenson et al. (1984), who successfully infected guinea pigs and monkeys with Lassa aerosols (Stephenson et al., 1984). Another demonstration of aerosol transmission was shown between mice infected by Lassa virus and healthy monkeys. Two of twelve monkeys died of Lassa fever after being held in the same room for 12 days after inoculation of the mice (Peters et al., 1987). It is likely that such a mechanism of transmission in rural areas occurs, especially during the dry season when the relative humidity is low, producing a lot of dust in the ambient atmosphere. At this time, many people remain at home because they have nothing to do in the fields. While waiting for the next plowing season, they relax, repair tools, sell crops, and spend a lot of time in the domestic area where the *Mastomys* are abundant. The risk of transmission could be then higher in this season, due to human behavior, and not due to Lassa virus prevalence in the rodent population.

Finally, hunting, trapping, butchering or eating rodents have been evoked as a possible risk behavior for contracting Lassa fever, but the relationship between rodent consumption and Lassa seroprevalence was not significant in Forest Guinea (ter Meulen et al., 1996; Kerneis et al., 2009).

TRANSMISSION: HUMAN TO HUMAN

As was seen during the first nosocomial outbreaks, Lassa virus is highly transmissible from human to human. This happens during patient care; the route of entry is usually a cut or a needle-stick injury when a healthy person comes into contact with the blood or secretions of the sick patient. It has often been the result of a lack of barrier nursing and monitoring a strict procedure for disinfection by hospital staff, especially by wearing gloves, gowns and masks (Fisher-Hoch et al., 1995). Many doctors and nurses have paid and are still paying with their lives for this lack of protection: Dr. J. Troup, 1970, Nigeria; Dr. Conteh, 2004, Sierra Leone and recently again in Nigeria (Ehichioya et al. 2010). Human-to-human transmission also occurs in the home when families take care of a patient, but also through sexual contact during recovery (McCormick, 1999). For example, a study of village communities in Sierra Leone showed that persons living in the same bedroom were infected more often than other unrelated members of the tribe (Fraser et al., 1974). Finally, Lassa fever could be a sexually transmitted disease, but we don't know how long the virus persists in semen or vaginal secretions. In urine and saliva, however, the virus persists intermittently for 2–4 days or up to two months after the acute phase. This represents a high risk of transmission during the recovery period (Monath et al., 1973; Monath et al., 1974b; Emond et al., 1982; Lunkenheimer et al., 1990). Anecdotally, Dr. Casals, the fourth case during the initial outbreak in the USA, was asked to avoid any contact with his family because he was shedding live virus in urine 32 days after the onset of fever. Virus shed by humans can also be aerosolized and several human-to-human transmissions by air were suspected such as those occurring in Jos, Nigeria in 1970 (Carey et al., 1972).

TRANSMISSION: HUMAN TO RODENT

This way of transmission is not documented in the literature, but this hypothesis is being developed to understand the epidemiology of the disease. Since humans can excrete the virus in urine and saliva during convalescence for up

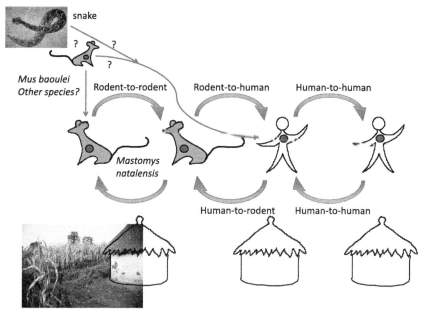

FIGURE 5.9   Wild and domestic cycle of transmission of Lassa virus.

to 2 months, and virus can survive outside the host for at least 12 days, they could become vectors by urinating in the surroundings of the village, or by spitting regularly on the soil. Rodents would be infected by digging the contaminated soil in some places around the houses, using the same paths as humans to enter and come out the houses. Figure 5.9 summarizes the cycle of transmission between rodents and humans.

DISTRIBUTION OF DISEASE

Lassa fever appears to have two geographically separate endemic areas: the Mano River region (Guinea, Sierra Leone and Liberia) in the West, and Nigeria in the East. In between these two areas, namely in Mali, Côte d'Ivoire, Ghana, Togo and Benin, no outbreak has ever been recorded, though isolated cases show evidence of viral circulation in these countries (Saltzmann, 1978; Günther et al., 2000; Akoua-Koffi et al., 2006; Atkin et al., 2009; Safronetz et al., 2010). Because virus was more prevalent in the rainy season than in the dry season, several environmental variables such as rainfall, temperature, vegetation index and elevation were entered in a computational model to produce a Lassa fever risk map (Figure 5.10, Fichet-Calvet & Rogers, 2009). The best predictors for risk of infection were rainfall between 1500 and 3000 mm, and annual mean temperature below 27°C. This was to some extent expected as simply superimposing the Lassa cases and the rainfall map demonstrated this correlation

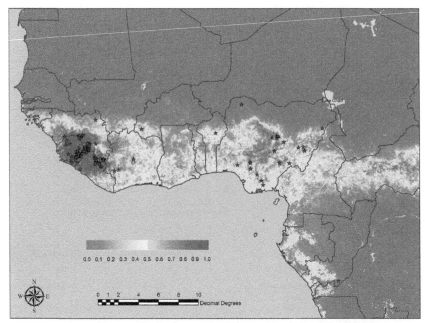

FIGURE 5.10   Lassa fever risk map in West Africa, with the higher risk located in the red zones (Fichet-Calvet & Rogers, 2009). Blue stars represent confirmed Lassa cases or studies demonstrating a seroprevalence superior to 10%.

(Figure 5.1). Nevertheless, the distribution of the reservoir *M. natalensis* is spread throughout Africa from East of Senegal to South Africa. But there is no Lassa fever in Central, Eastern and South Africa. One explanation could be that the host species is split into several genotypes with one genotype west to the Cameroon volcano chain, and another genotype east of this chain. This possibly explains why we have Lassa cases in Nigeria, and none in Cameroon, despite similar environmental conditions. Another explanation could be the presence of a more cryptic reservoir, which is not present in Central, Eastern and South Africa; this could be *M. baoulei.*

A persisting question is: why is the distribution of infected *M. natalensis* so patchy at the regional scale? Several spatial surveys in Guinea, Sierra Leone, Liberia, Cote d'Ivoire, Ghana and Nigeria indicated that very few localities are positive in West Africa. Investigations in high endemic zones showed some negative localities as well (Figure 5.11). This is a recurrent question for all zoonotic diseases, for which there is a larger distribution of the host, compared to those of the disease, for example *Myodes glareolus* / Puumala virus (*Bunyaviridae*), *Peromyscus maniculatus* / Sin Nombre virus (*Bunyaviridae*) and *Myotis daubentoni* / European bat lyssavirus type 2 (*Lyssavirus*). For Lassa in particular, we can suggest several explanations. The first could be local viral extinction due to demographic variations in host populations. In Forest Guinea, for

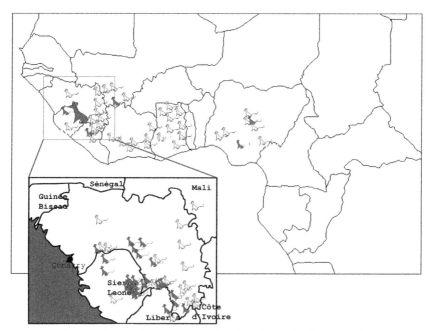

FIGURE 5.11   **Distribution of Lassa positive (in red) and negative (in green)** *M. natalensis* **in West Africa. The data are compiled from (Carey et al., 1972; Monath, 1975; Monath et al., 1974; Wulff et al., 1975; Robbins et al., 1983; McCormick et al., 1987b; Demby et al., 2001; Clegg, 2002; Lecompte et al., 2006; Fair et al., 2007; Coulibaly-N'Golo et al., 2011; Kronmann et al., in press; Safronetz et al., 2010; Olayemi, pers. com.).**

example, the village Bamakama was investigated twice, firstly in 1996–1997 by Demby et al. (2001), and secondly in 2005 by Lecompte et al. (2006). In 1996–1997, 9% of the *Mastomys* were positive (antigen or antibody positive), whereas no animal was found positive in 2005 (0/35). However, infected rodents are probably distributed in patches inside the village and it is also possible that the sampling did not cover the infected houses. In Sierra Leone, Keenlyside and co-workers (Keenlyside et al., 1983) demonstrated that the *Mastomys* in houses where infections had occurred were 10 times more infected than those trapped in control houses (39% versus 3.7%). A second explanation of the virus distribution could be due to human displacement. The classical hypothesis is the passive transportation of rats, including some infected ones, in trucks or in trains with food supplies along roads. This is often suggested because we are focused on a rodent-to-human transmission model, and not human-to-rodent transmission. But, in the latter scenario, there is a higher probability for humans to move and disperse the virus, than the rodent over long distances. Indeed, humans could be considered as vectors, shedding the virus in localities when they are traveling, such as visiting relatives, through trade or escaping conflicts. In this regard, it is interesting to note that Sierra Leone, where there was a civil war for

10 years between 1991 and 2002, is also the area with many Lassa fever cases. This was regularly reported by different nongovernmental organizations such as Merlin or the United Nations High Commission for Refugees (UNHCR), which worked in Kenema during the war. At the hospital, many cases came from different camps: Jimmi, Gerihun, Largo, Taiama, and Dauda (Bottineau & Kampudu, 2003). These camps were located in the distribution area of *M. natalensis*, but the most surprising was the case of Madina Oula camp, where human seroprevalence was 34% (59/171 in Lukaschevich et al., 1993), whereas there was no *M. natalensis* present in this area. Indeed, rodent surveillance in Guinea between 2002 and 2006 showed *M. natalensis* was absent from the coastal zone (Lecompte et al., 2006; Fichet-Calvet et al., 2009; Lalis et al., 2012). This case illustrates that a high human seroprevalence could be the consequence of human-to-human transmission only, without intervention of the reservoir at the beginning.

CO-EVOLUTION

It has often been suggested that the virus evolves with its host. This hypothesis has been demonstrated for the New World arenaviruses, then repeated for the first analyses of the Old World arenaviruses (Bowen et al., 1997; Mills et al., 1997; Salazar-Bravo et al., 2002; Gonzalez et al., 2007). But the more we discover new arenaviruses, the more the image of a co-evolution fades, because the comparison of phylogenetic trees of the virus on one hand, and their hosts on the other hand, do not match. In fact, the host switching would be frequent, as suggested by recent studies conducted on the viruses Gbagroube and Menekre in Côte d'Ivoire (Coulibaly-N'Golo et al., 2011) and on all the New World arenaviruses in the Americas (Irwin et al., 2012). The latest discovery of a Lassa-like virus in a pygmy mouse in Ghana points also in that direction. Figure 5.12 summarizes the state of knowledge of arenaviruses and their hosts in Africa.

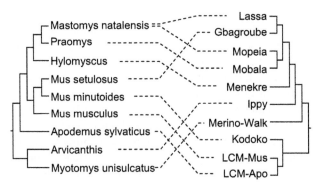

FIGURE 5.12   **Juxtaposition of rodent host and arenavirus tree.** *Reproduced from Coulibaly N'Golo et al., 2011.*

In the case of Lassa fever virus in particular, all studies show that it was a recent emergence of the order of 700–900 years, while its host *M. natalensis* emerged 2.3 million years ago (Brouat et al., 2009; Mouline et al., 2008). Both time scales do not match when analyzing the time of speciation in another arenavirus found in *M. natalensis*, Mopeia virus, which emerged 500 years ago in East Africa. Even the oldest arenavirus discovered in South Africa, Lujo, would be old at only 4500 years (Coulibaly-N'Golo et al., 2011).

But new opportunities are emerging thanks to the discovery of new arenaviruses in American snakes, the boa constrictor and the annulated tree boa (Stenglein et al., 2012). These arenaviruses are responsible for snake inclusion body disease, and can cause the loss of many animals in captivity. They are phylogenetically very divergent from the old and new world clades, and share similarities with filoviruses especially over the glycoprotein sequence. This is the first time that an arenavirus has been described in reptiles, sweeping away the strict association between rodents and this family of viruses. Given their basal position to all arenaviruses, we can assume that they are ancestral strains, possibly at the origin of a host switching between snakes and rodents, or between snakes and humans. Rodents may be contaminated by using the same habitats as snakes, but humans could also be contaminated when hunting snakes, snake consumption being common in Africa (Figure 5.13). Bites by snakes, whose deadly incidence is estimated at 0.01% in West Africa (Trape & Mané, 2006), would also be a route of transmission. Therefore, this finding is of huge importance because it opens the field for new investigations in evolutionary biology and epidemiology of Lassa virus.

CONCLUSIONS

The two Lassa fever foci, at a distance of 2000 km, remain an enigma because the reservoir is present everywhere in West Africa and also favorable climatic conditions exist between these regions. In fact, it will remain an enigma until we suggest a scenario involving the transmission of the virus from human to rodent, in a reversal of the normal route of zoonotic transmission as documented for other viruses (H1N1 influenza in swine, metapneumavirus in chimpanzees). Why not imagine that humans were transported with the virus from Nigeria to Sierra Leone? Historical investigations suggest that the Atlantic slave trade and its abolition in 1807 could be at the origin of the emergence of Lassa in Sierra Leone. Between 1808 and 1863, the Royal Navy had been very active in patrolling along the coast between Senegambia and Cameroon to intercept slave ships through its dedicated unit, the West African squadron. Thus, 96,000 captives were liberated in Freetown, with 69,000 of them originating from Nigeria and Benin (Eltis et al., 2008). This massive importation of potentially infected people in the first half of the 19th century could explain the emergence of Lassa fever 150–200 years ago in Sierra Leone. Furthermore, a unique lineage in the Sierra Leone foci in comparison to three lineages in the Nigerian foci could be a

FIGURE 5.13 **Hunter showing his game: a puff adder (*Bitis arietans*), Tanganya, Upper Guinea, 2005.** © *E. Fichet-Calvet*

signature of a genetic bottleneck, leading to a founder effect by a small virus population. Absence of virus positive *Mastomys* in Côte d'Ivoire and Ghana also argue towards a movement of the virus via direct transportation by sea rather than an overland translocation. Finally, traces of the Yoruba language in Sierra Leone argue towards a migration of Nigerians to Sierra Leone.

The human-to-rodent transmission allows us to understand the distribution of Lassa fever within each high endemic zone. Indeed, much evidence shows some positive and negative locations in the same area, either through observation of the human seroprevalence or the rodent virus prevalence. Why do some villages have Lassa virus positive *Mastomys*, and others do not? We could explain this situation by human displacements dispersing the virus randomly. The maintenance of each micro-focus would then be due to the presence of the reservoir, which can amplify the disease locally. As a commensal rodent, *Mastomys* fills the role of domestic animals, which are known to amplify the virus transmission of zoonosis towards humans (Karesh et al., 2012).

Beyond the biotic and abiotic conditions necessary for virus transmission, there is a large field almost uninvestigated: the social sciences. From Guinea to Nigeria, more than 100 different tribes share this territory, and their tolerance towards the rodents, the consumption of rodents and storage of food and crops vary. In Guinea, for example, Peul never "put a mouse into a pan," while Soussou, Malinke, Kissi and Guerze regularly eat mice (personal observations). The granaries are well developed in sub-Saharan areas, with the local population introducing an unexpected guard such as a snake that lives at the bottom of the granary (personal observations in Northern Cameroon). Therefore, some comparative studies in the cultural practices of local populations could help us to better understand the epidemiology of Lassa fever.

REFERENCES

Akoua-Koffi, C., ter Meulen, J., Legros, D., Akran, V., Aidara, M., Nahounou, N., Dogbo, P., & Ehouman, A. (2006). Detection of anti-Lassa antibodies in the Western Forest area of the Ivory Coast. *Méd. Trop.*, *66*, 465–468.

Albariño, C. G., Bird, B. H., Chakrabarti, A. K., Dodd, K. A., Erickson, B. R., & Nichol, S. T. (2011). Efficient rescue of recombinant Lassa virus reveals the influence of S segment noncoding regions on virus replication and virulence. *J. Virol.*, *85*, 4020–4024.

Asogun, D. A., Adomeh, D. I., Ehimuan, J., Odia, I., Hass, M., Gabriel, M., Olschläger, A., Becker-Ziaja, B., Folarin, O., Phelan, E., Ehiane, P. E., Ifeh, V. E., Uyigue, E. A., Oladapo, Y. T., Muoebonam, E. B., Osunde, O., Dongo, A., Okokhere, P. O., Okogbenin, S. A., Momoh, M., Alikah, S. O., Akhuemokhan, O. C., Imomeh, P., Odike, M. A., Gire, S., Anderson, K., Sabeti, P. C., Happi, C. T., Akpede, G. O., & Günther, S. (2012). Molecular diagnostics for Lassa fever at Irrua specialist teaching hospital, Nigeria: lessons learnt from two years of laboratory operation. *PLoS Neglected Tropical Diseases*, *6*, e1839.

Atkin, S., Anaraki, S., Gothard, P., Walsh, A., Brown, D., Gopal, R., Hand, J., & Morgan, D. (2009). The first case of Lassa fever imported from Mali to the United Kingdom, February 2009. *Euro Surveill*, *14*(10), pii=19145.

Auperin, D. D., & McCormick, J. B. (1989). Nucleotide sequence of the Lassa virus (Josiah strain) S genome RNA and amino acid sequence comparison of the N and GPC proteins to other arenaviruses. *Virology*, *168*, 421–425.

Auperin, D. D., Compans, R. W., & Bishop, D. H. (1982). Nucleotide sequence conservation at the 3′ termini of the virion RNA species of New World and Old World arenaviruses. *Virology*, *121*, 200–203.

Bausch, D. G., Demby, A. H., Coulibaly, M., Kanu, J., Goba, A., Bah, A., Condé, N., Wurtzel, H. L., Cavaliaro, K. F., Lloyd, E., Baldet, F. B., Cissé, S. D., Fofana, D., Savané, I. K., Tolno, R. T., Mahy, B., Wagoner, K. D., Ksiazek, T. G., Peters, C. J., & Rollin, P. E. (2001). Lassa fever in Guinea: I. Epidemiology of human disease and clinical observations. *Vector Borne and Zoonotic Dis.*, *1*, 269–281.

Boiro, I., Lomonossov, N. N., Sotsinski, V. A., Constantinov, O. K., Tkachenko, E. A., Inapogui, A. P., & Balde, C. (1987). Eléments de recherches clinico-épidémiologiques et de laboratoire sur les fièvres hémorragiques en Guinée. *Bulletin de la Société de Pathologie Exotique, 80,* 607–612.

Bottineau, M. C., & Kampudu, F. (2003). *Lassa Fever, summary of the situation in Sierra Leone focussing on refugee camps and way stations (1st of January to 30th of April 2003).* UNHCR.

Bowen, G. S., Tomori, O., Wulff, H., Casals, J., Noonan, A., & Downs, W. G. (1975). Lassa fever in Onitsha, East Central State, Nigeria, in 1974. *Bull. WHO., 52,* 599–603.

Bowen, M. D., Peters, C. J., & Nichol, S. T. (1997). Phylogenetic analysis of the Arenaviridae: patterns of virus evolution and evidence for cospeciation between arenaviruses and their rodent hosts. *Mol. Phylogenet. Evol., 8,* 301–316.

Bowen, M. D., Rollin, P. E., Ksiazek, T. G., Hustad, H. L., Bausch, D. G., Demby, A. H., Bajani, M. D., Peters, C. J., & Nichol, S. T. (2000). Genetic diversity among Lassa virus strains. *J. Virol., 74,* 6992–7004.

Briese, T., Paweska, J. T., McMullan, L. K., Hutchison, S. K., Street, C., Palacios, G., Kristova, M. l, Weyer, J., Swanepoel, R., Egholm, M., Nichol, S. T., & Lipkin, W. I. (2009). Genetic detection and characterization of Lujo virus, a new hemorrhagic fever-associated arenavirus from southern Africa. *PLoS Pathog., 5,* e1000455.

Britton-Davidian, J., Catalan, J., Granjon, L., & Duplantier, J. -M. (1995). Chromosomal phylogeny and evolution in the genus *Mastomys* (Mammalia, Rodentia). *J. Mammal, 76,* 248–262.

Brouat, C., Tatard, C., Ba, K., Cosson, J. F., Dobigny, G., Fichet-Calvet, E., Granjon, L., Lecompte, E., Loiseau, A., Mouline, K., Piry, S., & Duplantier, J. M. (2009). Phylogeography of the Guinea multimammate mouse (*Mastomys erythroleucus*): a case study for Sahelian species in West Africa. *J. Biogeog., 36,* 2237–2250.

Buckley, S. M., Casals, J., & Downs, W. G. (1970). Isolation and antigenic characterization of Lassa virus. *Nature, 227,* 174.

Carey, D. E., Kemp, G. E., White, H. A., Pinneo, L., Addy, R. F., Fom, A. L., Stroh, G., Casals, J., & Henderson, B. E. (1972). Lassa fever epidemiological aspects of the 1970 epidemic, Jos, Nigeria. *Trans. Roy. Soc. Trop. Med. Hyg., 66,* 402–408.

Clegg, J. C., Wilson, S. M., & Oram, J. D. (1991). Nucleotide sequence of the S RNA of Lassa virus (Nigerian strain) and comparative analysis of arenavirus gene products. *Virus Res., 18,* 151–164.

Clegg, J. C. S. (2002). Molecular phylogeny of the Arenaviruses. *Curr. Top. Microbiol. Immunol., 262,* 1–24.

Coulibaly-N'Golo, D., Allali, B., Kouassi, S. K., Fichet-Calvet, E., Becker-Ziaja, B., Rieger, T., Olschläger, S., Dosso, H., Denys, C., Ter Meulen, J., Akoue-Koffi, C., & Günther, S. (2011). Novel arenavirus sequences in *Hylomyscus* sp. and *Mus (Nannomys) setulosus* from Cote d'Ivoire: implications for evolution of arenaviruses in Africa. *PLoS ONE, 6,* e20893.

Cummins, D., McCormick, J. B., Bennett, D., Samba, J. A., Farrar, B., Machin, S. J., & Fisher-Hoch, S. P. (1990). Acute sensorineural deafness in Lassa fever. *JAMA, 264,* 2093–2096.

Demby, A. H., Inapogui, A., Kargbo, K., Koninga, J., Kourouma, K., Kanu, J., Coulibaly, M., Wagoner, K. D., Ksiazek, T. G., Peters, C. J., Rollin, P. E., & Bausch, D. G. (2001). Lassa fever in Guinea: II. Distribution and prevalence of Lassa virus infection in small mammals. *Vector Borne and Zoonotic Dis., 1,* 283–296.

Duplantier, J. M., Britton-Davidian, J., & Granjon, L. (1990). Chromosomal characterization of three species of the genus *Mastomys* in Senegal. *Zeitschrift für zoologische Systematik und Evolutionsforschung, 28,* 289–298.

Ehichioya, D. U., Hass, M., Becker-Ziaja, B., Ehimuan, J., Asogun, D. A., Fichet-Calvet, E., Kleinstuber, K., Lelke, M., ter Meulen, J., Akpede, G. O., Omilabu, S. A., Günther, S., & Olschläger, S. (2011). Current molecular epidemiology of Lassa virus in Nigeria. *J. Clin. Microbiol.*, *49*, 1157–1161.

Ehichioya, D. U., Asogun, D. A., Ehimuan, J., Okokhere, P. O., Pahlmann, M., Olschläger, S., Becker-Ziaja, B., Günther, S., & Omilabu, S. A. (2012). Hospital-based surveillance for Lassa fever in Edo State, Nigeria, 2005-2008. *Trop. Med. Int. Health*, *17*, 1001–1004.

Ehichioya, D. U., Hass, M., Olschläger, S., Becker-Ziaja, B., Onyebuchi Chukwu, C. O., Coker, J., Nasidi, A., Ogbu Ogugua, O., Günther, S., & Omilabu, S. A. (2010). Lassa fever, Nigeria, 2005-2008. *Emerging infectious diseases*, *16*(6), 1040–1041. doi: 10.3201/eid1606.100080 (June 2010).

Eltis, D., Richardson, D., Florentino, M., & Stephen, B. (2008). *The Trans-Atlantic slave trade database*. Available at www.slavevoyages.org.

Emond, R. T., Bannister, B., Lloyd, G., Southee, T. J., & Bowen, E. T. (1982). A case of Lassa fever: clinical and virological findings. *BMJ.*, *285*, 1001–1002.

Fair, J., Jentes, E., Inapogui, A., Kourouma, K., Goba, A., Bah, A., Tounkara, M., Coulibaly, M., Garry, R. F., & Bausch, D. G. (2007). Lassa virus-infected rodents in refugee camps in Guinea: a looming threat to public health in a politically unstable region. *Vector Borne Zoonotic Dis.*, *7*, 167–172.

Fichet-Calvet, E., Lecompte, E., Koivogui, L., Soropogui, B., Dore, A., Kourouma, F., Sylla, O., Daffis, S., Koulémou, K., & Ter Meulen, J. (2007). Fluctuation of abundance and Lassa virus prevalence in *Mastomys natalensis* in Guinea, West Africa. *Vector Borne Zoonotic Dis.*, *7*, 119–128.

Fichet-Calvet, E., Lecompte, E., Koivogui, L., Daffis, S., & Ter Meulen, J. (2008). Reproductive characteristics of *Mastomys natalensis* and Lassa virus prevalence in Guinea, West Africa. *Vector Borne Zoonotic Dis.*, *8*, 41–48.

Fichet-Calvet, E., & Rogers, D. J. (2009). Risk maps of Lassa fever in West Africa. *PLoS Neglected Tropical Diseases*, *3*, e388.

Fichet-Calvet, E., Lecompte, E., Veyrunes, F., Barriere, P., Nicolas, V., & Koulemou, K. (2009). Diversity and dynamics in a community of small mammals in coastal Guinea, West Africa. *Belg. J. Zool.*, *139*, 93–102.

Fisher-Hoch, S. P., Tomori, O., Nasidi, A., Perez-Oronoz, G. I., Fakile, Y., Hutwagner, L., & McCormick, J. B. (1995). Review of cases of nosocomial Lassa fever in Nigeria: the high price of poor medical practice. *BMJ.*, *311*, 857–859.

Frame, J. D. (1975). Surveillance of Lassa fever in missionaries stationed in West Africa. *Bull. WHO.*, *52*, 593–598.

Frame, J. D., Baldwin, J. M., Jr., Gocke, D. J., & Troup, J. M. (1970). Lassa fever, a new virus disease of man from West Africa. *Am. J. Trop. Med. Hyg.*, *19*, 670–676.

Fraser, D. W., Campbell, C. C., Monath, T. P., Goff, P. A., & Gregg, M. B. (1974). Lassa fever in the eastern province of Sierra Leone, 1970-1972. *Am. J. Trop. Med. Hyg.*, *23*, 1131–1139.

Georges, A. J., Gonzalez, J. P., Abdul-Wahid, S., Saluzzo, J. F., Meunier, D. M. Y., & McCormick, J. B. (1985). Antibodies to Lassa and Lassa-like viruses in man and mammals in the Central African Republic. *Trans. Royal Soc. Trop. Med. Hyg.*, *79*, 78–79.

Gonzalez, J. P., McCormick, J. B., Saluzzo, J. F., Herve, J. P., Georges, A. J., & Johnson, K. M. (1983). An arenavirus isolated from wild-caught rodents (*Praomys species*) in the Central African Republic. *Intervirology*, *19*, 105–112.

Gonzalez, J. P., Josse, R., Johnson, E. D., Merlin, M., Georges, A. J., Abandja, J., Danyod, M., Dupont, A., Ghogon, A., Koula-Bemba, D., Madelon, M. C., Sima, A., & Meunier, D. M. Y. (1989). Antibody prevalence against haemorrhagic fever viruses in randomized representative Central African populations. *Res. Virol.*, *140*, 319–331.

Gonzalez, J. P., Emonet, S., de Lamballerie, X., & Charrel, R. (2007). Arenaviruses. *Curr. Top. Microbiol. Immunol.*, *315*, 253–288.

Gouy de Bellocq, J. G., Borremans, B., Katakweba, A., Makundi, R., Baird, S. J., Becker-Ziaja, B., Günther, S., & Leirs, H. (2010). Sympatric occurrence of 3 arenaviruses, Tanzania. *Emerg. Infect. Dis.*, *16*, 692–695.

Granjon, L., Duplantier, J. M., Catalan, J., Britton-Davidian, J., & Bronner, G. N. (1996). Conspecificity of *Mastomys natalensis* (Rodentia: Muridae) from Senegal and South Africa: evidence from experimental crosses, karyology and biometry. *Mammalia*, *60*, 697–706.

Granjon, L., Duplantier, J. M., Catalan, J., & Britton-Davidian, J. (1997). Systematics of the genus *Mastomys* (Thomas, 1915) (Rodentia: Muridae): a review. *Belg. J. Zool.*, *127*, 7–18.

Günther, S., Emmerich, P., Laue, T., Kuhle, O., Asper, M., Jung, A., Grewing, T., ter Meulen, J., & Schmitz, H. (2000). Imported lassa fever in Germany: molecular characterization of a new lassa virus strain. *Emerg. Infect. Dis.*, *6*, 466–476.

Günther, S., & Lenz, O. (2004). Lassa virus. *Critical Reviews in Clinical Laboratory Sciences*, *41*, 339–390.

Henderson, B. E., Gary, G. W., Kissling, R. E., Frame, J. D., & Carey, D. E. (1972). Lassa fever virological and serological studies. *Trans. Royal Soc. Trop. Med. Hyg.*, *66*, 409–416.

Hubert, B., Meylan, A., Petter, F., Poulet, A., & Tranier, M. (1983). Different species in genus *Mastomys* from western, central and southern Africa (rodentia, Muridae). *Annales du Musée Royal pour l'Afrique Centrale, Sciences Zoologiques*, *237*, 143–148.

Irwin, N. R., Bayerlova, M., Missa, O., & Martinkova, N. (2012). Complex patterns of host switching in New World arenaviruses. *Molecular Ecology*, *21*, 4137–4150.

Ishii, A., Thomas, Y., Moonga, L., Nakamura, I., Ohnuma, A., Hang'ombe, B. M., Takada, A., Mweene, A. S., & Sawa, H. (2012). Molecular surveillance and phylogenetic analysis of Old World arenaviruses in Zambia. *J. Gen. Virol*, *93*, 2247–2251.

Karesh, W. B., Dobson, A. P., Llyod-Schmith, J. O., Lubroth, J., Dixon, M. A., Benett, M., Aldrich, S., Harrington, T., Formenty, P., Loh, E. H., Machalaba, C. C., Thomas, M. J., & Heymann, D. L. (2012). Ecology of zoonoses: natural and unnatural histories. *Lancet*, *380*, 1936–1945.

Keane, E., & Gilles, H. M. (1977). Lassa fever in Panguma hospital, Sierra Leone, 1973–6. *BMJ.*, *1*, 1399–1402.

Keenlyside, R. A., McCormick, J. B., Webb, P. A., Smith, E., Elliott, L., & Johnson, K. M. (1983). Case-control study of *Mastomys natalensis* and humans in Lassa virus-infected households in Sierra Leone. *Am. J. Trop. Med. Hyg.*, *32*, 829–837.

Kerneis, S., Koivogui, L., Magassouba, N., Koulemou, K., Lewis, R., Aplogan, A., Grais, R. F., Guerin, P. J., & Fichet-Calvet, E. (2009). Prevalence and risk factors of Lassa seropositivity in inhabitants of the forest region of Guinea: a cross-sectional study. *PLoS Neglected Tropical Diseases*, *3*, e548.

Khan, S. H., Goba, A., Chu, M., Roth, C., Healing, T., Marx, A., Fair, J., Guttieri, M. C., Ferro, P., Imes, T., Monagin, C., Garry, R. F., & Bausch, D. G. (2008). New opportunities for field research on the pathogenesis and treatment of Lassa fever. *Antiviral Research*, *78*, 103–115.

Klempa, B., Koulemou, K., Auste, B., Emmerich, P., Thome-Bolduan, C., Günther, S., Koivogui, L., Krüger, D. H., & Fichet-Calvet, E. (2013). Seroepidemiological study reveals regional co-occurrence of Lassa- and Hantavirus antibodies in Upper Guinea, West Africa. *Trop. Med. Int. Health*, *18*, 366–371.

Kronmann, K. C., Paintsil, S. N., Guirguis, F., Kronmann, L. C., Bonney, K., Obiri-Danso, K., Ampofo, W., & Fichet-Calvet, E. (in press). Two arenaviruses detected in pygmy mice in Ghana. *Emerg. Infect. Dis.*

Lalis, A., Leblois, R., Lecompte, E., Denys, C., Ter Meulen, J., & Wirth, T. (2012). The impact of human conflict on the genetics of *Mastomys natalensis* and Lassa virus in West Africa. *PLoS ONE, 7*, e37068.

Lecompte, E., Brouat, C., Duplantier, J. M., Galan, M., Granjon, L., Loiseau, A., Mouline, K., & Cosson, J. F. (2005). Molecular identification of four cryptic species of *Mastomys* (Rodentia, Murinae). *Biochemical Systematics and Ecology, 33*, 681–689.

Lecompte, E., Fichet-Calvet, E., Daffis, S., Koulemou, K., Sylla, O., Kourouma, F., Doré, A., Soropolui, B., Aniskin, V., Allali, B., Kouassi Kans, S., Lalis, A., Koivogui, L., Günther, S., Denys, C., & ter Meulen, J. (2006). *Mastomys natalensis* and Lassa fever. *West Africa. Emerg. Infect. Dis., 12*, 1971–1974.

Lecompte, E., ter Meulen, J., Emonet, S., Daffis, S., & Charrel, R. N. (2007). Genetic identification of Kodoko virus, a novel arenavirus of the African pigmy mouse (Mus *Nannomys minutoides*) in West Africa. *Virology, 364*, 178–183.

Leirs, H. (1994). *Population ecology of Mastomys natalensis (Smith, 1834). Implications for rodent control in Africa*. Bruxelles: Belgian Administration for Development Cooperation.

Leirs, H., Verhagen, R., & Verheyen, W. (1994). The basis of reproductive seasonality in *Mastomys* rats (Rodentia: Muridae) in Tanzania. *J. Trop. Ecol., 10*, 55–66.

Leirs, H., Verhagen, R., Verheyen, W., Mwanjabe, P., & Mbise, T. (1996). Forecasting rodent outbreaks in Africa: an ecological basis for *Mastomys* control in Tanzania. *J. App. Ecol, 33*, 937–943.

Leirs, H., Stenseth, N. C., Nichols, J. D., Hines, J. E., Verhagen, R., & Verheyen, W. (1997). Stochastic seasonality and non linear density-dependent factors regulate population size in an African rodent. *Nature, 389*, 176–180.

Lukashevich, L. S., Clegg, J. C., & Sidibe, K. (1993). Lassa virus activity in Guinea: distribution of human antiviral antibody defined using enzyme-linked immunosorbent assay with recombinant antigen. *J. Med. Virol., 40*, 210–217.

Lunkenheimer, K., Hufert, F. T., & Schmitz, H. (1990). Detection of Lassa virus RNA in specimens from patients with Lassa fever by using the polymerase chain reaction. *J. Clin. Microbiol., 28*, 2689–2692.

McCormick, J. B., King, I. J., Webb, P. A., Scribner, C. L., Craven, R. B., Johnson, K. M., Elliot, L. H., & Belmont-Williams, R. (1986). Lassa fever. Effective therapy with ribavirin. *N. Engl. J. Med., 314*, 20–26.

McCormick, J. B., King, I. J., Webb, P. A., Johnson, K. M., O'Sullivan, R., Smith, E. S., Trippel, S. ., & Tong, T. C. (1987a). A case-control study of the clinical diagnosis and course of Lassa fever. *J. Infect. Dis., 155*, 445–455.

McCormick, J. B., Webb, P. A., Krebs, J. W., Johnson, K. M., & Smith, E. S. (1987b). A prospective study of the epidemiology and ecology of Lassa fever. *J. Infect. Dis., 155*, 437–444.

McCormick, J. B. (1999). Lassa fever. In J. F. Saluzzo & B. Dodet (Eds.), *Emergence and control of rodent-borne viral diseases* (pp. 177–195). Elsevier.

McCormick, J. B., & Fisher-Hoch, S. P. (2002). Lassa fever. *Curr. Top. Microbiol. Immunol., 262*, 75–109.

Mills, J. N., Childs, J., Ksiazek, T. G., Peters, C. J., & Velleca, W. M. (1995). *Methods for trapping and sampling small mammals for virologic testing*. Atlanta: Centers dor Disease Control and Prevention.

Mills, J. N., Bowen, M. D., & Nichol, S. T. (1997). African arenaviruses—Coevolution between virus and murid host? *Belg. J. Zool., 127*, 19–28.

Monath, T. P., Mertens, P. E., Patton, R., Moser, C. R., Baum, J. J., Pinneo, L., Gary, G. W., & Kissling, R. E. (1973). A hospital epidemic of Lassa fever in Zorzor, Liberia, March-April 1972. *Am. J. Trop. Med. Hyg., 22*, 773–779.

Monath, T. P., Newhouse, V. F., Kemp, G. E., Setzer, H. W., & Cacciapuoti, A. (1974a). Lassa virus isolation from *Mastomys natalensis* rodents during an epidemic in Sierra Leone. *Science, 185,* 263–265.

Monath, T. P., Maher, M., Casals, J., Kissling, R. E., & Cacciapuoti, A. (1974b). Lassa fever in the eastern province of Sierra Leone, 1970–1972. II Clinical observations and virological studies on selected Hospital cases. *Am. J. Trop. Med. Hyg., 23,* 1140–1149.

Monath, T. P. (1975). Lassa fever: review of epidemiology and epizootiology. *Bull. WHO., 52,* 577–591.

Monson, M. H., Frame, J. D., Jahrling, P. B., & Alexander, K. (1984). Endemic Lassa fever in Liberia. I. Clinical and epidemiological aspects at Curran Lutheran Hospital, Zorzor, Liberia. *Trans. Royal Soc. Trop. Med. Hyg., 78,* 549–553.

Mouline, K., Granjon, L., Galan, M., Tatard, C., Abdoullaye, D., Ag Atteyine, S., Duplantier, J. M., & Cosson, J. F. (2008). Phylogeography of a Sahelian rodent species *Mastomys huberti*: a Plio-Pleistocene story of emergence and colonization of humid habitats. *Molecular Ecology, 17,* 1036–1053.

Murphy, F. A., Webb, P. A., Johnson, K. M., & Whitfield, S. G. (1969). Morphological comparison of Machupo with lymphocytic choriomeningitis virus: basis for a new taxonomic group. *J. Virol., 4,* 535–541.

Nakounne, E., Selekon, B., & Morvan, J. (2000). [Microbiological surveillance: viral hemorrhagic fever in Central African Republic: current serological data in man]. *Bulletin de la Société de Pathologie Exotique, 93,* 340–347.

Oldstone, M. B. A., & Campbell, K. P. (2011). Decoding arenaviruses pathogenesis: essential roles for alpha-Dystroglycan-virus interactions and the immune response. *Virology, 411,* 170–179.

Omilabu, S. A., Badaru, S. O., Okokhere, P., Asogun, D., Drosten, C., Emmerich, P., Becker-Ziaja, B., Schmitz, H., & Günther, S. (2005). Lassa fever, Nigeria, 2003 and 2004. *Emerg. Infect. Dis., 11,* 1642–1644.

Paweska, J. T., Sewlall, N. H., Ksiazek, T. G., Blumberg, L. H., Hale, M. J., Ian Lipkin, W., Weyer, J., Nichol, S. T., Rollin, P. E., McMullan, L. K., Paddock, C. D., Briese, T., Mnyaluza, J., Dinh, T. h., Mukonka, V., Ching, P., Duse, A., Richards, G., de Jong, G., Cohen, C., Ikalafeng, B., Mugero, C., Asomugha, C., Moltle, M. M., Nteo, D. M., Misia, E., Swanepoel, R., Zaki, S. R., & Outbreak Control and Investigation Teams (2009). Nosocomial outbreak of novel arenavirus infection, Southern Africa. *Emerg. Infect. Dis., 15,* 1598–1602.

Peters, C. J., Jahrling, P. B., Liu, C. T., Kenyon, R. H., McKee, K. T., & Barrera Oro, J. G. (1987). Experimental studies of arenaviral hemorrhagic fevers. *Curr. Top. Microbiol. Immunol., 134,* 5–65.

Robbins, C. B., Krebs, J. W., & Johnson, K. M. (1983). *Mastomys* (Rodentia: Muridae) species distinguished by hemoglobin pattern differences. *Am. J. Trop. Med. Hyg., 32,* 624–630.

Rowe, W. P., Murphy, F. A., Bergold, G. H., Casals, J., Hotchin, J., Johnson, K. M., Lehmann-Grube, F., Mims, C. A., Traub, E., & Webb, P. A. (1970). Arenoviruses: proposed name for a newly defined virus group. *J. Virol., 5,* 651–652.

Safronetz, D., Lopez, J. E., Sogoba, N., Traore, S. F., Raffel, S. J., Fischer, E. R., Egihara, H., Branco, L., Garry, R. F., Schwann, T. G., & Feldmann, H. (2010). Detection of Lassa virus. *Mali. Emerg. Infect. Dis., 16,* 1123–1126.

Sagripanti, J. L., Rom, A. M., & Holland, L. E. (2010). Persistence in darkness of virulent alphaviruses, Ebola virus, and Lassa virus deposited on solid surfaces. *Arch. Virol., 155,* 2035–2039.

Sagripanti, J. L., & Lytle, C. D. (2011). Sensitivity to ultraviolet radiation of Lassa, vaccinia, and Ebola viruses dried on surfaces. *Arch. Virol., 156,* 489–494.

Salazar-Bravo, J., Ruedas, L. A., & Yates, T. L. (2002). Mammalian reservoirs of Arenaviruses. *Curr. Top. Microbiol. Immunol., 262,* 25–63.

Saltzmann, S. (1978). *La fièvre de Lassa*. Neuchâtel: Editions des Groupes Missionnaires.

Stenglein, M. D., Sanders, C., Kistler, A. L., Ruby, J. G., Franco, J. Y., Reavill, D. R., Dunker, F., & Derisi, J. L. (2012). Identification, characterization, and in vitro culture of highly divergent arenaviruses from boa constrictors and annulated tree boas: candidate etiological agents for snake inclusion body disease. *mBio*, *3*, e00180–12.

Stephenson, E. H., Larson, E. W., & Dominik, J. W. (1984). Effect on environmental factors on aerosol-induced Lassa virus infection. *J. Med. Virol.*, *14*, 295–303.

Sylla, O. (2004). *Evaluation de l'audiométrie dans le diagnostic de l'ancienne infection de la Fièvre de Lassa à Bouramayah, dans la préfecture de Dubréka, République de Guinée*. (Medicine thesis), University Gamal Abdelnasser de Conakry.

Talani, P., Konongo, J. D., Gromyko, A. I., Nanga-Maniane, J., Yala, F., & Bodzongo, D. (1999). Prévalence des anticorps anti-fièvres hémorragiques d'origine virale dans la région du Pool (Congo-Brazzaville). *Médecine d'Afrique Noire*, *46*, 424–427.

ter Meulen, J., Lukashevich, I., Sidibe, K., Inapogui, A., Marx, M., Dorlemann, A., Yansane, M. L., Koulenou, K., Chang-Claude, J., & Schmitz, H. (1996). Hunting of peridomestic rodents and consumption of their meat as possible risk factors for rodent-to-human transmission of Lassa virus in the Republic of Guinea. *Am. J. Trop. Med. Hyg.*, *55*, 661–666.

Tomori, O., Fabiyi, A., Sorungbe, A., Smith, A., & McCormick, J. B. (1988). Viral hemorrhagic fever antibodies in Nigerian populations. *Am. J. Trop. Med. Hyg.*, *38*, 407–410.

Trape, J. F., & Mané, Y. (2006). *Guide des serpents d'Afrique occidentale : savane et désert*. IRD Editions.

Vezza, A. C., Cash, P., Jahrling, P., Eddy, G., & Bishop, D. H. (1980). Arenavirus recombination: the formation of recombinants between prototype pichinde and pichinde munchique viruses and evidence that arenavirus S RNA codes for N polypeptide. *Virology*, *106*, 250–260.

Walker, D. H., Wulff, H., Lange, J. V., & Murphy, F. A. (1975). Comparative pathology of Lassa virus infection in monkeys, guinea-pigs, and *Mastomys natalensis*. *Bull. WHO.*, *52*, 523–534.

Wulff, H., Fabiyi, A., & Monath, T. P. (1975). Recent isolation of Lassa virus from Nigerian rodents. *Bull. WHO.*, *52*, 609–613.

Yalley-Ogunro, J. E., Frame, J. D., & Hanson, A. P. (1984). Endemic Lassa fever in Liberia. VI. Village serological surveys for evidence of Lassa virus activity in Lofa County, Liberia. *Trans. Royal Soc. Trop. Med. Hyg.*, *78*, 764–770.

Chapter 6

Henipaviruses

Deadly Zoonotic Paramyxoviruses of Bat Origin

Glenn A. Marsh[1] and Lin-Fa Wang[1,2]
[1]CSIRO Australian Animal Health Laboratory, Geelong, Australia, [2]Duke-NUS Graduate Medical School, Singapore

INTRODUCTION

The genus *Henipavirus* within the family *Paramyxoviridae* consists of emerging paramyxoviruses that are some of the deadliest of known human pathogens. Hendra (HeV) and Nipah (NiV) viruses are zoonotic pathogens that can cause respiratory and encephalitic illness in humans with mortality rates that exceed 70%. Discovered in the 1990s, HeV and NiV challenged the view that paramyxoviruses contained genomes of uniform size and organization, being highly species restricted and in humans causing mostly nonlethal or preventable disease.

The henipaviruses are bat-borne viruses and, in addition to HeV and NiV which are both present in Southeast Asia and have caused human disease, include viruses that have been shown to be present in bats, with an almost global distribution. Cedar virus has recently been described as a nonpathogenic henipavirus present in pteropid bats in Australia.

GLOBAL DISTRIBUTION

Serological and molecular evidence for henipavirus-related pathogens in bat species has been documented in many places in the absence of identifiable disease (Figure 6.1) including Thailand (Wacharapluesadee et al., 2005), China (Li et al., 2008), Indonesia (Sendow et al., 2006; Sendow et al., 2009), Papua New Guinea (Breed et al., 2010), Madagascar (Lehle et al., 2007) and West Africa (Hayman et al., 2008; Drexler et al., 2009; Drexler et al., 2012), and Central and South America (Drexler et al., 2012).

CLASSIFICATION

HeV and NiV belong to the family *Paramyxoviridae*, subfamily *Paramyxovirinae* (Figure 6.2). Following the isolation and characterization of HeV, the

The Role of Animals in Emerging Viral Diseases. http://dx.doi.org/10.1016/B978-0-12-405191-1.00006-5

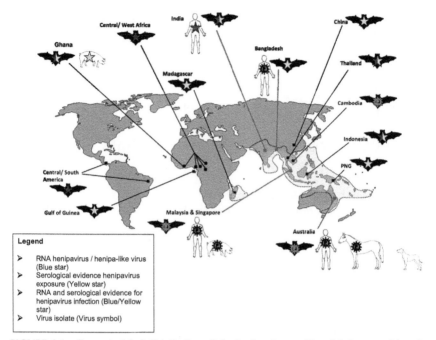

FIGURE 6.1 **Current global distribution of the henipaviruses.** The global range of bats in the genus *Pteropus*, described by Hall and Richards (2000), is depicted as shaded areas within the dotted line.

genome size and lack of conservation with other known paramyxoviruses led to the suggestion of a new virus genus within the *Paramyxovirinae* (Wang 2000). Following the identification of NiV and the demonstration of its close similarity to HeV, this led to the creation of the genus *Henipavirus* (Eaton et al., 2007).

Paramyxoviruses are lipid-enveloped viruses and contain a negative-sense RNA genome and are classified into several genera. These genera contain human pathogens such as measles virus, mumps virus, the parainfluenza viruses, and respiratory syncytial virus (RSV). Paramyxoviruses affecting nonhuman hosts include canine distemper virus that primarily infects dogs, rinderpest virus that infects cattle, and cetacean morbillivirus that is found in marine mammals such as dolphins and porpoises. For many paramyxoviruses, the host range is limited and cross-species transmission events are rare; however, the recently emerged henipaviruses prove to be an exception, displaying high virulence and a wide host range (Virtue et al., 2009).

GENOME ORGANIZATION

The henipavirus genome consists of six genes, encoding at least nine proteins, with a genome organization that is consistent with other paramyxoviruses (Figure 6.3). The HeV genome is 18,234 nucleotides (nt) and NiV 18,246 nt long, making them significantly larger than most other paramyxoviruses, with

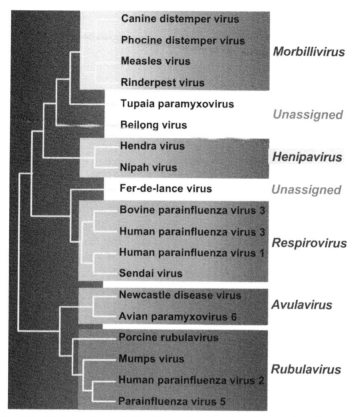

FIGURE 6.2 **Phylogenetic relationship of selected viruses in the subfamily** *Paramyxoviridae.* The phylogenetic tree was based on protein sequences of nucleocapsid (N) proteins of viruses representing the five existing genera and several unclassified viruses.

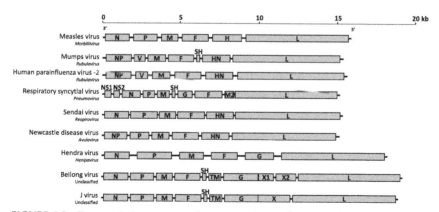

FIGURE 6.3 **Representative genomes of paramyxoviruses, demonstrating gene order.** The gene sizes and untranslated regions are drawn to approximate scale. The P/V genes of paramyxoviruses may encode up to 5 genes including P, V, W, C, C' SB, Y1, Y2, I and D (reviewed in Lamb and Parks, 2007).

the exception of the rodent-borne J-virus and Beilong virus (Li et al., 2006; Magoffin et al., 2007; Jack et al., 2008). The increased genome size of the henipaviruses is due to a longer open reading frame encoding the phosphoprotein (P) and to the large 3' untranslated regions that flank each gene with the exception of the large polymerase gene (Yu et al., 1998; Harcourt et al., 2000; Wang et al., 2000; Harcourt et al., 2001).

REPLICATION CYCLE

HeV and NiV virions attach to host cell surfaces through the identified receptor molecules ephrin-B2 and ephrin-B3, which are expressed on a large number of different cell types (Bonaparte et al., 2005; Negrete et al., 2005, 2006). After G binds to the cell surface receptor, the F protein then mediates fusion between the viral envelope and the host cell membrane, resulting in the release of the viral ribonucleocapsid into the cytoplasm (Lamb and Parks, 2007). Following release into the host cell cytoplasm, transcription of viral mRNA begins in a manner consistent with other paramyxoviruses, with the encapsidated genome serving as the template for mRNA synthesis. Following the switch from mRNA transcription, full length anti-genomes are produced which act as templates for synthesis of new, negative-sense genomic RNA. Newly synthesized genomes are encapsidated with nucleoprotein and interact with the other structural proteins to form new virions, which bud from the cell surface.

HENDRA VIRUS

HeV was first identified during a disease investigation of acute respiratory disease in thoroughbred horses in Hendra, a suburb of Brisbane, Australia (Murray et al., 1995a, 1995b; Selvey et al., 1995). During this outbreak, 13 horses succumbed to disease; in addition, a horse trainer and a stable worker also became infected, and the trainer subsequently died. In October 1995, a third person was diagnosed with a HeV infection in Mackay, approximately 1000 km north of Brisbane. Retrospective analysis demonstrated that an outbreak had occurred in Mackay in August 1994, with the infected individual helping with the autopsy on two horses that died. At the time, he had a mild undiagnosed infection from which he recovered. Fourteen months later, fatal encephalitis developed, suggesting either virus persistence or recrudescence (Hooper et al., 1996).

Including the initial events of 1994, there have been 41 events identified to date (Table 6.1). Following the initial events of 1994 up until 2004, HeV was considered a rare event with only one equine infection being identified in 1999. From 2004 until 2010, an increase in HeV cases was observed with most years seeing two disease events occurring. In 2011, an unprecedented number of HeV cases was observed, with 18 disease events being identified. Intensive surveillance of bat colonies during this period demonstrated an increase in the amount of HeV being shed by colonies. The reason for this large increase in observed

TABLE 6.1 Outbreaks of Hendra virus in Australia, 1994–March 2013

Date	Location	No. of horses infected	Human infections/deaths
August, 1994	Mackay, QLD	2	1/1
September, 1994	Hendra, QLD	20	2/1
January, 1999	Trinity Beach, QLD	1	–
October, 2004	Gordonvale, QLD	1	1/0
December, 2004	Townsville, QLD	1	–
June, 2006	Peachester, QLD	1	–
October, 2006	Murwillumbah, NSW	1	–
June, 2007	Peachester, QLD	1	–
July, 2007	Clifton Beach, QLD	1	–
July, 2008	Redlands, QLD	5	2/1
July, 2008	Proserpine, QLD	3	–
July, 2009	Rockhampton, QLD	3	1/1
September, 2009	Bowen, QLD	3	–
May, 2010	Tewantin, QLD	1	–
June, 2011	Kerry, QLD	1	–
June, 2011	Wollongbar, NSW	2	–
June, 2011	Macksville, NSW	1	–
June, 2011	Mt Alford, QLD	3	–
July, 2011	Park Ridge, QLD	1	–
July, 2011	Kuranda, QLD	1	–
July, 2011	Hervey Bay, QLD	1	–
July, 2011	Boondall, QLD	1	–
July, 2011	Lismore, NSW	1	–
July, 2011	Logan, QLD	1	–
July, 2011	Chinchilla, QLD	1	–
July, 2011	Mullumbimby, NSW	1	–
August, 2011	Ballina, NSW	1	–
August, 2011	South Ballina, NSW	2	–

Continued

TABLE 6.1 Outbreaks of Hendra virus in Australia, 1994–March 2013—cont'd

Date	Location	No. of horses infected	Human infections/deaths
August, 2011	Mullumbimby, NSW	1	-
August, 2011	Currumbin, QLD	1	-
August, 2011	North Ballina, NSW	1	-
September, 2011	Beachmere, QLD	2	-
January, 2012	Townsville, QLD	1	-
May, 2012	Rockhampton, QLD	1	-
May, 2012	Ingham, QLD	1	-
June, 2012	Rockhampton, QLD	3	-
July, 2012	Mackay, QLD	1	-
July, 2012	Cairns, QLD	1	-
September, 2012	Port Douglas, QLD	1	-
November, 2012	Ingham, QLD	1	-
January, 2013	Mackay, QLD	1	-
February, 2013	Atherton tablelands, QLD	1	-

QLD – Queensland, state of Australia, NSW – New South Wales, state of Australia

cases is unknown, but it is believed to be multifactorial with the health/nutritional status of bats and weather events possibly being involved (Field et al., 2011). The year 2012 continued the trend of an increased number of identified events, with eight outbreaks occurring. All of these cases have occurred in Queensland or northern NSW, despite seropositive bats from all four species of fruit bats in Australia and a geographic range extending down the east coast of Australia to Melbourne (Young et al., 1996).

Horse disease

In horses, HeV causes a severe, often fatal, febrile illness associated with respiratory and neurological signs. HeV in both naturally and experimentally infected horses is a rapidly progressing disease, with death occurring usually within 48 hours after the onset of clinical signs. The incubation period for horses is believed to range from 5 to 10 days following infection (Murray et al.,

1995a; Williamson et al., 1998; Marsh et al., 2011). Clinical signs observed in infected horses including fever, tachycardia, inappetence, depression, dyspnea, and restlessness. As the disease progresses frothy nasal discharge has been observed. Neurological signs such as ataxia, head pressing and myoclonic twitches can develop (Williamson, et al., 1998; Marsh et al., 2011). Signs of disease in experimentally challenged animals are independent on the route of exposure (intranasal, intravenous or subcutaneous) (Murray, et al., 1995a; Williamson et al., 1998; Marsh et al., 2011). HeV infection in horses has a case fatality rate of approximately 75% with some horses observed to seroconvert following asymptomatic infection.

HeV spreads systemically in infected horses and therefore at post mortem examination virus can be detected in all tissues of animals, with virus isolation possible from many of these tissues. In experimentally challenged animals, viral RNA can be detected in nasal swabs, oral swabs, urine and rectal swabs prior to the onset of clinical signs. Despite virus being shed in many fluids, transmission of HeV between horses is an unlikely event (Williamson, et al., 1998; Marsh et al., 2011) and seems to require very close contact under natural conditions (Field et al., 2010).

Human disease

HeV is a zoonotic pathogen of significant concern to human health. From seven humans who had confirmed HeV infection, four have resulted in death. All human infections have resulted from close physical contact with infected horses (Selvey et al., 1995; O'Sullivan et al., 1997; Playford et al., 2010). Symptoms have varied between patients, with an estimated incubation period of 7–10 days. Initial disease signs are influenza-like, with fever, lethargy, headaches, muscle pain, nausea and vomiting. Disease then progresses to a fulminating encephalitis with multi-organ failure (Selvey et al., 1995; Playford, et al., 2010). In one case, late onset encephalitis has been observed 14 months after an unidentified exposure to infected horses. At the time of exposure, the individual suffered from an influenza-like illness and aseptic meningitis (Baldock et al., 1996; O'Sullivan, et al., 1997).

HeV infections in other animals

Until 2011, natural infection of animals other than horses with HeV had not been observed. However, during the cluster of cases that occurred in 2011, neutralizing antibodies to HeV were detected in a dog on a property with three equine cases. The dog was not observed to display any clinical signs of disease, although the presence of high levels of antibodies indicated exposure to HeV infection. Early laboratory studies had suggested that dogs could become subclinically infected (Westbury et al., 1996); the risk posed by these infections to the health of humans and other animal is not known. Experimental challenge studies are underway to characterize the disease in dogs and to assess the risk associated with these infections.

NATURAL RESERVOIRS OF HENIPAVIRUSES

Following the identification of HeV, a comprehensive sero-survey was carried out to find the natural reservoir of HeV. Serological evidence identified flying foxes (suborder *Megachiroptera,* genus *Pteropus*) as the likely reservoir host (Young, et al., 1996). HeV has subsequently isolated on several occasions from two species of pteropid bats (Halpin et al., 2000; Smith et al., 2011). Using this knowledge from HeV, the natural reservoir for NiV was also rapidly identified as pteropid bats in Malaysia. NiV was isolated from urine collected from under a colony of *Pteropus hypomelanus* on Tioman Island situated off the east coast of peninsular Malaysia (Chua et al., 2002). In addition, NiV has also been isolated from *P. lylei* in Cambodia (Reynes et al., 2005). Field studies have shown the seroprevalence of antibodies to henipaviruses in pteropid bats varies from 10% to 50% in numerous countries, including Australia (Young, et al., 1996; Halpin et al., 2000), Malaysia (Yob et al., 2001), India (Yadav et al., 2012), Cambodia (Reynes, et al., 2005), Indonesia (Sendow et al., 2006; Sendow, et al., 2009), Papua New Guinea (Breed, et al., 2010) and Thailand (Wacharapluesadee et al., 2010). This high level of sero-positivity in bats suggests that henipaviruses transmit readily within colonies. Perhaps not surprisingly as the natural reservoir, henipaviruses have never been associated with any disease in pteropid bats.

Experimental challenge has also failed to produce clinical diseases in bat species (Williamson, et al., 1998, 2000; Halpin et al., 2011). Fruit bats have been challenged with high viral doses of HeV and NiV virus, resulting in only some bats seroconverting at low levels. On histopathological examination of challenged bats viral antigen was only occasionally observed, a striking difference to that observed in other animals. Additionally, despite vigorous sampling of bats, re-isolation of virus from challenged bats was a very rare event, occurring only from urine of one HeV challenged nonpregnant bat and from a few tissues including fetus of two pregnant bats challenged with HeV. NiV has never been re-isolated from experimentally challenged animals. Together, this data demonstrates that spill-over of henipaviruses from bats should be a rare event, which is confirmed by the relatively small number of outbreaks that occur.

TRANSMISSION OF HENDRA VIRUS

The mechanisms of transmission of HeV from bats to horses are not known, although a number of possible scenarios have been described. The successful isolation of HeV from uterine fluid and fetal tissues of pregnant bats suggests that incidental exposure of horses to aborted fetuses, placental tissues or birthing membranes and fluids through grazing is a potential mode of transmission (Halpin, et al., 2000). This hypothesis is supported by the seasonal clustering of HeV outbreaks in a June to September period that coincides with late-stage pregnancy and the birthing period. However, not all outbreaks have occurred during this seasonal period. The ingestion of saliva-laden spats, generated by flying foxes during feeding, by horses has also been hypothesised as a mechanism for transmission, along with ingestion of flying fox urine or feces from contaminated pastures, feed or water (Field, et al., 2001, 2007).

Of seven cases of human infection with HeV, transmission has been attributed to close contact with infected horses in six of seven cases. This transmission of virus most likely involves direct exposure to respiratory secretions, blood and/or saliva from sick horses, or exposure to fluids and tissues during post-mortem examination of horses (O'Sullivan, et al., 1997; Hanna et al., 2006; Playford, et al., 2010). The potential of virus transmission via urine has also been demonstrated experimentally in a cat model for HeV infection (Westbury, et al., 1996) and was the likely mechanism of cat-to-horse transmission in an experimental setting (Williamson, et al., 1998). The shedding of HeV by experimentally challenged horses in nasal secretions prior to the onset of fever and clinical disease provides evidence that asymptomatically infected animals pose a transmission risk to humans, although this risk is low compared to horses showing clinical disease, in which viral loads in secretions are significantly higher (Marsh et al., 2011).

There is no clear evidence that horse-to-horse transmission occurs with HeV, with many outbreaks consisting of disease in single horses on properties with multiple horses. The few occasions when horse-to-horse transmission has been suggested, this may have been due to indirect transmission through environmental contamination or mechanical transfer of secretions during stabling or treatment and care of animals (Baldock, et al., 1996; Field et al., 2001, 2010).

There are no reports of human-to-human transmission of HeV, although evidence of virus in urine and nasopharyngeal aspirates by PCR and isolation of HeV from nasopharyngeal secretions have been described (Playford, et al., 2010). There is also no evidence for bat-to-human transmission of HeV. Following the discovery of bats as the HeV reservoir species a serological survey of Queensland bat carers was carried out. This study screened 128 individuals for HeV infection and found no evidence of infections, despite many carers reporting prolonged contact with flying foxes, including being scratched or bitten (Selvey et al., 1996).

TRANSMISSION OF NiV VIRUS IN MALAYSIA, BANGLADESH AND INDIA

Outbreaks of Nipah virus have been reported from Malaysia, Bangladesh and India (Table 6.2). Clear differences have been observed in regards to transmission of NiV in the outbreaks in Bangladesh and India compared to the Malaysian outbreak. The most significant of these is the capacity for NiV to transmit from human-to-human in Bangladesh and India. Human infections in the Malaysia outbreak were almost all associated with transmission from infected pigs (Tan et al., 1999). By contrast, in Bangladesh, no intermediate animal host has been identified, with human infections being associated with direct bat-to-human transmission and more worryingly human-to-human transmission (Chadha, et al., 2006; Luby et al., 2009a). Mortality rates of outbreaks in India and Bangladesh (43–100%) have also been higher than the 38.5% rate reported for Malaysia (Chua et al., 2003; Chong et al., 2008).

Bat-to-human transmission of NiV

Following the discovery of bats as the natural reservoir in Malaysia, a serological survey was carried out on 153 individuals living in close proximity to a *P. hypomelanus* colony on Tioman Island (the location of the Malaysian bat NiV isolate). No evidence of human infection through contact with bats was identified, matching that observed with HeV in Australia (Chong et al., 2003). However, in contrast, in Bangladesh and India, bat-to-human transmission is an important route of human infection. Sequence analyses of NiV strains associated with outbreaks indicate that numerous introductions of NiV into the human population have occurred (Harcourt et al., 2005). It has been estimated that over 20 such bat-to-human transmission events have occurred in Bangladesh since 2001 (Luby, et al., 2009b). Epidemiological evidence for zoonotic transmission appears to primarily implicate indirect bat-to-human transmission; contact with food or surfaces contaminated with infectious material from the saliva, urine or feces of bats has been identified as the primary source of infection. For example, in an outbreak in Goalanda, Bangladesh in January 2004, a statistically significant association between illness and tree-climbing was identified for teenage boys. The trees identified are known roost sites for *P. giganteus* suggesting exposure to virus via material such as feces, urine or saliva left by bats (Montgomery et al., 2008).

An additional source of NiV infection in Bangladesh has been identified as consumption of locally harvested fresh date palm juice (Luby et al., 2006). Date palm sap is a national delicacy of Bangladesh, harvested overnight from date palms and in most cases consumed raw (Nahar et al., 2010). Using infrared photography, investigators have observed that bats are frequent visitors to trees from which date palm sap is collected, with bats frequently observed feeding directly from tree taps and collection pots (ICDDRB, 2005). This data highlights the potential of NiV to be a food-borne pathogen, with experimental evidence demonstrating virus remains viable for up to three days in fruit juice (Fogarty et al., 2008).

TABLE 6.2 Outbreaks of Nipah virus 1999–March 2013

Country	Location	Date	Number of cases	Fatalities (%)	References
Malaysia/	-	1999	268	106 (40%)	Chua et al., 2000
Singapore	-	1999	11	1 (9%)	Paton et al., 1999
India	Siliguri	Jan–Feb 2001	92	67 (74%)	Chadha et al., 2006
Bangladesh	Meherpur	April–May 2001	13	9 (69%)	Hsu et al., 2004
Bangladesh	Naogaon	January 2003	12	8 (67%)	Hsu et al., 2004
Bangladesh	Manikganj and Rajbari	Jan–Feb 2004	62	46 (74%)	ICDDRB, 2004
Bangladesh	Faridpur	Feb–April 2004	36	27 (75%)	Hossain et al., 2008
Bangladesh	Tangail	January 2005	12	11 (92%)	Luby et al., 2006
India	Nadia	Feb–May 2007	Approx. 50	5 (10%)	Arankalle et al., 2011
Bangladesh	Thakurgaon	Jan–Feb 2007	7	3 (43%)	ICDDRB, 2007
Bangladesh	Manikganj and Rajbari	February 2008	9	8 (89%)	Rahman et al., 2012
Bangladesh	Fardipur and Gopalganj	Dec 2009–Mar2010	17	15 (88%)	ICDDRB, 2010
Bangladesh	Lalmonirhat	February 2011	unknown	35	Anonymous, 2011
Bangladesh	Joypurhat	January 2012	6	6 (100%)	Anonymous, 2012
Bangladesh	Rajshahi	January 2013	1	1 (100%)	Anonymous, 2013
Bangladesh	Natore	January 2013	2	1 (50%)	Anonymous, 2013

Continued

TABLE 6.2 Outbreaks of Nipah virus 1999–March 2013—cont'd

Country	Location	Date	Number of cases	Fatalities (%)	References
Bangladesh	Mymenshingh	January 2013	2	2 (100%)	Anonymous, 2013
Bangladesh	Pabna	January 2013	2	1 (50%)	Anonymous, 2013
Bangladesh	Rajbari	January 2013	1	1 (100%)	Anonymous, 2013
Bangladesh	Chittagong	January 2013	1	1 (100%)	Anonymous, 2013
Bangladesh	Nilphamari	January 2013	3	3 (100%)	Anonymous, 2013
Bangladesh	Gaibandha	January 2013	1	1 (100%)	Anonymous, 2013
Bangladesh	Nougaon	January 2013	2	2 (100%)	Anonymous, 2013
Bangladesh	Jhenaidah	January 2013	1	1 (100%)	Anonymous, 2013
Bangladesh	Dhaka	January 2013	2	2 (100%)	Anonymous, 2013
Bangladesh	Kurigram	February 2013	1	1 (100%)	Anonymous, 2013

RISK FACTORS FOR INCREASED HENIPAVIRUS OUTBREAKS

Flying fox populations are under ever-growing pressure because of environmental change such as deforestation. The loss of natural habitat has led to bat colonies forming in urban areas and encroaching into agricultural areas. This has been suggested as driving forces in the emergence of the henipaviruses from their natural ecological niche (Field, et al., 2001; Chua et al., 2002; Plowright et al., 2011). Prevention of henipavirus spill-over from bats to livestock animals or humans is complicated by a lack of understanding of the mechanisms by which henipavirus infection is maintained within bat populations or the mecha nisms of transmission to spill-over hosts. With an increasing global distribution of henipaviruses, and increased bat–human and bat–livestock interactions, spill-over of henipaviruses to humans and animals is likely to continue.

DEVELOPMENT OF VACCINE AND THERAPEUTICS

There are no licensed or approved therapeutics for treating henipavirus disease in humans. Several antiviral approaches have been reported and tested in animal models and show promise as treatment against disease (reviewed in Broder, 2012). Among the approaches for development of a therapy, passive immunotherapy with polyclonal or monoclonal antibody against the virus envelope glycoproteins has been the most successful (Guillaume et al., 2004, 2006; Zhu et al., 2008). From these studies, a clinical grade batch of a monoclonal antibody, m102.4 (Zhu et al., 2008), has been manufactured in Australia for future use in people in the event of a high risk exposure.

The most promising approach to prevent human disease from HeV in Australia is a recently licensed equine vaccine, Equivac®HeV. This vaccine is a recombinant form of the HeV G glycoprotein. The preliminary data show seroconversion following vaccination of horses and prevents disease following exposure to an otherwise lethal HeV challenge. In addition, vaccination prevented viral shedding and replication of HeV in tissues (Middleton and Wang, unpublished results).This vaccine is the first vaccine licensed and commercially deployed against a BSL-4 agent.

ACKNOWLEDGMENTS

We thank Dr. Bronwyn Clayton for assistance with the production of Figure 6.1. Recent work conducted by the authors is supported in part by an OCE Postdoctoral Fellowship (GAM) and a CEO Science Leader Award (LFW) from the CSIRO Office of the Chief Executive.

REFERENCES

Anonymous, 2011. Nipah Encephalitis, Human - Bangladesh (05): (Rangpur). ProMED-mail, March 8, 2011 archive no. 20110308.0756(20110308.0756).

Anonymous, 2012. Nipah virus, fatal - Bangladesh: (Faridpur). ProMED-mail, February 4, 2012 archive no. 20110204.0402(20110204.0402).

Anonymous, 2013. Nipah outbreak. Retrieved 22 March 2013, from http://www.iedcr.org/index. php?option=com_content&view=article&id=106.

Arankalle, V. A., Bandyopadhyay, B. T., Ramdasi, A. Y., Jadi, R., Patil, D. R., Rahman, M., Majumadar, M., Bannerjee, P. S., Hati, A. K., Goswami, R. P., Neogi, D. K., & Mishra, A. C. (2011). Genomic characterization of Nipah virus, West Bengal, India. *Emerg. Infect. Dis.*, *17*, 907–909.

Baldock, F. C., Douglas, I. C., Halpin, K., Field, H., Young, P. L., & Black, P. F. (1996). Epidemiological investigations into the 1994 Equine morbillivirus outbreaks in Queensland, Australia. *Sing. Vet. J.*, *20*, 57–61.

Bonaparte, M. I., Dimitrov, A. S., Bossart, K. N., Crameri, G., Mungall, B. A., Bishop, K. A., Choudry, V., Dimitrov, D. S., Wang, L. F., Eaton, B. T., & Broder, C. C. (2005). Ephrin-B2 ligand is a functional receptor for Hendra virus and Nipah virus. *Proc Natl Acad Sci U S A*, *102*, 10652–10657.

Breed, A. C., Yu, M., Barr, J. A., Crameri, G., Thalmann, C. M., & Fa Wang, L. (2010). Prevalence of henipavirus and rubulavirus antibodies in pteropid bats, Papua New Guinea. *Emerg. Infect. Dis.*, *16*, 1997–1999.

Broder, C. C. (2012). Henipavirus outbreaks to antivirals: the current status of potential therapeutics. *Curr. Opinion Virol*, *2*, 176–187.

Chadha, M. S., Comer, J. A., Lowe, L., Rota, P. A., Rollin, P. E., Bellini, W. J., Ksiazek, T. G., Mishra, A. M. (2006). Nipah virus-associated encephalitis outbreak, Siliguri, India. *Emerg. Infect. Dis.*, *12*, 235–240.

Chong, H. T., Tan, C. T., Goh, K. J., Lam, S. K., & Chua, K. B. (2003). The risk of human Nipah virus infection directly from bats (*Pteropus hypomelanus*) is low. *Neurol. J. Southeast Asia*, *8*, 31–34.

Chong, H. T., Hossain, M. J., & Tan, C. T. (2008). Differences in epidemiologic and clinical features of Nipah virus encephalitis between the Malaysian and Bangladesh outbreaks. *Neurology Asia*, *13*, 23–26.

Chua, K. B., Bellini, W. J., Rota, P. A., Harcourt, B. H., Tamin, A., Lam, S. K., Kziazek, T. G., Rollin, P. E., Zaki, S. R., Shieh, W., Goldsmith, C. S., Gubler, D. T., Roehrig, J. T., Eaton, B., Gould, A. R., Olson, J., Field, H., Danieli, P., Ling, A. E., Peters, C. J., Anderson, L. J., & Mahy, B. W. (2000). Nipah virus: a recently emergent deadly paramyxovirus. *Science*, *288*(5470), 1432–1435.

Chua, K. B., Koh, C. L., Hooi, P. S., Wee, K. F., Khong, J. H., Chua, B. H., Chan, Y. P., Lim, M. E., & Lam, S. K. (2002). Isolation of Nipah virus from Malaysian Island flying-foxes. *Microbes and Infection*, *4*, 145–151.

Chua, K. B. (2003). Nipah virus outbreak in Malaysia. *J. Clin. Virol.*, *26*, 265–275.

Drexler, J. F., Corman, V. M., Gloza-Rausch, F., Seebens, A., Annan, A., Ipsen, A., Kruppa, T., Müller, M. A., Kalko, E. K., Adu-Sarkodie, Y., Oppong, S., & Drosten, C. (2009). Henipavirus RNA in African bats. *PLoS One*, *4*, e6367.

Drexler, J. F., Corman, V. M., Müller, M. A., Maganga, G. D., Vallo, P., Binger, T., Gloza-Rausch, F., Rasche, A., Yordanov, S., Seebens, A., Oppong, S., Sarkodie, Y. A., Pongombo, C., Lukashev, A. N., Schmidt-Chanasit, J., Stöcker, A., Carneiro, A. J. B., Erbar, S., Maisner, A., Fronhoffs, F., Buettner, R., Kalko, E. K. V., Kruppa, T., Franke, C. R., Kallies, R., Yandoko, E. R. N., Herrler, G., Reusken, C., Hassanin, A., Krüger, D. H., Matthee, S., Ulrich, R. G., Leroy, E. M., & Drosten, C. (2012). Bats host major mammalian paramyxoviruses. *Nature Communications*, *3*, 796.

Eaton, B. T., Mackenzie, J. S., & Wang, L. F. (2007). Henipaviruses. In D. M. Knipe & P. M. Howley (Eds.), *Fields Virology* (5th edition, pp. 1587–1600). Philadelphia: Lippincott, Williams and Wilkins.

Field, H., Young, P., Yob, J. M., Mills, J., Hall, L., & Mackenzie, J. (2001). The natural history of Hendra and Nipah viruses. *Microbes Infect.*, *3*, 307–314.

Field, H. E., Mackenzie, J. S., & Daszak, P. (2007). Henipaviruses: emerging paramyxoviruses associated with fruit bats. *Curr. Top. Microbiol. Immunol.*, *315*, 133–159.

Field, H., Schaaf, K., Kung, N., Simon, C., Waltisbuhl, D., Hobert, H., Moore, F., Middleton, D., Crook, A., Smith, G., Daniels, P., Glanville, R., & Lovell, D. (2010). Hendra virus outbreak with novel clinical features, Australia. *Emerg. Infect. Dis.*, *16*, 338–340.

Field, H., de Jong, C., Melville, D., Smith, C., Smith, I., Broos, A., Kung, Y. H., McLaughlin, A., & Zeddeman, A. (2011). Hendra virus infection dynamics in Australian fruit bats. *PLoS ONE*, *6*, e28678.

Fogarty, R., Halpin, K., Hyatt, A. D., Daszak, P., & Mungall, B. A. (2008). Henipavirus susceptibility to environmental variables. *Virus Res.*, *132*, 140–144.

Guillaume, V., Contamin, H., Loth, P., Georges-Courbot, M. C., Lefeuvre, A., Marianneau, P., Chua, K. B., Lam, S. K., Buckland, R., Deubel, V., & Wild, T. F. (2004). Nipah virus: vaccination and passive protection studies in a hamster model. *J. Virol.*, *78*, 834–840.

Guillaume, V., Contamin, H., Loth, P., Grosjean, I., Courbot, M. C., Deubel, V., Buckland, R., & Wild, T. F. (2006). Antibody prophylaxis and therapy against Nipah virus infection in hamsters. *J. Virol.*, *80*, 1972–1978.

Hall, L., & Richards, G. (2000). *Flying Foxes, Fruit and Blossom Bats of Australia.* Sydney: UNSW Press.

Halpin, K., Young, P. L., Field, H. E., & Mackenzie, J. S. (2000). Isolation of Hendra virus from pteropid bats: a natural reservoir of Hendra virus. *J. Gen. Virol.*, *81*, 1927–1932.

Halpin, K., Hyatt, A. D., Fogarty, R., Middleton, D., Bingham, J., Epstein, J. H., Rahman, S. A., Hughes, T., Smith, C., Field, H. E., Daszak, P., & Grp, H. E. R. (2011). Pteropid Bats are Confirmed as the Reservoir Hosts of Henipaviruses: A Comprehensive Experimental Study of Virus Transmission. *Am. J. Trop. Med. Hyg.*, *85*, 946–951.

Hanna, J. N., McBride, W. J., Brookes, D. L., Shield, J., Taylor, C. T., Smith, I. L., Craig, S. B., & Smith, G. A. (2006). Hendra virus infection in a veterinarian. *Med. J. Aust.*, *185*, 562–564.

Harcourt, B. H., Tamin, A., Ksiazek, T. G., Rollin, P. E., Anderson, L. J., Bellini, W. J., & Rota, P. A. (2000). Molecular characterization of Nipah virus, a newly emergent paramyxovirus. *Virology*, *271*, 334–349.

Harcourt, B. H., Tamin, A., Halpin, K., Ksiazek, T. G., Rollin, P. E., Bellini, W. J., & Rota, P. A. (2001). Molecular characterization of the polymerase gene and genomic termini of Nipah virus. *Virology*, *287*, 192–201.

Harcourt, B. H., Lowe, L., Tamin, A., Liu, X., Bankamp, B., Bowden, N., Rollin, P. E., Comer, J. A., Ksiazek, T. G., Hossain, M. J., Gurley, E. S., Breiman, R. F., Bellini, W. J., & Rota, P. A. (2005). Genetic characterization of Nipah virus, Bangladesh, 2004. *Emerg. Infect. Dis.*, *11*, 1594–1597.

Hayman, D., Suu-Ire, R., Breed, A., McEachern, J., Wang, L., Wood, J., & Cunningham, A. (2008). Evidence of henipavirus infection in West African fruit bats. *PLoS ONE*, *3*, e2739.

Hooper, P. T., Gould, A. R., Russell, G. M., Kattenbelt, J. A., & Mitchell, G. (1996). The retrospective diagnosis of a second outbreak of equine morbillivirus infection. *Aust. Vet. J.*, *74*, 244–245.

Hossain, M. J., Gurley, E. S., Montgomery, J. M., Bell, M., Carroll, D. S., Hsu, V. P., Formenty, P., Croisier, A., Bertherat, E., Faiz, M. A., Azad, A. K., Islam, R., Molla, M. A., Ksiazek, T. G., Rota, P. A., Comer, J. A., Rollin, P. E., Luby, S. P., & Breiman, R. F. (2008). Clinical presentation of nipah virus infection in Bangladesh. *Clin. Infect. Dis*, *46*, 977–984.

Hsu, V. P., Hossain, M. J., Partashar, U. D., Ali, M. M., Ksiazek, T. G., Kuzmin, I., Niezgoda, M., Rupprecht, C., Bresee, J., & Breiman, R. F. (2004). Nipah virus encephalitis reemergence. *Bangladesh. Emerg. Infect. Dis*, *10*, 2082–2087.

ICDDRB. (2004). Nipah Encephalitis Outbreak Over Wide Area of Western Bangladesh, *2004 Health Sci. Bull.*, *2*, 7–11.

ICDDRB. (2005). Nipah virus outbreak from date palm juice. *Health Sci. Bull.*, *3*, 1–5.

ICDDRB. (2007). Person-to-person transmission of Nipah infection in Bangladesh. *Health Sci. Bull.*, *5*, 1–6.

ICDDRB. (2010). Bangladesh. Nipah outbreak in Faridpur District, Bangladesh, 2010. *Health Sci. Bull.*, *8*, 6–11.

Jack, P. J., Anderson, D. E., Bossart, K. N., Marsh, G. A., Yu, M., & Wang, L. F. (2008). Expression of novel genes encoded by the paramyxovirus J virus. *J. Gen. Virol.*, *89*, 1434–1441.

Lamb, R. A., & Parks, G. D. (2007). Paramyxoviruses. In D. M. Knipe & P. M. Howley (Eds.), *Fields Virology* (5th ed., Vol. 2). Philadelphia: Lippincott Williams and Wilkins.

Lehle, C., Razafitrimo, G., Razainirina, J., Andriaholinirina, N., Goodman, S. M., Faure, C., Georges-Courbot, M. -C., Rousset, D., & Reynes, J. -M. (2007). Henipavirus and Tioman virus antibodies in pteropodid bats. *Madagascar. Emerg. Infect. Dis*, *13*, 159–161.

Li, Z., Yu, M., Zhang, H., Magoffin, D. E., Jack, P. J., Hyatt, A., Wang, H. Y., & Wang, L. F. (2006). Beilong virus, a novel paramyxovirus with the largest genome of non-segmented negative-stranded RNA viruses. *Virology*, *346*, 219–228.

Li, Y., Wang, J., Hickey, A. C., Zhang, Y., Li, Y., Wu, Y., Zhang, H., Yuan, J., Han, Z., McEachern, J., Broder, C. C., Wang, L. -F., & Shi, Z. (2008). Antibodies to Nipah or Nipah-like viruses in bats, China. *Emerg. Infect. Dis*, *14*, 1974–1976.

Luby, S. P., Rahman, M., Hossain, M. J., Blum, L. S., Husain, M. M., Gurley, E., Khan, R., Ahmed, B. -A., Rahman, S., Nahar, N., Kenah, E., Comer, J. A., & Ksiazek, T. G. (2006). Foodborne transmission of Nipah virus. *Bangladesh. Emerg. Infect. Dis*, *12*, 1888–1894.

Luby, S. P., Gurley, E. S., & Hossain, M. J. (2009a). Transmission of human infection with Nipah virus. *Clin. Infect. Dis.*, *49*, 1743–1748.

Luby, S. P., Hossain, M. J., Gurley, E. S., Ahmed, B. N., Banu, S., Khan, S. U., Homaira, N., Rota, P. A., Rollin, P. E., Comer, J. A., Kenah, E., Ksiazek, T. G., & Rahman, M. (2009b). Recurrent zoonotic transmission of Nipah virus into humans, Bangladesh, 2001–2007. *Emerg. Infect. Dis.*, *15*, 1229–1235.

Magoffin, D. E., Mackenzie, J. S., & Wang, L. F. (2007). Genetic analysis of J-virus and Beilong virus using minireplicons. *Virology*, *364*, 103–111.

Marsh, G. A., Haining, J., Hancock, T. J., Robinson, R., Foord, A. J., Barr, J. A., Riddell, A., Heine, H. G., White, J. R., Crameri, G., Field, H. E., Wang, L. F., & Middleton, D. (2011). Experimental infection of horses with Hendra virus/Australia/horse/2008/Redlands. *Emerg. Infect. Dis.*, *17*, 2232–2238.

Montgomery, J. M., Hossain, M. J., Gurley, E., Carroll, G. D., Croisier, A., Bertherat, E., Asgeri, N., Formenty, P., Keeler, N., Comer, J., Bell, A. R., Akram, K., Molla, A. R., Zomari, K., Islam, M. R., Wagoner, K., Mills, J. N., Rollin, P. E., Ksiazek, T. G., & Breiman, R. F. (2008). Risk factors for Nipah virus encephalitis in Bangladesh. *Emerg. Infect. Dis.*, *14*, 1526–1532.

Murray, K., Rogers, R., Selvey, L., Selleck, P., Hyatt, A., Gould, A., Gleeson, L., Hooper, P., & Westbury, H. (1995a). A novel morbillivirus pneumonia of horses and its transmission to humans. *Emerg. Infect. Dis.*, *1*, 31–33.

Murray, K., Selleck, P., Hooper, P., Hyatt, A., Gould, A., Gleeson, L., Westbury, H., Hiley, L., Selvey, L., Rodwell, B., & Ketterer, P. (1995b). A morbillivirus that caused fatal disease in horses and humans. *Science*, *268*, 94–97.

Nahar, N., Sultana, R., Gurley, E. S., Hossain, M. J., & Luby, S. P. (2010). Date palm sap collection: exploring opportunities to prevent Nipah transmission. *Ecohealth*, *7*, 196–203.

Negrete, O. A., Levroney, E. L., Aguilar, H. C., Bertolotti-Ciarlet, A., Nazarian, R., Tajyar, S., & Lee, B. (2005). EphrinB2 is the entry receptor for Nipah virus, an emergent deadly paramyxovirus. *Nature*, *436*, 401–405.

Negrete, O. A., Wolf, M. C., Aguilar, H. C., Enterlein, S., Wang, W., Muhlberger, E., Su, S. V., Bertolotti-Ciarlet, A., Flick, R., & Lee, B. (2006). Two key residues in ephrinB3 are critical for its use as an alternative receptor for Nipah virus. *PLoS pathogens*, *2*, e7.

O'Sullivan, J. D., Allworth, A. M., Paterson, D. L., Snow, T. M., Boots, R., Gleeson, L. J., Gould, A. R., Hyatt, A. D., & Bradfield, J. (1997). Fatal encephalitis due to novel paramyxovirus transmitted from horses. *The Lancet*, *349*, 93–95.

Paton, N. I., Leo, Y. S., Zaki, S. R., Auchus, A. P., Lee, K. E., Ling, A. E., Chew, S. K., Ang, B., Rollin, P. E., Umpathi, T., Sng, I., Lee, C. C., Lim, E., & Ksiazek, T. G. (1999). Outbreak of Nipah-virus infection among abattoir workers in Singapore. *Lancet*, *354*, 1253–1256.

Playford, E. G., McCall, B., Smith, G., Slinko, V., Allen, G., Smith, I., Moore, F., Taylor, C., Kung, Y. H., & Field, H. (2010). Human Hendra virus encephalitis associated with equine outbreak, Australia, 2008. *Emerg. Infect. Dis.*, *16*(2), 219–223.

Plowright, R. K., Foley, P., Field, H. E., Dobson, A. P., Foley, J. E., Eby, P., & Daszak, P. (2011). Urban habituation, ecological connectivity and epidemic dampening: the emergence of Hendra virus from flying foxes (*Pteropus* spp.). *Proc. Biol. Sci. Royal Soc.*, *278*, 3703–3712.

Rahman, M. A., Hossain, M. J., Sultana, S., Homaira, N., Khan, S. U., Rahman, M., Gurley, E. S., Rollin, P. E., Lo, M. K., Comer, J. A., Lowe, L., Rota, P. A., Ksiazek, T. G., Kench, E., Sharker, Y., & Luby, S. P. (2012). Date palm sap linked to Nipah virus outbreak in Bangladesh, 2008. *Vector Borne Zoonotic Dis*, *12*, 65–72.

Reynes, J. M., Counor, D., Ong, S., Faure, C., Seng, V., Molia, S., Walston, J., Georges-Courbot, M. C., Deubel, V., & Sarthou, J. L. (2005). Nipah virus in Lyle's flying foxes, Cambodia. *Emerg Infect Dis*, *11*, 1042–1047.

Selvey, L. A., Wells, R. M., McCormack, J. G., Ansford, A. J., Murray, K., Rogers, R. J., Lavercombe, P. S., Selleck, P., & Sheridan, J. W. (1995). Infection of humans and horses by a newly described morbillivirus. *Med. J. Aust.*, *162*, 642–645.

Selvey, L., Taylor, R., Arklay, A., & Gerrard, J. (1996). Screening of bat carers for antibodies to equine morbillivirus. *Commun. Dis. Intell.*, *20*, 477–478.

Sendow, I., Field, H. E., Curran, J., Darminto, C., Morrissy, Meehan, G., Buick, T., & Daniels, P. (2006). Henipavirus in *Pteropus vampyrus* bats, Indonesia. *Emerg. Infect. Dis.*, *12*, 711–712.

Sendow, I., Field, H. E., Adjid, A., Ratnawati, A., Breed, A. C., Morrisey, C., Darminto, & Daniels, P. (2009). Screening for Nipah Virus Infection in West Kalimantan Province, Indonesia. *Zoonoses Public Health*, *57*, 499–503.

Smith, I., Broos, A., de Jong, C., Zeddeman, A., Smith, C., Smith, G., Moore, F., Barr, J., Crameri, G., Marsh, G., Tachedjian, M., Yu, M., Kung, Y. H., Wang, L. F., & Field, H. (2011). Identifying Hendra virus diversity in pteropid bats. *PLoS One*, *6*, e25275.

Tan, K. S., Tan, C. T., & Goh, K. J. (1999). Epidemiological aspects of Nipah virus infection. *Neurol. J. Southeast Asia*, *4*, 77–81.

Virtue, E. R., Marsh, G. A., & Wang, L. F. (2009). Paramyxoviruses infecting humans: the old, the new and the unknown. *Future Microbiol*, *4*, 537–554.

Wacharapluesadee, S., Lumlertdacha, B., Boongird, K., Wanghongasa, S., Chanhome, L., Rollin, P., Stockton, P., Rupprecht, C. E., Ksiazek, T. G., & Hemachudha, T. (2005). Bat Nipah virus. *Thailand. Emerg. Infect. Dis.*, *11*, 1949–1951.

Wacharapluesadee, S., Boongird, K., Wanghongsa, S., Ratanasetyuth, N., Supavonwong, P., Saengsen, D., Gongal, G. N., & Hemachudha, T. (2010). A longitudinal study of the prevalence of Nipah virus in *Pteropus lylei* bats in Thailand: evidence for seasonal preference in disease transmission. *Vector Borne Zoonotic Dis.*, *10*, 183–190.

Wang, L. F., Yu, M., Hansson, E., Pritchard, L. I., Shiell, B., Michalski, W. P., & Eaton, B. T. (2000). The exceptionally large genome of Hendra virus: support for creation of a new genus within the family Paramyxoviridae. *J. Virol.*, *74*, 9972–9979.

Westbury, H. A., Hooper, P. T., Brouwer, S. L., & Selleck, P. W. (1996). Susceptibility of cats to equine morbillivirus. *Aust. Vet. J.*, *74*, 132–134.

Williamson, M. M., Hooper, P. T., Selleck, P. W., Gleeson, L. J., Daniels, P. W., Westbury, H. A., & Murray, P. K. (1998). Transmission studies of Hendra virus (equine morbillivirus) in fruit bats, horses and cats. *Aust. Vet. J.*, *76*, 813–818.

Williamson, M. M., Hooper, P. T., Selleck, P. W., Westbury, H. A., & Slocombe, R. F. (2000). Experimental Hendra virus infection in pregnant guinea-pigs and fruit bats (*Pteropus poliocephalus*). *J. Comp. Pathol.*, *122*, 201–207.

Yadav, P. D., Raut, C. G., Shete, A. M., Mishra, A. C., Towner, J. S., Nichol, S. T., & Mourya, D. T. (2012). Detection of Nipah virus RNA in fruit bat (*Pteropus giganteus*) from India. *Am. J. Trop. Med. Hyg.*, *87*, 576–578.

Yob, J. M., Field, H., Rashdi, A. M., Morrissy, C., van der Heide, B., Rota, P., bin Adzhor, A., White, J., Daniels, P., Jamaluddin, A., & Ksiazek, T. (2001). Nipah virus infection in bats (order Chiroptera) in peninsular Malaysia. *Emerg. Infect. Dis.*, *7*, 439–441.

Young, P. L., Halpin, K., Selleck, P. W., Field, H., Gravel, J. L., Kelly, M. A., & Mackenzie, J. S. (1996). Serologic evidence for the presence in Pteropus bats of a paramyxovirus related to equine morbillivirus. *Emerg. Infect. Dis.*, *2*, 239–240.

Yu, M., Hansson, E., Shiell, B., Michalski, W., Eaton, B. T., & Wang, L. F. (1998). Sequence analysis of the Hendra virus nucleoprotein gene: comparison with other members of the subfamily Paramyxovirinae. *J. Gen. Virol.*, *79*, 1775–1780.

Zhu, Z., Bossart, K. N., Bishop, K. A., Crameri, G., Dimitrov, A. S., McEachern, J. A., Feng, Y., Middleton, D., Wang, L. F., Broder, C. C., & Dimitrov, D. S. (2008). Exceptionally potent cross-reactive neutralization of Nipah and Hendra viruses by a human monoclonal antibody. *J. Infect. Dis.*, *197*, 846–853.

Chapter 7

The Role of Birds in the Spread of West Nile Virus

Paul Gale and Nicholas Johnson

Animal Health and Veterinary Laboratories Agency, Surrey, United Kingdom

INTRODUCTION

West Nile virus (WNV) belongs to the *Flavivirus* genus that contains a growing number of zoonotic viruses transmitted by mosquitoes or ticks (Gould and Solomon, 2005). Table 7.1 provides a summary of the main mosquito-borne flaviviruses and illustrates the range of human diseases associated with infection. Key viruses within this group are the yellow fever and Dengue viruses, which between them cause millions of human infections annually. While most of these viruses are principally found in the tropics and subtropics, reflecting the abundance and diversity of the mosquito vector, it is clear from the past two decades that flaviviruses are capable of establishing foci of infection in temperate regions during peak vector periods of the year and on occasion over-wintering to become endemic. Indeed WNV has dramatically expanded its geographic range in the past ten years (Brault, 2009) achieving endemicity across North and South America with continued incidence of human mortality and morbidity. Indeed evidence for introduction and establishment of WNV was reported in Argentina as early as January 2006 (Diaz et al., 2008). Increases in global commerce, climate change, ecological factors and the emergence of novel viral genotypes play significant roles in the emergence of this virus although the exact mechanisms and relative importance of each is uncertain (Brault, 2009). This chapter reviews the evidence for the significance of different mechanisms of virus translocation with a focus on the role of birds in the emergence of disease.

VIRUS CLASSIFICATION AND STRUCTURE

WNV is classified within the family *Flaviviridae* and genus *Flavivirus*. In addition, the *Flaviviridae* contains two other genera. Firstly, the genus *Pestivirus*, whose members are not transmitted by arthropod vectors but through direct

TABLE 7.1 An Overview of Zoonotic Mosquito-borne Flaviviruses

Virus	Geographic distribution	Principal vector	Livestock affected	Wildlife affected	Human disease
Dengue virus	Africa, Asia, Americas, Europe	*Aedes* spp.	None	None	Systemic febrile illness; Hemorrhagic fever
Japanese encephalitis virus	Asia	*Culex* spp.	Pigs, horses	Avian spp.	Encephalitis
Murray Valley encephalitis virus	Australasia, Indonesia	*Culex annulorostris*	Horse, cattle	None	Encephalitis
St. Louis encephalitis virus	Americas	*Culex* spp.	None	Avian spp.	Encephalitis
Usutu virus	Africa, Europe	Not determined	None	Avian spp.	Encephalitis
Wesselsbron virus	Africa, Asia	*Aedes* spp.	Sheep, cattle	None	Hemorrhagic fever
West Nile virus	Worldwide	*Culex* spp.	Horse	Avian spp.	Encephalitis
Yellow fever	Africa, Americas	*Aedes and Haemogogus* spp.	None	None	Hemorrhagic fever
Zika virus	Africa, Asia	Not determined	None	Not determined	Systemic febrile illness

contact and aerosols, and includes significant pathogens of livestock such as bovine viral diarrhea virus and classical swine fever virus. The second is the *Hepacivirus* genus, whose principal member is hepatitis C virus. This virus is transmitted sexually and through blood products in humans.

Mosquitoes are not the only vector of flaviviruses. Phylogeny of the genus has revealed the intimate co-evolution of particular virus species with particular arthropod vectors including mosquitoes and ticks (Gould et al., 2001; Randolph and Rogers, 2006). Indeed, the tick-borne clade of flaviviruses include tick-borne encephalitis virus, which infects a range of vertebrate hosts (Randolph and Rogers, 2006). Phylogenetically, WNV clusters with viruses that are transmitted by *Culex* species of mosquitoes with birds

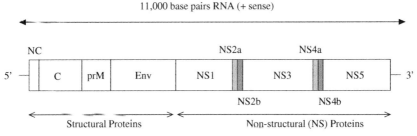

FIGURE 7.1   **Schematic diagram of the West Nile virus genome.**

as the main vertebrate host. This relates WNV to other zoonotic viruses such as Japanese encephalitis virus (JEV) and confirms the long established antigenic relationship that placed WNV within the Japanese encephalitis virus serocomplex (Calisher et al., 1989). Other zoonotic viruses within this group include St. Louis encephalitis virus (SLEV) and Murray Valley encephalitis virus (MVEV). As the names suggest, the key clinical manifestation of these viruses is the ability to cause encephalitis that in a small proportion of cases can result in death.

The WNV genome is approximately 11,000 base pairs in length. For example, the New York 99 (NY99) strain's genome is 11,029 base pairs long (Lanciotti et al., 2002). The genome is a single-stranded ribonucleic acid (RNA) molecule in the positive sense that is immediately translated on entry into a susceptible cell. The genome includes a single open reading frame that encodes a large precursor polyprotein that is proteolytically cleaved after translation into three structural proteins, that form progeny nucleocapsids and envelope proteins, and seven nonstructural (NS) proteins that mediate replication of new genomic copies (Figure 7.1). The RNA dependent RNA polymerase (RdRp) activity of the nonstructural protein 5 (NS5) of WNV is critical for virus replication through copying complementary negative strands and using these as templates for generation of new positive strand genomes (Malet et al., 2007). The genome also includes short noncoding or untranslated regions (UTRs) at both the 5' and 3' termini that are required for encapsidation and replication by the RdRp. The virion capsid is enclosed by a host-derived phospholipid bilayer membrane into which the envelope (E) glycoprotein is inserted. The final virion is approximately 50 nm in diameter. The E protein is the main virus component of the viral envelope and mediates binding and entry to the host cell. The crystal structure of the E glycoprotein has been determined and shows that epitopes targeted by neutralizing WNV-specific antibodies map to a region of domain III that is exposed on the virus surface and has been implicated in receptor binding (Kanai et al., 2006).

A further feature of WNV is its ability, in common with most viruses, to subvert the mammalian host interferon system. A number of mechanisms have been identified through which the virus achieves this. An early observation was a WNV-induced delay in Interferon Response Factor 3 (IRF3) activation providing an early block in interferon stimulated gene (ISG) responses (Fredricksen et al., 2004). Subsequently, specific interferon antagonist functions have been attributed to particular proteins, particularly inhibition of interferon transcription by proteins NS4B (Munoz-Jordan et al., 2005), NS1 (Schole and Mason, 2005) and NS2 (Liu et al., 2006). This ability enables the virus to overcome the initial interferon response although it is not thought to influence neuroinvasion. There is also evidence for subversion of the mosquito antiviral response. RNA interference (RNAi) is the predominant antiviral response against invading RNA viruses in insects (Schnettler et al., 2012) and flavivirus replication is enhanced in mosquitoes depleted for RNAi factors. Recently it has been shown that the 3'-UTR region of WNV and DENV efficiently suppressed RNAi pathways in both mammalian and insect cells (Schnettler et al., 2012).

DISCOVERY AND ARTHROPOD-BORNE VIRUS PERSISTENCE

West Nile virus was first isolated from the blood of a woman suffering a febrile illness in the West Nile district of Uganda (Smithburn et al., 1940). Subsequent studies made further isolations of WNV from human sera in Egypt (Melnick et al., 1951), and from birds and mosquitoes (Work et al., 1953). This established that mosquitoes were the likely virus vector through blood-feeding and that birds played a role in the persistence of the virus within the environment. Until the discovery of flaviviruses unable to replicate in mammalian cells (Cook et al., 2012), it was assumed that all flaviviruses needed to cycle between an arthropod vector and a vertebrate host to allow persistence and horizontal transmission of the virus. Transovarial transmission (TOT) of virus to mosquito progeny has been demonstrated for WNV in several mosquito species including *Culex vishnui*, *Culex salinarius*, *Aedes triseriatus* and *Culex pipiens* (Mishra and Mourya, 2001; Dohm et al., 2002; Unlu et al., 2010). Fechter-Leggett et al. (2012) reported vertical transmission rates from *Culex pipiens* complex females to egg rafts and larvae of 50% and 40%, respectively. Although the estimated minimal filial infection rate of 2.0 infected females per 1000 WNV-positive females was low, this could contribute to epizootic summertime transmission in very large mosquito populations (Fechter-Leggett et al., 2012). Furthermore, the important contribution of WNV TOT is the potential to create an overwintering population that reinitiates enzootic transmission the following spring (Fechter-Leggett et al., 2012). In the absence of other amplifying mechanisms, TOT together with transstadial transmission and sexual transmission between adult mosquitoes are not sufficient to sustain WNV leading to its extinction. Thus, WNV is maintained in nature in a cycle between mosquitoes and vertebrate hosts with the predominant and preferred host reservoir being birds (Colpitts et al., 2012). For example, in the northeastern United States of America (USA), *Cx. pipiens pipiens* is recognized as the primary enzootic vector responsible for transmission of virus among wild bird populations (Andreadis, 2012). Successive transmission cycles of arbovirus between an arthropod and a vertebrate host enhance virus persistence and amplification, and such viruses need to replicate efficiently in both hosts.

The Basic Reproductive Number (R_0) defines the number of new cases of infection that arise at some time in the future from one case of the infection at the present time when introduced into a population of totally susceptible hosts (Rogers and Randolph, 2006). The R_0 for mosquito-borne diseases is defined mathematically as:

$$R_0 = \frac{mbca^2 e^{-\mu T}}{\mu r} \tag{7.1}$$

where m = the ratio of vector numbers to host numbers, b = the transmission coefficient from vertebrate to vector, c = the transmission coefficient from vector to vertebrate, a = the biting rate of the vectors on the hosts of interest, μ = the mortality rate of the vectors, T = the extrinsic incubation period (EIP, see below) of the infection and r = rate of recovery of the host from infection. If the value of R_0 lies below 1.0 the disease will decline and eventually be eradicated. If R_0 is above 1.0, the disease will increase from very low levels to reach an equilibrium prevalence. Thus, from Eqn 7.1, it can be seen that increases in magnitudes of m, b, c and a and decreases in μ, r and T (EIP) all serve to increase the value R_0.

This brief overview of the ecology of arthropod-borne viruses, together with the mathematical definition of R_0, identify a number of key criteria that are needed to drive transmission of viruses such as WNV. Firstly, the vector must be present and in such abundance that numerous vertebrate hosts are bitten. Thus R_0 increases with the vector/host abundance ratio, m, and the square of the biting rate, a (Eqn 7.1). In temperate regions, high mosquito numbers and biting rates usually only occur during the summer months when temperatures are sufficiently high to promote rapid progression of the mosquito life cycle. Furthermore mosquito activity and hence biting rate, a, are dependent on temperature (Rogers and Randolph, 2006). Mosquitoes need access to water to lay eggs and enable larvae to develop. Mosquito species differ in their preferences for breeding sites. Some species are flood plain mosquitoes and others use car tires or water containers in urban areas. In California there has been an interesting association between neglected swimming pools in homes repossessed in the financial crisis and breeding sites for mosquitoes giving rise to WNV outbreaks (Reisen et al., 2008). Predicting such scenarios before they occur would be a goal of horizon scanning techniques (Gale and Breed, 2012). Finally, the transmission coefficient from bird host to mosquito vector (b in Eqn 7.1) must be sufficiently high. Bird species differ in their reservoir competence (Komar et al., 2003) and it has been demonstrated that the percentage of *Cx. tarsalis* females infected with WNV increases with the log of the titer of WNV in the donor birds at the time of blood-feeding (Reisen et al., 2006). In the northeastern USA,

Cx pipiens pipiens is strongly ornithophilic. In other states by contrast it shows an increased affinity for human hosts and has been proposed as a key bridge vector (Andreadis, 2012). Unlike ticks, which take only one blood meal per stage, adult female mosquitoes may take many blood meals and are infectious for their entire lives. For WNV, humans and horses fall into the category of "dead-end" hosts—i.e., they are susceptible to infection and clinical disease but play no role in the ecology and persistence of the virus (Kramer et al., 2008). This is because the low level of viremia in mammals is usually not sufficient to be transmitted to mosquitoes such that $b = 0$ in Eqn 7.1, thereby ending the transmission cycle (Colpitts et al., 2012). As shown by Eqn 7.1 a combination of some or all these factors contribute to increasing R_0 and maintaining an outbreak.

DISTRIBUTION OF WEST NILE VIRUS

Since its first isolation WNV has been detected widely in Africa from countries such as Nigeria, Ethiopia and South Africa (Jupp, 2001), and Asia, including Malaysia and Pakistan (Reisen et al., 1982). A closely related virus, Kunjin virus, was isolated in Australia (Mackenzie et al., 1994) and is now considered a subtype of WNV. This suggested that WNV had the widest distribution of any known flavivirus, even before its emergence in North America.

Currently, there are five WNV lineages based on genomic phylogeny. Lineage 1 dominates and contains viruses from outbreaks around the world and includes Kunjin virus. Lineage 2 was originally thought to be restricted to sub-Saharan Africa, but has been responsible for outbreaks of West Nile fever in Hungary and Greece. Lineage 3 was isolated in the Czech Republic (Bakonyi et al., 2005) and lineage 4 was isolated in Russia (Pripilov et al., 2002). Lineage 5 was isolated in India (Bondre et al., 2007) and has been responsible for a number of human infections.

DESCRIPTION OF DISEASE IN HUMANS

Transmission follows a bite from an infected mosquito. Other, less frequent means of transmission include transfusion of blood or blood products, and organ transplantation from an infected donor. Mother to child infection has been observed through both transplacental transmission and through breast feeding. The incubation period following exposure ranges from 2–6 days. Humans have a short, low-level viremia lasting 2–4 days, which is insufficient to infect mosquitoes. Seroconversion, often as early as 5 days post infection, leads to rapid clearance of virus from the blood following the development of antibodies (Solomon et al., 2003).

Infection in humans can have variable outcomes. In most cases, infection is asymptomatic and can only be demonstrated by seroconversion. Up to 30% of infections lead to a febrile episode, referred to as West Nile fever, with disease in the form of a flu-like illness that includes symptoms of fever, back pain, fatigue and myalgia. This can last from a couple of days to weeks (Hayes et al., 2005). Less commonly observed is a maculopapular rash, particularly in children, and occasionally severe cases of myocarditis, pancreatitis and hepatitis (Hollidge et al., 2010). In less than 1% of cases, infection leads to West Nile (WN) neuroinvasive disease. Of these, 60% present with encephalitis and 40% with meningitis, and up to 10% die. The factors that make individuals susceptible to WN neuroinvasive disease are poorly understood but the risk is higher in the elderly and immunocompromised. In those that survive neuroinvasive disease, up to 70% have permanent neurological sequelae, including flaccid paralysis (Gyure, 2009). In a small number of infected patients, persistent infection has been observed with virus being shed in urine for years following initial disease (Murray et al., 2010).

There is no treatment for WNV infection and no vaccine has been approved for human use to date. However, there is evidence that immune responses elicited by other flavivirus vaccines, including JEV, tick-borne encephalitis virus and yellow fever virus, may neutralize WNV suggesting that a low level of protection is possible (Mansfield et al., 2011). Currently a large number of candidate vaccines are undergoing human phase II clinical trials for efficacy (De Filette et al., 2012).

DISEASE IN HORSES AND OTHER DOMESTIC ANIMALS

For reasons that are not clearly understood, equids are the one group of domestic animals that develop disease following infection with WNV from a mosquito bite. Other species such as dogs and chickens can become infected. Indeed chickens seroconvert after exposure by mosquito bite and survive showing no signs of disease, but would not contribute to the local WNV transmission cycle because they have low virus titers (Langevin et al., 2001). These are positive attributes of a sentinel species for the emergence of WNV in an area.

As in humans, a minority of infected horses, approximately 10%, develop neurological disease. Signs of disease include ataxia, paresis or hind limb paralysis, skin fasciculations, muscle tremors and rigidity (Castillo-Olivares and Wood, 2004). Rare signs include dysmetria, hyperexcitability and hyperaesthesia. Experimental infection of horses has confirmed that a low level of viremia of short duration develops that is unable to act as an amplifying host for WNV (Bunning et al., 2002).

Vaccines for the protection of horses from WNV infection were introduced rapidly after the emergence of the virus in North America. These were formalin inactivated virus with adjuvant (Dauphin and Zientara, 2007). Recent developments have included recombinant envelope protein, virus-like particles and DNA vaccines (reviewed recently by De Filette et al., 2012).

DIAGNOSIS OF WEST NILE VIRUS

Detection of WNV in humans is challenging due to the low level transient viremia. Virus isolation can be achieved through growth in tissue culture with Vero cells causing a lytic cytopathic effect. This can be controlled by addition of a gel overlay to the cell monolayer that leads to the development of defined plaques after a number of days. An alternative to virus isolation from solid organ tissues is that of virus antigen capture. Assays such as the Vec Test® are rapid and have been applied to the detection of WNV from oral and tissue swabs from bird samples (Stone et al., 2004). Detection of virus genome can be achieved through sensitive molecular methods such as reverse transcription polymerase chain reaction (RT-PCR) (Lanciotti et al., 1999) and reverse transcription loop mediated isothermal amplification (RT-LAMP) (Parida et al., 2004). In the absence of virus detection, diagnosis is dependent on serological methods including plaque reduction neutralization (PRN) and enzyme linked immunosorbent assays (ELISAs) specific for IgM and IgG (Dauphin and Zientara, 2007). Similar approaches can be applied to diagnosis of WNV in horses (Kleiboeker et al., 2004).

MOSQUITO VECTOR

The ability of different mosquito species to acquire and transmit WNV is highly variable (Colpitts et al., 2012). The earliest studies on the ecology of WNV detected the virus predominantly from *Culex* species of mosquito (Taylor et al., 1953) and *Culex* mosquitoes are accepted as the primary global transmission vector (Colpitts et al., 2012). For example, *Culex tarsalis* is the main mosquito vector of WNV in the western United States and can feed on a variety of avian and mammalian species. As birds are the main amplifying vertebrate species, it is unsurprising that an ornithophilic group of mosquito species is the main vector for WNV. In North America many *Culex* species have been shown to be competent for WNV transmission (Sardelis et al., 2001). These include *Culex erythrothorax*, *Culex nigripalpus*, *Culex pipiens pipiens*, *Culex pipiens quinquifasciatus*, *Culex restuans*, *Culex salinarius* and *Culex tarsalis*. In Europe, *Culex pipiens pipiens* and *Culex modestus* are important vector species, and in Australia, *Culex annulirostris* is considered the major vector of Kunjin virus (Hall et al., 2002). In South Africa *Culex univittatus* has been shown to efficiently transmit WNV to birds (Jupp, 1974). In addition to *Culex* species, WNV has been detected in a further ten genera of mosquitoes including *Ochlerotatus*, *Aedes*, *Anopheles*, *Coquilletidia*, *Aedemmyia*, *Mansonia*, *Mimomyia*, *Psorophora*, *Culiseta* and *Uranoteania* (Zeller and Schuffenecker, 2004). It is likely that some of these act as bridge vectors for transmission to humans and equines.

A further aspect associated with vector competence of the mosquito in predicting the emergence and spread of arboviruses is the extrinsic incubation period (EIP). This is the time period from ingestion by the mosquito of a blood meal containing virus to the point when that vector is then able to transmit virus to a new vertebrate host. During the intervening time, usually measured in days, the virus must bind to receptors on and replicate in the mosquito midgut cells, overcome the mosquito immune system and spread through the mosquito to infect the salivary glands to be in a position to infect a new host when the vector takes another blood meal. Two critical factors that influence the EIP are the environmental temperature and the vector competence for the given virus strain. Thus the EIP of WNV in *Cx. tarsalis* decreased from 30 days at 18°C to 7 days at 30°C (Reisen et al., 2006). The temperature of the mosquito reflects the environmental temperature. A key observation during the WNV epidemic in North America was the emergence of a variant that had a number of sequence changes within its genome compared to the genome of the original virus introduced in 1999 (Ebel et al., 2004; Davis et al., 2005). Thus, while replication of the strain of WNV originally introduced into North America, NY99, was more efficient at warmer temperatures than that of a South African strain (Reisen et al., 2006), the genotype WN02, which was first detected in 2001 and spread across the USA, was more efficient than the genotype NY99. Kilpatrick et al. (2008) demonstrated that the proportion of *Cx. pipiens* mosquitoes transmitting virus could be modeled with a degree-day term with temperature raised to the fourth power, and that almost double the proportion of *Cx. pipiens* mosquitoes transmit WN02 compared to NY99. Thus warmer temperatures increased the advantage of the WN02 genotype over the NY99 genotype virus with transmission by *Culex* mosquitoes accelerating sharply with increasing temperature. This may have facilitated further spread across USA and this variant, WN02, rapidly displaced the original virus (NY99) from the entire geographical range of the epidemic. The phenotypic change that the relatively small number of genomic changes bestowed on the virus was to reduce the EIP within competent mosquitoes by up to four days (Moudy et al., 2007). Evidence against RdRp being a rate-limiting factor for the WNV EIP is that after intrathoracic inoculation there was no difference in EIP between the WN02 and NY99 strains in *Cx. pipiens*, suggesting that differences in their interactions with the mosquito midgut are important (Moudy et al., 2007). A shorter EIP at elevated temperature could reflect disruption of the cell junctions in the midgut epithelium which would allow leakage of the virus and hence more rapid infection of the salivary glands (Kilpatrick et al., 2008). This however, would not necessarily explain the observed differences between WNV genotypes in transmission rates in *Cx. pipiens* at elevated temperature. Laperriere et al. (2011) have simulated the seasonal cycles of bird, equine and human cases of WNV in Minnesota. The model is based on an SEIR (Susceptible, Exposed, Infectious, Removed) model involving transmission between birds, humans, horses and mosquitoes. The model runs for the entire period 2002–2009 and is the first model for WNV that simulates the seasonal cycle by explicitly considering the environmental temperature (daily air-temperature measurements from six climate stations located in the Minneapolis metropolitan area). All the major peaks in the observed time series were caught by the simulations. This is consistent with temperature being a major driver of WNV in the USA.

EMERGENCE OF WEST NILE VIRUS IN THE MEDITERRANEAN BASIN

West Nile virus has been responsible for sporadic outbreaks of disease in countries around the Mediterranean Sea since the 1960s (Hubálek and Halouzka, 1999). These have involved infections in humans and/or horses (reviewed by Zeller and Schuffenecker, 2004). All outbreaks were reported during the late summer months between July and September. Since 2000, there have been further outbreaks that have been documented in detail (Table 7.2). These have included outbreaks in Russia (Lvov et al., 2000), Morocco (Schuffenecker et al., 2005), Portugal (Esteves et al., 2005), Italy

TABLE 7.2 Outbreaks of WNV Around the Mediterranean Basin (2000 to Present)

Year	Year	Species infected	Reference
2000	France	Horses	Murgue et al., 2001
	Israel	Humans, birds, horses	Hindiyeh et al., 2001; Steinmann et al., 2002; Malkinson et al., 2002
2001	Russia	Humans	Lvov et al., 2002
2002	Russia	Humans	Lvov et al., 2002
2003	Hungary	Birds	Bakonyi et al., 2006
	Morocco	Horses	Schuffenecker et al., 2005
2004	Hungary	Birds	Bakonyi et al., 2006
	Portugal	Mosquitoes	Esteves et al., 2005
2005	Hungary	Birds	Erdélyi et al., 2007
2008	Italy	Equines, Birds	Savini et al., 2008
	Austria	Birds	Wodak et al., 2011
2009	Austria	Birds	Wodak et al., 2011
2010	Romania	Humans	Neghina and Neghina, 2011
	Greece	Humans	Papa et al., 2011
2010/2011	Spain	Horses	Gárcia-Bocanegra et al., 2012

(Savini et al., 2008) and Greece (Papa et al., 2011). The outbreak in Greece is distinct in that it has been dominated by human infections caused by WNV lineage 2. Most outbreaks, with the exception of a previous emergence in Hungary, were attributed to lineage 1. Many outbreaks were reported over consecutive years and, through phylogenetic analysis of isolated virus, there is strong evidence to suggest that the virus has overwintered in the vector rather than been repeatedly introduced.

Some evidence for the role of migrating birds in the introduction of WNV to new regions comes from the 1999 outbreak in Israel (Malkinson et al., 2002). In the preceding year, WNV was detected in brain tissue of juvenile white storks (*Ciconia ciconia*) migrating southwards through Israel for the first time from breeding grounds in Europe. This species migrates between sub-Saharan Africa and the Iberian Peninsula using the East Atlantic flyway and between southern Africa and Central Europe using the East Africa West Asian flyway. Although Malkinson et al. (2002) suggested that strong winds had blown the birds off their usual migration route through Jordan, it should be noted that a satellite tracking investigation (Berthold et al., 2004) showed one individual migrating southwards through Israel in each of six years. That bird bred in Germany each year and migrated southeast through Poland, Romania and Bulgaria crossing the Bosphorus into Turkey (Berthold et al., 2004). The WNV-infected birds in the study of Malkinson et al. (2002) were migrating southwards on the East Africa West Asian flyway and had not flown over Israel before (being juveniles), but had presumably become infected with WNV at some point along their route of migration in Europe. WNV was also isolated from domestic geese (*Anser anser domesticus*) in Israel in 1998 and 1999 and from a captive white-eyed gull (*Larus leucophthalmus*) in 1999 suggesting vector transmission to nonmigratory birds (Malkinson et al., 2002). The virus persisted in Israel, overwintered and emerged in subsequent years with cases in humans (Hindiyeh et al., 2001) and horses (Steinman et al., 2002). Climatic conditions were also important in the outbreak in Israel in 2000 (Paz, 2006). Thus, each of those epidemics appeared after a long heatwave with extreme heat being more significant than high air humidity for increasing WNV cases (Paz, 2006). Indeed, the minimum temperature was the most important climatic factor. This example will be discussed in more detail later in this chapter as a focus of discussion on the role of migratory birds and WNV spread.

EMERGENCE OF WEST NILE VIRUS IN THE WESTERN HEMISPHERE

The most dramatic and long-lasting incursion of WNV with a well-documented impact on a human population was the introduction of WNV into North America in 1999. The virus was identified as the agent causing the deaths of captive and wild avian species on September 23, 1999 by RT-PCR and genome sequencing (Lanciotti et al., 1999). Retrospective serological investigation of human and bird samples identified seropositive cases from early August, suggesting that infection had occurred in July implying that this was the likely time of introduction. Autopsy investigation confirmed that bird deaths were due to encephalitis with brain lesions being readily identified. The final seropositive human case in the New York area was reported on September 5, one of 62 laboratory-confirmed cases that year. The final bird death was reported on November 5, 1999. It has been estimated that as many as 3,000 American crows (*Corvus brachyrhynchos*) were infected in the New York City area during the first year of the outbreak. During 1999, reports of dead birds occurred in the neighboring states of Connecticut and New Jersey, suggesting dispersal of virus up to 250 kilometers (km) from the supposed point of introduction (Rappole and Hubálek, 2003).

Speculation on the method of virus introduction has included long-range migration of an infected bird from Africa. This might have been ship-assisted. The Northern Wheatear (*Oenanthe oenanthe*) is known to migrate from the Canadian Arctic to Africa and back through satellite tracking studies. However, most migrants cannot attempt the journey from Africa to the USA because of the prevailing direction of the jet stream. Indeed, Rappole et al. (2000) cite relatively few species of bird that are known to complete trans-Atlantic migration annually and note that the numbers are so small that the risk of establishment of WNV seems low through this route. Alternative theories for the introduction of WNV to North America included the importation of an infected mosquito, bird or human, or an act of bioterrorism. Prior to summer 1999, the virus had never been detected in the Americas, although the related flavivirus St. Louis encephalitis virus was endemic in southern USA. The large number of avian mortalities, both in wild indigenous species, particularly of the genus *Corvidae*, and exotic species in urban zoos (Steele et al., 2000) provided a stark contrast to outbreaks of WNV in Europe where avian deaths were rare. This was followed by incidence of fever and encephalitis in humans. It was initially hoped that the virus would not survive the North American winter but this early optimism was unfounded, with the relentless expansion of the virus to new areas in subsequent years. The virus re-emerged in May of 2000 approximately 50 km north of New York City, again in an American crow. Further spread occurred both to the north and south. By the end of 2000, WNV had been detected almost 600 km in all directions from New York. However, the appearance of cases appeared to show a steady movement rather than sudden large distance jumps that might be predicted by spread by long-distance migratory birds. By 2001 the virus was detected throughout the eastern United States, particularly in the southern states of Florida and Louisiana, perhaps reflecting the high abundance of vectors in these states. Bird migration could have contributed to WNV emergence in this region. In support of this, seroprevalence studies on birds undergoing spring and autumn migration have shown seropositivity levels ranging from 3–10% (Dusek et al., 2009). Viremia was also reported in a small number of birds, in particular the gray catbird (*Dumetella carolensis*).

A possible explanation for the gradual dispersal observed for WNV spread in the western hemisphere might be through spread by nonmigratory or sedentary birds such as the house sparrow (*Passer domesticus*). This species, like many indigenous birds of America, has been shown to support high WNV viremia (Komar et al., 2003) and thus is likely to enable WNV transmission to mosquitoes. The species is also capable of flying up to 50 km (Rappole and Hubálek, 2003). Another species that may have enabled rapid WNV spread is the American robin (*Turdus migratorius*), which is a common target for ornithophilic *Culex* species mosquitoes such as *Cx. plpiens*, *Cx. cinquifaciatus*, *C. restuans*, *Cx. tarsalis* and *Cx. nigripalpus*. Canadian populations of the American robin are long-range migrants between Canada and Mexico. Locally in suburban Chicago (USA), Hamer et al. (2011) estimated that two avian species, the American robin and the house sparrow, produced 95.8% of the infectious *Cx. pipiens* mosquitoes and showed a significant positive association with WNV infection in *Culex* spp. mosquitoes. Introduction of WNV into Argentina by migrating birds is a popular hypothesis, although relatively few North American breeding birds migrate to Argentina, and austral migrants number fewer than boreal migrants (Diaz et al., 2008). Of 211 migratory birds tested in Argentina, all were seronegative (Diaz et al., 2008). There is some evidence for lateral transmission of WNV in American crows at roost sites through fecal contamination (Dawson et al., 2007). This could explain transmission during the winter when mosquitoes are not active.

The ecological impact of WNV introduction into North America was manifest as a dramatic decline in the population of many indigenous bird species, estimated in some cases to have dropped by 50% (LaDeau et al., 2007). This manifestation of WNV spread in the US has been used as an early warning system to indicate an increase in virus activity (Mostashari et al., 2003). This gives advanced warning to public health services, laboratory facilities and can lead to the initiation of vector control.

A further factor that is thought to have been influential in the transmission of WNV across America is the modification of the environment by humans. The conversion of virgin landscape for both agricultural and urban land use promoted certain avian species over others, making these species more abundant. Some of these were important for WNV transmission (Kilpatrick, 2011) and may have accelerated the spread of virus through the landscape. However, this does not appear to be the case in Europe where the landscape has been even more heavily modified, although a related virus, Usutu, is now spreading in many countries of Europe and may be followed by WNV in the future. Anthropogenic modifications of ecosystems, in particular expansion of rice cultivation and changes in pest-management strategies as directed by EU regulations, in the Camargue region of France have been implicated in changes in the abundance of *Culex modestus*, an important vector of WNV (Poncon et al., 2007). The importance of anthropogenic changes in vector-borne disease recrudescence should not be underestimated.

Associated with the spread of WNV through North America were clinical cases of infection in humans and increasing numbers of cases in horses as the epidemic spread to rural areas. In 2002 alone there were 4156 human cases of WNV infection resulting in 284 deaths (Granwehr et al., 2004). By 2008 the virus had reached the Western seaboard of the United States, been translocated into Canada and south into Latin America (Kramer et al., 2008). The virus is now considered endemic in the Western Hemisphere with limited options for control or elimination. The introduction of vaccination for horses has ameliorated the impact on equids but the lack of a human vaccine or effective treatment means that WNV remains a significant public health issue throughout the Americas.

POTENTIAL ROLE OF BIRDS IN THE INTRODUCTION AND SPREAD OF WNV

Migratory birds have been implicated in the spread of numerous zoonotic diseases, including viruses such as WNV, Crimean-Congo hemorrhagic fever virus (CCHFV) and avian influenza virus (Hubálek, 2004). However, there is little direct evidence to confirm this, although CCHFV has been detected recently in ticks on migrant birds in Morocco (Palomar et al., 2013). In order to consider the role of birds in the spread of WNV, we will focus on the species believed to have been responsible for the introduction of the virus into Israel in 1998, the white stork (*Ciconia ciconia*) (Malkinson et al., 2002). The case for those white storks introducing WNV into Israel was strengthened by the fact that sequence analysis of the envelope gene of the stork isolate showed almost complete identity with isolates from the geese and nonmigrating white-eyed gull. High vertebrate biodiversity appears to be an important factor in reducing transmission of some zoonotic pathogens among wildlife hosts. Thus, Swaddle and Calos (2008) found a lower incidence of WNV in humans in eastern US counties that have greater avian diversity when controlled for confounding factors.

The biology and ecology of the white stork (*Ciconia ciconia*)

The white stork is a large, distinct European migrant species. Thus, a flock of debilitated or dying birds is more obvious than smaller, cryptic migrants which may migrate individually. White storks are predominantly white with black secondary and primary feathers (Figure 7.2A and 7.2B). Adults can measure up to 115 cm from the tip of the beak to tail and have a wingspan of over 2 m. The species breeds locally in Europe, either in the Iberian Peninsula or eastern European countries but also Germany, France and the Netherlands. It is also a passage migrant to more northerly European countries including the UK. There is also a population that migrates to the Middle East from Africa. They overwinter in sub-Saharan Africa as far south as South Africa itself. Being a large bird, long distance migration is partly dependent on nonpowered flight,

FIGURE 7.2   **Images of the white stork (*Ciconia ciconia*) feeding (A), in flight (B) and a group of birds soaring on a thermal prior to migration from Spain (C).** *(Paul Gale).*

which requires considerably less energy than powered flight (Figure 7.2C). This is achieved by soaring up on the warm air thermal currents that rise over land and then gliding, assisted by the prevailing winds. Such thermals do not develop over large water bodies so large migratory birds such as storks and raptors that use this method are restricted to predominantly land routes, either crossing the Strait of Gibraltar on the Eastern Atlantic flyway or around the eastern edge of the Mediterranean Sea travelling through the Middle East (Leskem and Yom-Tov, 1998). Satellite tracking has established a migration route for the white stork from its breeding grounds in Germany to Africa through Israel via Slovakia, Hungary, Romania and Bulgaria (Berthold et al., 2004). The northerly migration takes place in spring between February and April, with birds arriving at the breeding grounds from March onwards. The southerly migration occurs in autumn with birds travelling over the Middle East during late August and September. White storks are marshland and wetland birds and so spend considerable time in habitats with high mosquito abundance. This makes them a potential species to become infected with WNV and transfer infectious agents across long distances, although the duration of viremia may only last for a few days (Komar et al., 2003), thus limiting long-distance spread to those birds that are infected immediately prior to embarking on migration. In this respect, birds differ from arthropods in their vector capability, with arthropods being infectious for their entire lives.

Based on the assumption that the white storks were responsible for the introduction of disease, the surprising observation is that contrary to the assertion that WNV is endemic in Africa and is transported north, the outbreak in Israel in 1998 appear to have its origins in areas north of the country where the juvenile birds originated, probably in Europe. This implies that the storks were infected either in their breeding grounds immediately prior to departure or at stopover sites as they migrated south. More detailed analysis of satellite tracking data, together with information on the location and length of stopovers, will give additional information with which to assess the range over which migrating birds could be viremic. Furthermore, there is some evidence that since the mid-1980s increasing numbers of storks have stopped their annual migration from Europe to Africa for the winter with many living in Spain and Portugal the whole year round (Anon, 2013). This reflects the availability of food in rubbish dumps throughout the year. Thus global anthropogenic changes, including climatic factors, may have an impact on the risk of introduction of exotic viruses by migrating birds (Gale et al., 2009).

Evidence for West Nile virus infection in white storks and their potential role as a host reservoir

The three factors that are critical to the ability of a bird species to act as a host reservoir for an arthropod-borne virus are firstly the ability of the virus to replicate in that bird species to give a viremia, including the duration and titer resulting from infection, secondly the generation of immunologically naive individuals that could become infected and hence serve as hosts, and thirdly a high survival rate after infection (Carrara et al., 2005). There is little information on the WNV viremia in the white stork, although there is extensive data from experimental infection of other avian species. Thus, for example, Reisen et al. (2006) reported a variable viremic response to WNV in house sparrows (*Passer domesticus*) and house finches (*Carpodacus mexicanus*) compared to white-crowned sparrows (*Zonotrichia leucophys*) and that infection rates of *Culex tarsalis* females were low at bird titers $< 6.5 \log_{10}$ plaque forming units (pfu) / ml. Although not susceptible to disease, inoculation of chickens using both infected mosquitoes and subcutaneous infection resulted in a transient viremia but at a low level of $< 4.0 \log_{10}$ pfu/ml, insufficient to infect mosquitoes (Langevin et al., 2001; Komar et al., 2003). A wide-ranging study inoculating a range of captive-bred and wild-caught avian species with WNV demonstrated that infection in many species, but particularly *Passeriformes*, resulted in a four to five day viremia with titers as high as $10 \log_{10}$ pfu/ml (Komar et al., 2003).

Such levels would be sufficient to infect mosquitoes. These and other studies suggest that a species susceptible to WNV infection would have a transient viremia of $> 7 \log_{10}$ pfu/ml lasting between one and six days from infection. The second factor, the appearance of susceptible individuals, is driven primarily from bird breeding. White storks have up to four chicks per brood, which if they survive to fledging will provide a major component of the southerly migrating bird flocks. As immunologically naïve individuals, it is these that are most likely to become infected. However, the birth rates of large birds such as storks is much lower than those of rodents, which serve as host reservoirs for Venezuelan equine encephalitis virus (VEEV) for example (Carrara et al., 2005). Thus birds are unlikely to represent a constant reservoir of infection unlike rodents which breed all year in tropical regions.

The adaptation of arboviruses to replicate at elevated temperatures facilitates utilization of new host species, for example birds (Brault, 2009), which may be important in range expansion. Of considerable interest is the increased replication efficiency of different WNV strains in vertebrate hosts at elevated temperatures. Thus, replication of a WNV strain from Kenya in cell culture was reduced 6,500-fold at 44°C relative to levels at 37°C, while replication of the strain of WNV originally introduced into North America (NY99) was only reduced by 17-fold (Kinney et al., 2006). Kinney et al. (2006) suggest that the ability of the natural temperature-resistant strain, NY99, to replicate at high temperatures could be important in the increased avian virulence of the NY99 genotype. Thus, mean body temperatures of WNV NY99-infected American crows ranged between 40.5 and 44.5°C (Kinney et al., 2006) potentially giving the NY99 strain a clear replicative advantage in transmission within that bird population.

Where does exposure occur?

The next question is where do white storks become infected? Two areas of Europe where WNV seroconversion in wild birds has been demonstrated and where white storks breed are southern Spain and southern France. Thus, Figuerola et al. (2007) reported high prevalence of WNV antibodies in juvenile common coots (*Fulica atra*) on the Donana, southern Spain, in September–October 2003 (37.5%) and 2004 (28.8%), falling to less than 20% in January–February 2006. Of 95 birds captured in 2 consecutive years, 59% had no detectable antibodies in either year, 21% seroreverted, 6.3% seroconverted, and 13.7% had antibodies in both years. Seroconversion confirms that WNV circulation is present in the Donana region. Similarly, of 271 black-billed magpies (*Pica pica*) captured in the Camargue (southern France), 10.7% had neutralizing antibodies, adults being more frequently seropositive than juveniles (Jourdain et al., 2008). In addition to Spain and France, outbreaks of WNV in horses and/or humans have been reported in Romania, Hungary, Portugal, Italy and Greece (Zeller et al., 2010). It is conceivable that white storks migrating through Israel on the East Africa/West Asian flyway were infected in Romania, Hungary or Greece.

Serological and virological testing in northern latitude countries such as Germany (Linke et al., 2007; Seidowski et al., 2010) and Poland (Hubálek et al., 2008) has provided evidence that there are a small number of seropositive storks found in northern Europe. Depending on the serology assay used, levels range from less than 1% to over 6%. In the absence of demonstration of virus, this was at least partly attributed to antibody transferred in the egg. The converse of this observation is that as many as 95% of the new generation of white storks leaving the breeding grounds in Germany and Poland are immunologically naïve for WNV and thus potentially susceptible to infection should it occur on the southerly migration. Locations such as the Hungarian plain would be ideal stopover points, with numerous rivers such as the Danube and Tisza providing wetland areas for wading birds to rest before flying over the Balkan Mountains to the south. Other wetland areas found around the Mediterranean Sea include the Camargue, Ebro and Po Deltas in the western Mediterranean in France, Spain and Italy respectively or the Amvrakiko Gulf and Aliakmonas Delta in Greece and the Gediz and Göksu Deltas in Turkey (Jourdain et al., 2007). These sites attract large numbers of migrant birds in addition to providing an all-year-round location for resident birds. It is possible that infection could occur here, leading to viremic birds leaving central and southern Europe as they progress around the edge of the eastern Mediterranean.

Such a scenario could have been responsible for the emergence of lineage 2 WNV in Greece in 2010. Six years previously, this lineage was reported in Hungary, some 800 miles to the north of Greece, in 2004 (Bakonyi et al., 2006). During subsequent years this lineage was reported in birds of prey in 2005 (Erdélyi et al., 2007), in horses with neurological signs (Kutasi et al., 2011) and in birds of prey in the neighboring country of Austria in 2008/09 (Wodak et al., 2011). This suggests that WNV lineage 2 was endemic in this region in the years leading up to 2010. Migrating birds leaving northern Europe would pass through this region during the time of peak vector activity in late summer, perhaps making brief overnight rest stops in wetland areas. During these stops they would be exposed to biting mosquitoes and could become infected. Over the following days as they continued their migration they would become viremic and potentially infectious to biting mosquitoes at rest stops further south and east. This could have been the method of introduction to Greece. However, the manifestation of WNV lineage 2 infection differed in these two locations. In Central Europe, birds were affected without any evidence for human involvement (Bakonyi et al., 2006) while in Greece infected humans were the first indication in July of 2010 that an outbreak had begun (Papa et al., 2011). It is likely that the virus had been present for some time prior to this, possibly introduced into the region during the previous year. Further investigations have shown that virus is present in the mosquito population (Papa et al., 2011) and within the indigenous bird population (Valiakos et al., 2011).

These observations suggest that WNV can be transported in a north–south axis during both migration periods. However, it is likely that movement is restricted to shorter distances than the full migration due to the limited period of viremia which is typically from 1.3 to 6.0 days (Komar et al., 2003). An alternative hypothesis might be persistent infection with continuous shedding, although there is little evidence for this from infection studies in birds. There is also strong evidence that both WNV lineages 1 and 2 have over-wintered in Europe and are persistently emerging in late summer in a similar manner to that in North America. The detection of the virus is more a function of the type and level of surveillance taking place within individual countries. It should be remembered that bird migration has been taking place for thousands of years, and the question of "Why now?" has to be addressed, particularly for viruses such as CCHFV and WNV which have been endemic in Africa for many years. In this respect, other routes such as global air transport of infected mosquitoes, both adults and larvae, need further consideration. Indeed the probability of at least one WNV-infected mosquito being introduced into the UK aboard aircraft from the United States each summer was 0.99 with a mean of 5.2 infected mosquitoes between May to October each year (Brown et al., 2012).

CONCLUSIONS

A growing body of evidence suggests that the movement of infected birds is a driver for West Nile virus spread although other routes, in particular the transportation of infected mosquitoes, need greater consideration. Spread by birds can range from short distance translocations by both local movement and through long-distance migration. The critical component of this is the ability of the bird to sustain a viremia of sufficient magnitude for a sufficient length of time to infect biting mosquitoes while retaining the capacity to fly. Adaptation of the virus has been shown to be a key factor in the spread of WNV to new areas. Current outbreaks are dominated by WNV lineage 1 and 2 viruses, and this trend is likely to continue. Options for control are currently limited for WNV in the absence of a human vaccine and surveillance for virus emergence is critical. Many countries around the Mediterranean Basin have instigated extensive virological surveillance in mosquitoes, wildlife, domestic animals and in human populations, in order to monitor current WNV activity. Vector control and public information on the dangers of mosquito bites remain the main options for reducing the risk of transmission and infection. However, the effects of potential anthropogenic changes on mosquito populations and their interactions with humans should also be considered. The emergence of WNV is highly unpredictable. Countries around the Mediterranean Sea have and will continue to host repeated outbreaks of WNV infection that will result in varying levels of disease in humans and horses for the foreseeable future.

REFERENCES

Andreadis, T. G. (2012). The contribution of *Culex pipiens* complex mosquitoes to transmission and persistence of West Nile virus in North America. *J. Am. Mosq. Control Assoc.*, *28*(4S), 137–151.

Anon, (2013). White storks have stopped migrating—New project to discover why. http://www.wildlifeextra.com/go/news/white-stork-migration.html#cr. Access 14 March 2013.

Bakonyi, T., Hubálek, Z., Rudolf, I., & Nowotny, N. (2005). Novel flavivirus or new lineage of West Nile virus, Central Europe. *Emerg. Infect. Dis.*, *11*, 225–331.

Bakonyi, T., Ivanics, E., Erdélyi, K., Ursu, K., Ferenczi, E., Weissenböck, H., & Nowotny, N. (2006). Lineage 1 and 2 strains of encephalitis West Nile virus, central Europe. *Emerg. Infect. Dis.*, *12*, 618–623.

Berthold, P., Kaatz, M., & Querner, U. (2004). Long-term satellite tracking of white stork (*Ciconia ciconia*) migration: constancy versus variability. *J. Ornithol.*, *145*, 356–359.

Bondre, V. P., Jadi, R. S., Mishra, A. C., Yergollear, P. N., & Anranhalle, V. A. (2007). West Nile virus isolates from India: evidence for a distinct genetic lineage. *J. Gen. Virol.*, *88*, 875–884.

Brault, A. C. (2009). Changing patterns of West Nile virus transmission: altered vector competence and host susceptibility. *Vet. Res.*, *40*, 40–43.

Bunning, M. L., Bowen, R. A., Cropp, C. B., Sullivan, K. G., Davis, B. S., Komar, N., Godsey, M. S., Baker, D., Hettler, D. L., Holmes, D. A., Biggerstaff, B. J., & Mitchell, C. J. (2002). Experimental infection of horses with West Nile virus. *Emerg. Infect. Dis.*, *8*, 380–386.

Brown, E. B. E., Adkin, A., Fooks, A. R., Stephenson, B., Medlock, J. M., & Snary, E. L. (2012). Assessing the risk of West Nile virus-infected mosquitoes from transatlantic aircraft: Implications for disease emergence in the United Kingdom. *Vector-borne and Zoonotic Diseases*, *12*, 310–320.

Calisher, C. H., Karabatsos, N., Dalrymple, J. M., Shope, R. E., Porterfield, J. S., Westaway, E. G., & Brandt, W. E. (1989). Antigenic relationships between flaviviruses as determined by cross-neutralization tests with polyclonal antisera. *J. Gen. Virol.*, *70*, 37–43.

Carrara, A. -S., Gonzales, M., Ferro, C., Tamayo, M., Aronson, J., Paessler, S., Anishchenko, M., Boshell, J., & Weaver, S. C. (2005). Venezuelan equine encephalitis virus infection of spiny rats. *Emerg. Infect. Dis.*, *11*, 664–669.

Castillo-Olivares, J., & Wood, J. N. (2004). West Nile infection of horses. *Vet. Res.*, *35*, 467–483.

Colpitts, T. M., Conway, M. J., Montgomery, R. R., & Fikrig, E. (2012). West Nile virus: biology, transmission and human infection. *Clin. Microbiol. Rev.*, *25*, 635–648.

Cook, S., Moureau, G., Kitchen, A., Gould, E. A., de Lamballerie, X., Holmes, E. C., & Harbach, R. E. (2012). Molecular evolution of the insect-specific flaviviruses. *J. Gen. Virol.*, *93*, 223–224.

Dauphin, G., & Zientara, S. (2007). West Nile virus: recent trends in diagnosis and vaccine development. *Vaccine*, *25*, 5573–5576.

Davis, C. T., Ebel, G. D., Lanciotti, R. S., Brault, A. C., Guzman, H., Siirin, M., Lambert, A., Parsons, R. E., Beasley, D. W., Novak, R. J., Elizondo-Quiroga, D., Green, E. N., Young, D. S., Stark, L. M., Drebot, M. A., Artsob, H., Tesh, R. B., Kramer, L. D., & Barrett, A. D. (2005). Phylogenetic analysis of North American West Nile virus isolates, 2001-2004: evidence for the emergence of a dominant genotype. *Virology*, *9*, 860–863.

Dawson, J. R., Stone, W. B., Ebel, G. D., Young, D. S., Galinski, D. S., Pensabene, J. P., Franke, M. A., Eidson, M., & Kramer, L. D. (2007). Crow deaths caused by West Nile virus during winter. *Emerging Infectious Diseases*, *13*, 1912–1914.

De Filette, M., Ulbert, S., Diamond, M. S., & Sanders, N. N. (2012). Recent progress in West Nile virus diagnosis and vaccination. *Vet. Res.*, *43*, 16.

Diaz, L. A., Komar, N., Visintin, A., Juri, M. J. D., Stein, M., Allende, R. L., Spinsanti, L., Konigheim, B., Aguilar, J., Laurito, M., Almiron, W., & Contigiani, M. (2008). West Nile virus in birds, Argentina. *Emerg. Infect. Dis.*, *14*, 689–690.

Dohm, D. J., Sardelis, M. R., & Turell, M. J. (2002). Experimental vertical transmission of West Nile virus by *Culex pipiens* (Diptera: Culicidae). *J. Med. Entomol*, *39*, 640–644.

Dusek, R. J., McLean, R. G., Kramer, L. D., Ubico, S. R., Dupuis, A. P., Jr., Ebel, G. D., & Guptill, S. C. (2009). Prevalence of West Nile virus in migratory birds during spring and fall migration. *Am. J. Trop. Med. Hyg*, *81*, 1151–1158.

Ebel, G. D., Carricaburu, J., Young, D., Bernard, K. A., & Kramer, L. D. (2004). Genetic and phenotypic variation of West Nile virus in New York, 2000-2003. *Am. J. Trop. Med. Hyg, 71*, 493–500.

Erdélyi, K., Ursu, K., Ferenczi, E., Szeredi, L., Rátz, F., Skáre, J., & Bakonyi, T. (2007). Clinical and pathologic features of lineage 2 West Nile virus infections in birds of prey in Hungary. *Vector Borne and Zoonotic Diseases*, *7*, 181–188.

Esteves, A., Almeida, A. P., Galão, R. P., Parreira, R., Rodrigues, J. C., Sousa, C. A., & Novo, M. T. (2005). West Nile virus in Southern Portugal, 2004. *Vector Borne and Zoonotic Diseases*, *5*, 410–413.

Fechter-Leggett, E., Nelms, B. M., Barker, C. M., & Reisen, W. K. (2012). West Nile virus cluster analysis and vertical transmission in *Culex pipiens* complex mosquitoes in Sacramento and Yolo counties, California, 2011. *J. Vector Ecol.*, *37*, 442–449.

Figuerola, J., Soriguer, R., Rojo, G., Tejedor, C. G., & Jimenez-Clavero, M. A. (2007). Seroconversion in wild birds and local circulation of West Nile virus, Spain. *Emerg. Infect. Dis.*, *13*, 1915–1917.

Fredricksen, B. L., Smith, M., Katze, M. G., Shi, P. -Y., & Gale, M., Jr. (2004). The host response to West Nile virus infection limits viral spread through the activation of the interferon regulatory factor 3 pathway. *J. Virol.*, *78*, 7737–7747.

Gale, P., Drew, T., Phipps, L. P., David, G., & Wooldridge, M. (2009). The effect of climate change on the occurrence and prevalence of livestock diseases in Great Britain: a review. *J. Appl. Microbiol.*, *106*, 1409–1423.

Gale, P., & Breed, A. C. (2012). Horizon scanning for emergence of new viruses: From constructing complex scenarios to online games. *Transboundary and Emerging Diseases*. Epub doi:10.1111/j.1865-1682.2012.01356.x.

Gárcia-Bocanegra, I., Jaén-Téllez, J. A., Napp, S., Arenas-Montes, A., Fernández-Molera, V., & Arenas, A. (2012). Monitoring of the West Nile virus epidemic in Spain between 2010 and 2011. *Transbound. Emerg. Dis*, *59*, 448–455.

Gould, E. A., de Lamballerie, X., Zanotto, P. M., & Holmes, E. C. (2001). Evolution, epidemiology, and dispersal of flaviviruses revealed by molecular phylogenies. *Adv. Vir. Res.*, *57*, 71–103.

Gould, E. A., & Solomon, T. (2005). Pathogenic flaviviruses. *The Lancet*, *371*, 500–509.

Granwehr, B. P., Lillibridge, K. M., Higgs, S., Mason, P. W., Aronson, J. F., Campbell, G. A., & Barrett, A. D. T. (2004). West Nile virus: where are we now? *Lancet Infect. Dis.*, *4*, 547–556.

Gyure, K. A. (2009). West Nile virus infections. *J. Neuropath. Exp. Neurol.*, *68*, 1053–1060.

Hall, R. A., Broom, A. K., Smith, D. W., & Mackenzie, J. S. (2003). The ecology and epidemiology of Kunjin virus. *Curr Top. Microbiol. Immunol.*, *267*, 253–269.

Hamer, G. L., Chaves, L. F., Anderson, T. K., Kitron, U. D., Brawn, J. D., Ruiz, M. O., Loss, E. D., Walker, E. D., & Goldberg, T. L. (2011). Fine-Scale Variation in vector host use and force of infection drive localized patterns of West Nile virus transmission. *PLoS ONE*, *6*, e23767.

Hayes, E. B., Komoar, N., Nasci, R. S., Montgomery, S. P., O'Leary, D. R., & Campbell, G. L. (2005). Epidemiology and transmission of West Nile virus disease. *Emerg. Infect. Dis.*, *11*, 1167–1173.

Hindiyeh, M., Shulman, L. M., Mendelson, E., Weiss, L., Grossman, Z., & Bin, H. (2001). Isolation and characterization of West Nile virus from the blood of viraemic patients during the 2000 outbreak in Israel. *Emerg. Infect. Dis.*, *7*, 748–750.

Hollidge, B. S., González-Scarano, F., & Soldan, S. S. (2010). Arboviral encephalitides: transmission, emergence and pathogenesis. *J. Neuroimmunol. Pharmacol.*, *5*, 428–442.

Hubálek, Z., & Halouzka, J. (1999). West Nile fever—a reemerging mosquito-borne viral disease in Europe. *Emerg. Infect. Dis.*, *5*, 643–650.

Hubálek, Z. (2004). An annotated checklist of pathogenic microorganisms associated with migratory birds. *J. Wild. Dis.*, *40*, 639–659.

Hubálek, Z., Wegner, E., Halouzka, J., Tryjanowski, P., Jerzak, L., Sikutová, S., Rudolf, I., Kruszewicz, A. G., Jaworski, Z., & Wlodarczyk, R. (2008). Serologic survey of potential vertebrate hosts for West Nile virus in Poland. *Virol. Immunol.*, *21*, 247–253.

Jourdain, E., Gauthier-Clerc, M., Bicout, D. J., & Sabatier, P. (2007). Bird migration routes and risk of pathogen dispersion into western Mediterranean wetlands. *Emerg. Infect. Dis.*, *13*, 365–372.

Jourdain, E., Gauthier-clerc, M., Sabatier, P., Grege, O., Greenland, T., Leblond, A., Lafaye, M., & Zeller, H. G. (2008). Magpies as hosts for West Nile virus, southern France. *Emerg. Infect. Dis.*, *14*, 158–160.

Jupp, P. G. (1974). Laboratory studies on the transmission of West Nile virus by Culex (*Culex univittatus*) Theobald; factors influencing transmission rate. *J. Med. Entomol.*, *11*, 455–458.

Jupp, P. G. (2001). The ecology of West Nile virus in South Africa and the occurrence of outbreaks in humans. *Annals of the New York Academy of Science*, *951*, 143–152.

Kanai, R., Kar, K., Anthony, K., Gould, L. H., Ledizet, M., Fikrig, E., Marasco, W. A., Koski, R. A., & Modis, Y. (2006). Crystal structure of West Nile virus envelope glycoprotein reveals viral surface epitopes. *J. Virol.*, *80*, 11000–11008.

Kilpatrick, A. M., Meola, M. A., Moudy, R. M., & Kramer, L. D. (2008). Temperature, viral genetics, and the transmission of West Nile virus by *Culex pipiens* mosquitoes. *PLoS Pathog.*, *4*, e1000092.

Kilpatrick, A. M. (2011). Globalisation, land use and the invasion of West Nile virus. *Science*, *334*, 323–327.

Kinney, R. M., Huang, C. Y., Whiteman, M. C., Bowen, R. A., Langevin, S. A., Miller, B. R., & Brault, A. C. (2006). Avian virulence and thermostable replication of the North American strain of West Nile virus. *J. Gen. Virol.*, *87*, 3611–3622.

Kleiboeker, S. B., Loiacono, C. M., Rottinghaus, A., Pue, H. L., & Johnson, G. C. (2004). Diagnosis of West Nile virus infections in horses. *J. Vet. Diag. Inv.*, *16*, 2–10.

Komar, N., Langevin, S., Hinten, S., Nemeth, N., Edwards, E., Hettler, D., Davis, B., Bowen, R., & Bunning, M. (2003). Experimental infection of North American birds with the New York 1999 strain of West Nile virus. *Emerg. Infect. Dis.*, *9*, 311–322.

Kramer, L. D., Styer, L. M., & Ebel, G. D. (2008). A global perspective on the epidemiology of West Nile virus. *Ann. Rev. Entomol.*, *53*, 61–81.

Kutasi, O., Bakony, T., Lecollinet, S., Biksi, I., Ferenczi, E., Bahoun, C., Sardi, S., Zientara, S., & Szenci, O. (2011). Equine encephalomyelitis outbreak caused by a genetic lineage 2 West Nile virus in Hungary. *J. Vet. Int. Med.*, *25*, 586–591.

LaDeau, S. L., Kilpatrick, A. M., & Marra, P. P. (2007). West Nile virus emergence and large-scale declines of North American bird populations. *Nature*, *447*, 710–713.

Lanciotti, R. S., Roehrig, J. T., Deubel, V., Smith, J., Parker, M., Steele, Crise, B., Volpe, K. E., Crabtree, M. B., Scherret, J. H., Hall, R. A., Mackenzie, J. S., Cropp, C. B., Panigrahy, B., Ostlund, E., Schmitt, B., Malkinson, M., Banet, C., Weissman, J., Komar, N., Savage, H. M., Stone, W., McNamara, T., & Gubler, D. (1999). Origins of the West Nile virus responsible for an outbreak of encephalitis in the northeastern United States. *Science*, *286*, 2333–2337.

Lanciotti, R. S., Ebel, G. D., Deubel, V., Kerst, A. J., Murri, S., Meyer, R., Bowen, M., McKinney, N., Morrill, W. E., Crabtree, M. B., Kramer, L. D., & Roehrig, J. T. (2002). Complete genome sequences and phylogenetic analysis of West Nile virus strains isolated from the United States, Europe and Middle East. *Virology, 298,* 96–105.

Langevin, S. A., Bunning, M., Davis, B., & Komar, N. (2001). Experimental infection of chickens as candidate sentinels for West Nile virus. *Emerg. Infect. Dis., 7,* 726–729.

Laperriere, V., Brugger, K., & Rubel, F. (2011). Simulation of the seasonal cycles of bird, equine and human West Nile virus cases. *Prev. Vet. Med., 98,* 99–110.

Leskem, Y., & Yom Tov, Y. (1998). Routes of migrating soaring birds. *IBIS, 140,* 41–52.

Liu, W. J., Wang, X. J., Clark, D. C., Lobigs, M., Hall, R. A., & Khromykh, A. A. (2006). A single amino acid substitution in the West Nile virus non-structural protein NS2A disables its ability to inhibit alpha/beta interferon induction and attenuates virus virulence in mice. *J. Virol., 80,* 2396–2404.

Linke, S., Niedrig, M., Kaiser, A., Ellerbrok, H., Müller, K., Müller, T., Conraths, F. J., Mühle, R. U., Schmidt, D., Köppen, U., Bairlen, F., Berthold, P., & Pauli, G. (2007). Serologic evidence of West Nile virus infections in wild birds captured in Germany. *Am. J. Trop. Med. Hyg., 77,* 358–364.

Lvov, D. K., Butenko, A. M., Gromashevsky, V. L., Larichev, V. P., Gaidamovich, S. Y., Vyshmirsky, O. I., Zhukov, A. N., Lazorenko, V. V., Salko, V. N., Kovtunov, A. I., Galimzyanov, K. M., Platonov, A. E., Morozova, T. N., Khutoretskaya, N. V., Shishkino, E. O., & Skvortsova, T. M. (2000). Isolation of two strains of West Nile virus during an outbreak in southern Russia, 1999. *Emerg. Infect. Dis., 6,* 373–376.

Lvov, D., Lvov, D. K., Kovtunov, A. I., Butenko, A. M., & Zhukov, A. N. (2002). West Nile fever in southern Russia—epidemiological, clinical, genetic peculiarities (1999–2001). *Proceedings of the XIIth International Congress of Virology,* Paris, 27 July – 1 August 2002, p46.

Mackenzie, J. S., Lindsay, M. D., Coelen, R. J., Broom, A. K., Hall, R. A., & Smith, D. W. (1994). Arboviruses causing human disease in the Australasian zoogeographic region. *Arch. Virol., 136,* 447–467.

Malet, H., Egloff, M. P., Selisko, B., Butcher, R. E., Wright, P. J., Roberts, M., Gruez, A., Sulzenbacher, G., Vonrhein, C., Bricogne, G., Mackenzie, J. M., Khromykh, A. A., Davidson, A. D., & Canard, B. (2007). Crystal structure of the RNA polymerase domain of the West Nile virus non-structural protein 5. *J. Biol. Chem., 282,* 10678–10689.

Malkinson, K., Banet, C., Weissman, Y., Pokamunski, S., King, R., Drouet, M. T., & Deubel, V. (2002). Introduction of West Nile virus in the Middle East by migrating white storks. *Emerg. Infect. Dis., 8,* 392–397.

Mansfield, K. L., Horton, D. L., Johnson, N., Li, L., Barrett, A. D. T., Smith, D. J., Galbraith, S., Solomon, T., & Fooks, A. R. (2011). Flavivirus-induced antibody cross-reactivity. *J. Gen. Virol., 92,* 2821–2829

Melnick, J. L., Paul, J. R., Riordan, J. T., Barnett, V. H. H., Coldblum, N., & Zabin, E. (1951). Isolation from human sera in Egypt of a virus apparently identical to West Nile virus. *Proc. Soc. Exp. Biol. Med., 77,* 661–665.

Mishra, A. C., & Mourya, D. T. (2001). Transovarial transmission of West Nile virus in *Culex vishnui* mosquito. *Indian Journal of Medical Research, 114,* 212–214.

Mostashari, F., Kulldorff, M., Hartman, J. J., Miller, J. R., & Kulasekera, V. (2003). Dead bird clusters as an early warning system for West Nile virus activity. *Emerg. Infect. Dis., 9,* 641–646.

Moudy, R. M., Meola, M. A., Morin, L. L., Ebel, G. D., & Kramer, L. D. (2007). A newly emergent genotype of West Nile virus is transmitted earlier and more efficiently in Culex mosquitoes. Am. J. Trop. Med. *Hyg., 77,* 365–370.

Munoz-Jordan, J. L., Laurent-Rolle, M., Ashur, J., Martinez-Sobrido, L., Ashok, M., Lipkin, W. I., & Garcia-Sastre, A. (2005). Inhibition of alpha/beta interferon signalling by the NS4B protein of flaviviruses. *J. Virol., 79*, 8004–8013.

Murgue, B., Murri, S., Zientarra, S., Labie, J., Durrand, B., Durand, J. P., & Zeller, H. (2001). West Nile outbreak in horses in sourthern France, 2000: the return after 35 years. *Emerg. Infect. Dis., 7*, 692–696.

Murray, K., Walker, C., Herrington, E., Lewis, J. A., McCormick, J., Beasley, D. W. C., Tesh, R. B., & Fischer-Hoch, S. (2010). Persistent infection with West Nile virus years after initial infection. *J. Infect. Dis., 201*, 2–4.

Neghina, A. M., & Neghina, R. (2011). Reemergence of human infections with West Nile virus in Romania, 2010: an epidemiological study and brief review of the past situation. *Vector Borne and Zoonotic Diseases, 11*, 1289–1292.

Palomar, A. M., Portillo, A., Santibanez, P., Mazuelas, D., Arizaga, J., Crespo, A., Gutierrez, O., Cuadrado, J. F., & Oteo, J. A. (2013). Crimean-Congo hemorrhagic fever virus in ticks from migratory birds, Morocco. *Emerg. Infect. Dis., 19*, 260–263.

Papa, A., Bakonyi, T., Zanthopoulou, K., Vázquez, A., Tenorio, A., & Nowotny, N. (2011). Genetic characterization of West Nile virus lineage 2, Greece 2010. *Emerg. Infect. Dis., 17*, 920–922.

Parida, M., Posadas, G., Inoue, S., Hasebe, F., & Morita, K. (2004). Real-time reverse transcription loop-mediated isothermal amplification for rapid detection of West Nile virus. *J. Clin. Microbiol., 42*, 257–263.

Paz, S. (2006). The West Nile Virus outbreak in Israel (2000) from a new perspective: The regional impact of climate. *International Journal of Environmental Health Research, 16*, 1–13.

Poncon, N., Balenghien, T., Toty, C., Ferre, J. B., Thomas, C., Dervieux, A., L'ambert, G., Schaffner, F., Bardin, O., & Fontenille, D. (2007). Effects of local antropogenic changes on potential malaria vector *Anopheles hyrcanus* and West Nile virus vector *Culex modestus*, Camargue, France. *Emerg. Infect. Dis., 12*, 1810–1815.

Pripilov, A. G., Kinney, R. M., Sanokhvalov, E. I., Savage, H. M., Al'kjovskii, S. V., Tsuchiya, K. R., Gramashevshii, V. L., Sadykova, G. K., Shatalov, A. G., Vyshemirskii, O. I., Usachev, E. V., Moknonov, V. V., Voronina, A. G., Butenko, A. M., Larichev, V. F., Zhukov, A. N., Kovtunov, A. I., Gubler, D. J., & L'vov, D. K. (2002). Analysis of new variants of West Nile fever. *Vopr. Virusol., 47*, 36–41.

Randolph, S. E., & Rogers, D. J. (2006). Tick-borne disease systems: Mapping geographic and phylogenetic space. *Adv. Parasitol., 62*, 263–291.

Rappole, J. H., Derrickson, S. R., & Hubálek, Z. (2000). Migratory birds and spread of West Nile virus in the Western Hemisphere. *Emerg. Infect. Dis., 6*, 319–328.

Rappole, J. H., & Hubálek, Z. (2003). Migratory birds and West Nile virus. *J. App. Microbiol., 94*, 47S–58S.

Reisen, W. K., Hayes, C. G., Azra, K., Niaz, S., Mahmood, F., Parveen, T., & Boreham, P. F. (1982). West Nile virus in Pakistan. II. Entomological studies at Changa Manga National Forest, Punjab Province. *Trans. Roy. Soc. Trop. Med. Hyg., 76*, 437–448.

Reisen, W. K., Fang, Y., & Martinez, V. M. (2006). Effects of temperature on the transmission of West Nile virus by *Culex tarsalis* (Diptera: Culicidae). *J. Med. Entomol., 43*, 309–317.

Reisen, W. K., Takahashi, R. M., Carroll, B. D., & Quiring, R. (2008). Delinquent mortgages, neglected swimming pools, and West Nile virus, California. *Emerg. Infect. Dis., 14*, 1747–1749.

Rogers, D. J., & Randolph, S. E. (2006). Climate change and vector-borne diseases. *Advances in Parasitology, 62*, 345–381.

Sardelis, M. R., Turrell, M. J., Dohm, D. J., & O'Guinn, M. L. (2001). Vector competence of selected North American *Culex* and *Coquillettidia* mosquitoes for West Nile virus. *Emerg. Infect. Dis., 7*, 1018–1022.

Savini, G., Monaco, F., Calistri, P., & Lelli, R. (2008). Phylogenetic analysis of West Nile virus isolated in Italy in 2008. *Eurosurveillance, 13*, pii: 19048.

Schnettler, E., Sterken, M. G., Leung, J. Y., Metz, S. W., Geertsema, C., Goldbach, R. W., Vlak, J. M., Kohl, A., Khromykh, A. A., & Pijlman, G. P. (2012). Noncoding flavivirus RNA displays RNA interference suppressor activity in insect and mammalian cells. *J. Virol., 96*, 13486–13500.

Schuffenecker, I., Peyfritte, C. N., el Harrak, M., Murri, S., Leblond, A., & Zeller, H. G. (2005). West Nile virus in Morocco, 2003. *Emerg. Infect. Dis., 11*, 306–309.

Schole, F., & Mason, P. W. (2005). West Nile virus replication interferes with both poly I: C-induction and attenuates virus virulence in mice. *J. Virol., 80*, 2396–2404.

Seidowski, D., Ziegler, U., von Rönn, J. A., Müller, K., Hüppop, K., Müller, T., Freuling, C., Mühle, R. U., Nowotny, N., Ulrich, R. G., Niedrig, M., & Groschup, M. H. (2010). West Nile virus monitoring of migratory and resident birds in Germany. *Vector Borne Zoonotic Dis., 10*, 639–647.

Smithburn, K. C., Hughes, T. P., Burke, A. W., & Paul, J. H. (1940). A neurotropic virus isolated from the blood of a native of Uganda. *Am. J. Trop. Med. Hyg., 20*, 471–492.

Solomon, T., Ooi, M. M., Beasley, D. W. C., & Mallewa, M. (2003). West Nile encephalitis. *British Med. J., 326*, 865–869.

Steele, K. E., Linn, M. J., Schoepp, R. J., Komar, N., Geisbert, T. W., Manduca, R. M., Calle, P. P., Raphael, B. L., Clippinger, T. L., Larsen, T., Smith, J., Lanicotti, R. S., Panella, N. A., & McNamara, T. (2000). Pathology of fatal West Nile virus infections in native and exotic birds during 1999 outbreak in New York City, New York. *Vet. Pathol., 37*, 208–224.

Steinmann, A., Banet, C., Sutton, G. A., Yadin, H., Hadar, S., & Brill, A. (2002). Clinical signs of West Nile virus during encephalomyelitis in horses during the outbreak in Israel in 2000. *Vet. Rec., 13*, 47–49.

Stone, W. B., Okoniewski, J. C., Therrien, J. E., Kramer, L. D., Kauffman, E. B., & Eidson, M. (2004). VecTest as diagnostic and surveillance tool for West Nile virus in dead birds. *Emerg. Infect. Dis., 10*, 2175–2181.

Swaddle, J. P., & Calos, S. E. (2008). Increased avian diversity is associated with lower incidence of human West Nile infection: observation of the dilution effect. *PLoS ONE, 3*, e2488.

Taylor, R. M., Hurlbut, H. S., Dressler, H. R., Spangler, E. W., & Thrasher, D. (1953). Isolation of West Nile virus from *Culex* mosquitoes. *J. Egypt. Med. Assoc., 36*, 199–208.

Unlu, I., Mackay, A. J., Roy, A., Yates, M. M., & Foil, L. D. (2010). Evidence of vertical transmission of West Nile virus in field-collected mosquitoes. *J. Vector Ecol., 35*, 95–99.

Valiakos, G., Touloudi, A., Iacovakis, C., Athanasiou, L., Birtsas, P., Spyrou, V., & Billinis, C. (2011). Molecular detection and phylogenetic analysis of West Nile virus lineage 2 in sedentary wild birds (Eurasian Magpie), Greece, 2010. *Eurosurveillance, 16*, pii: 19862.

Wodak, E., Richter, S., Bagó, A., Revilla-Fernández, S., Weissenböck, H., Nowotny, N., & Winter, P. (2011). Detection and molecular analysis of West Nile virus infections in birds of prey in the eastern part of Austria in 2008 and 2009. *Vet. Microbiol., 149*, 358–366.

Work, T. H., Hurlbut, H. S., & Taylor, R. M. (1953). Isolation of West Nile virus from hooded crow and rock pigeon in the Nile delta. *Proc. Soc. Exp. Biol. Med., 84*, 719–722.

Zeller, H. G., & Schuffenecker, I. (2004). West Nile virus: an overview of its spread in Europe and the Mediterranean Basin in contrast to its spread in the Americas. *Eur. J. Clin. Microbiol. Infect. Dis., 23*, 147–156.

Zeller, H., Lenglet, A., & van Bortel, W. (2010). West Nile virus: the need to strengthen preparedness in Europe. *Euro Surveill, 15*(34), pii=19647.

Chapter 8

Rift Valley Fever Virus

A Virus with Potential for Global Emergence

Janusz T. Paweska[1,2] and Petrus Jansen van Vuren[1]

[1]Center for Emerging and Zoonotic Diseases, National Institute for Communicable Diseases of the National Health Laboratory Service, South Africa, [2]Division Virology and Communicable Diseases Surveillance, School of Pathology, University of the Witwatersrand, Johannesburg, South Africa

INTRODUCTION

Rift Valley fever virus (RVFV) is a zoonotic RNA arthropod-borne virus of the genus *Phlebovirus* in the family *Bunyaviridae* (Schmaljohn and Nichol, 2007) affecting primarily domestic ruminants and humans (Swanepoel & Paweska, 2011). The virus was first identified in 1930, during an investigation of an "enzootic hepatitis" associated with high rates of abortions among pregnant ewes and acute deaths of newborn lambs on a farm near Lake Naivasha in the Rift Valley of Kenya (Daubney et al., 1931). Subsequently periodic epizootics occurred in Kenya until the first major outbreak was recognized in South Africa in 1950–51, which caused an estimated 500,000 abortions and 100,000 deaths of sheep. Further epizootics were subsequently confirmed in Namibia, Mozambique, Zimbabwe, Zambia, the Sudan, and other East African countries. A second major and more widespread outbreak in South Africa caused extensive losses of sheep and cattle in 1974–76. During this epizootic the potential lethality of the virus for man associated with encephalitis and/or hemorrhagic fever was first recognized. By the late 1980s the geographic distribution of the virus had further extended, with outbreaks of the disease or evidence of RVFV activity having been reported in many countries of sub-Saharan and central Africa (Swanepoel & Coetzer, 2004). The virus was isolated for the first time outside of continental Africa in 1979 across the Indian Ocean in Madagascar (Morvan et al., 1991). The first occurrence of disease in Egypt in 1977–78, resulted in estimated 200,000 human infections, and some 18,000 cases of illness, of which at least 598 were fatal due to encephalitis and/or hemorrhagic fever (Meegan, 1981). Outbreaks of the disease in West Africa were first recorded in Mauritania and Senegal in 1987–88 (Jouan et al., 1988; Ksiazek

The Role of Animals in Emerging Viral Diseases. http://dx.doi.org/10.1016/B978-0-12-405191-1.00008-9

et al., 1989). In 2000, RVFV spread across the Red Sea into the Arabian Peninsula (Jupp et al., 2002). Between 2006 and 2011, a resurgence of severe outbreaks of RVF was reported from East Africa (Mohamed et al., 2010; Nguku et al., 2010; Shieh et al., 2010), Madagascar (Andriamandimby et al., 2010), South Africa (Archer et al., 2008; Métras et al., 2012), and for the first time virus activity was reported from the Archipelago of Comoros, on the French Island of Mayotte (Cêtre-Sossah et al., 2012). An increasing sequence database significantly contributed to studies of the evolution and the spread of the virus and clearly demonstrates the high potential of the virus to be introduced and accomplish endemic status in previously naïve areas. Results of whole and/or partial genome sequence and Bayesian analysis suggest that RVFV lineages in current circulation are likely to have all descended from a relatively recent ancestor (Grobbelaar et al., 2011). Despite extensive geographic dispersion, and wide range of susceptible arthropod vectors and vertebrate hosts, RVFV displays low genetic diversity (Bird et al., 2007). The existence of only one serotype of RVFV and a high degree of conservation of genes encoding the surface glycoproteins indicate that a single vaccine should protect against all currently circulating RVFV genetic variants. Likewise, the high preservation of the RNA-dependent RNA polymerase sequence suggests that antiviral drugs should effectively inhibit replication of all known genetic lineages of the virus. Increased knowledge and better understanding of virulence determinants and host defense mechanisms (Ikegami & Makino, 2011; Nfon et al., 2012) has stimulated in recent years a significant progress in the development of RVF vaccines (Bird et al., 2011; Brennan et al., 2011; Kortekaas et al., 2011; Morrill & Peters, 2011; Dodd et al., 2012; Lihoradowa and Ikegami, 2012) and antiviral drugs (Narayanan et al., 2012; Scott et al., 2012). However, at present immunization of humans can only be accomplished through the use of an experimental inactivated RVF vaccine with limited availability to laboratory workers (Rusnak et al., 2011). Likewise, except for restricted animal use of the live attenuated Smithburn or inactivated vaccines in Africa (Kamal, 2011; Lagerqvist et al., 2012), specific therapies or safe, efficacious, and affordable vaccines are not yet licensed for public or veterinary applications elsewhere. Most recently an animal vaccination program based on the use of a naturally occurring RVFV mutant (Clone-13) was introduced in South Africa (von Teichman et al., 2011).

The renewed interest in RVFV as a significant veterinary and public health threat has been a subject of numerous comprehensive reviews describing current knowledge on various aspects of the disease and its etiological agent, including epidemiology, pathogenesis, control and prevention, development of new diagnostic tools, therapeutics and vaccines (Bouloy & Flick, 2009; Pepin et al., 2010; Boshra et al., 2011; Swanepoel and Paweska, 2011), surveillence, prediction and control strategies (Bird and Nichol, 2012); other reviews are devoted to aspects of molecular biology and genetic diversity (Bouloy and Weber, 2010; Ikegami, 2012), viral and host determinants of virulence (Ikegami and Makino, 2011) or have included RVF as a part of a broader discussion on emerging diseases, climate change and anthropogenic factors (Martin et al., 2008; Clements and Pfeiffer, 2009; Chevalier et al., 2010). Most recently Olive at al. (2012) presented a detailed review of studies related to the role of wild mammals in the maintenance of RVFV. However, in spite of recent advances in research, a better understanding of the epidemiology and ecology of the virus, and particularly the enigma surrounding natural transmission cycles, remains a challenge for science.

RVFV TAXONOMY AND MOLECULAR BIOLOGY

RVFV is a member of the genus *Phlebovirus* in the family Bunyaviridae and like all bunyaviruses it is characterized by a tri-segmented single-stranded RNA genome of negative or ambisense polarity. All the replication steps occur in the cytoplasm of infected cells and virions mature by budding in the Golgi compartment (Ikegami & Makino, 2011). RVFV particles are 90–110 nm in diameter and consist of an envelope and a ribonucleocapsid (RNP). The envelope is composed of a lipid bilayer containing heterodimers of Gn and Gc glycoproteins which are the building blocks of 122 capsomers (110 hexamers and 12 pentamers) arranged in T=12 lattice. In contrast to other RNA viruses, bunyaviruses lack the matrix protein that connects the glycoproteins with the viral core. However, the cytoplasmic tail of the Gn seems to directly interact with RNPs, which would compensate for the absence of matrix protein (Piper et al., 2011). The viral ribonucleoproteins corresponding to each of the three genomic segments associated with numerous copies of the nucleoprotein N and the RNA dependent RNA polymerase L are packaged into the virion (Sherman et al., 2009) (Figure 8.1A). The RVFV genome (~12kb) consists of three single-stranded virion RNA (vRNA) molecules or segments designated large (L), medium (M) and small (S) which are used as templates to generate complementary RNA (cRNA) and messenger RNA (mRNA). The L- and M-segments are of negative polarity. The L-segment encodes for the viral RNA-dependent RNA polymerase (L protein). The M-segment encodes four proteins in a single open reading frame (ORF): the precursor to the glycoproteins Gn and Gc, and two nonstructural proteins, designated Nsm1 and NSm2 (Gerrard and Nichol, 2007). The S-segment utilizes an ambisense strategy to code for two proteins, the nucleoprotein N and a nonstructural protein NSs. The NSs forms a ribbon-like filament in the nucleus (Struthers & Swanepoel, 1982). This feature is rather unique for a virus replicating in the cytoplasm. Moreover, it is particular to RVFV and not shared with NSs proteins of other bunyaviruses. The RVFV genome coding strategy is illustrated in Figure 8.1B.

A schematic diagram of RVFV replication cycle is shown in Figure 8.2. The RVFV replication cycle is similar to that of other negative-stranded RNA viruses (Bouloy and Weber, 2010) and can be divided into three major events: (a) attachment, uptake and fusion, (b) primary transcription, translation and replication, and (c) virus assembly and release. The virus–cell interaction is initiated by

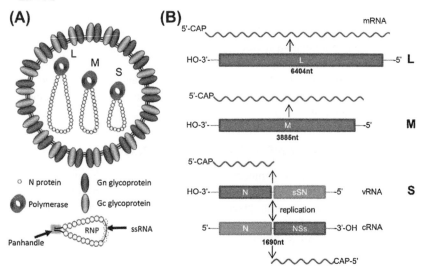

FIGURE 8.1 **The RVFV virion and transcription strategy. (A) Schematic representation of an RVFV particle with the Gn and Gc glycoproteins anchored in the viral envelope. The three genome segments S, M, and L are encapsidated by the nucleoprotein (N) into ribonucleo-proteins (RNP) and associate with the viral polymerase. The terminal ends of the genomic RNA segments are conserved in sequence and are invertedly complementary, permitting the ends of the RNAs to base-pair with a panhandle structure. (B) The RVF genome consists of three single-stranded RNA segments of the L- and M-segments are of negative and the S-segment of ambisense polarity. The viral polymerase is responsible for primary transcription of the negative-sense RNAs and subsequent replication of the three genomic segments.** Blue = genomic sense, orange = antigenomic-sense. *Courtesy of Steffen Mattjijn de Boer, Central Veterinary Institute of Wageningen University and Research Center, Lelystad, The Netherlands.*

binding of the virions to specific cell surface receptor(s). Based on the available data it is assumed that after binding, bunyavirus uptake occurs through receptor-mediated endocytosis. Entry is predicted to employ a class II fusion mechanism that is activated by low pH following endocytosis of the virion (Filone et al., 2006). Recently it was shown that phleboviruses exploit DC-SIGN as a receptor (Lozach et al., 2011). After uptake, the virus is trafficked along the endocytic pathway towards the perinuclear-localized lysosomes (Schmaljohn & Nichol, 2007), but little is known regarding the early phases of phlebovirus infection that precede the release of RNPs into the cytosol.

Primary transcription starts after release of the viral genome into the cytosol where RNA replication and RNP assembly take place in so-called viral facto-ries composed of Golgi membranes, actin and viral proteins associated with the rough endoplasmic reticulum and mitochondria (Fontana et al., 2008). Virions are transported within vacuoles to the cell surface where fusion of the vacu-ole and plasma membranes allows for the release of virus. During the replica-tion cycle, each vRNA (segment) is transcribed into mRNA and is replicated through a process that involves the synthesis of the exact copy of the genome,

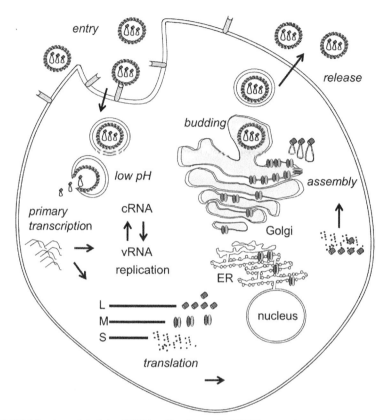

FIGURE 8.2   **A model of the RVFV replication cycle. RVFV enters host cells by a receptor-mediated endocytosis. The low pH in endosomal compartments triggers fusion of the viral envelope and endosomal membrane, followed by genome release into the cytosol, where primary transcription of the genomic-sense RNA (vRNA) into mRNA is initiated by the ribonucleoproteins (RNP)-associated RNA polymerases. mRNAs transcribed from the L-, M- and S-segments are translated and viral proteins accumulate. Complementary RNA (cRNA) serves as a template for vRNA replication and contrariwise. In Golgi-associated structures newly formed vRNAs are encapsidated by the nucleocapsid protein and associate with the polymerase protein, resulting in the formation of RNPs. Upon synthesis the glycoproteins Gn and Gc exit the endoplasmic reticulum (ER) and accumulate in the membranes of the Golgi network where virions are assembled through budding. Newly formed RVFV virions subsequently travel from Golgi complex to the plasma membrane where they are released.**
Courtesy of Steffen Mattjijn de Boer, Central Veterinary Institute of Wageningen University and Research Center, Lelystad, The Netherlands.

called complementary RNA (cRNA) or antigenome. The cRNA representing the copy of the S ambisense segment serves as a template for the synthesis of the NSs mRNA. The S cRNA is present in the input virus, allowing for expression of this protein at an early time of replication, an indication that it has an important role during infection. Messenger (mRNA) synthesis is initiated through a cap-snatching mechanism whereas the synthesis of cRNA is initiated with 5′ nucleoside triphosphates. Furthermore, cRNA is the complete copy of the vRNA whereas mRNAs terminate in the noncoding region before the 5′ end of the template for the L- and M-segments or in the intergenic region for the S-segment. The switch between the two activities remains unknown and although several polymerase consensus motifs were found in the L protein (Muller et al., 1994), the different domains responsible for the activities of cap-snatching or transcription termination have still to be determined. Systems to manipulate the genome of RVFV, including construction of minigenomes, have now been well established advancing further research on various steps in RNA synthesis, i.e., transcription, replication, transcription termination and packaging, and developing of multivalent live-attenuated vaccines (Brennan et al., 2011).

THE FUNCTIONS AND ROLE OF RVFV PROTEINS IN DISEASE PATHOLOGY

Bunyaviruses utilize the endonuclease activity associated with L protein to obtain 5′ –capped RNA, which can be used as primers for viral mRNA transcription (Reguera et al., 2010). The glycoproteins are essential for the penetration of the virus and virions escape from infected cells. As the most exposed structural viral components, they are recognized by the immune system and induce the production of neutralizing antibodies which play an important role in protection (Mandel et al., 2010). The glycoproteins mediate virus entry into many cell types through specific receptors which remain to be identified. Both Gn and Gc likely contribute to the virion assembly process and interact with N protein (Habjan et al., 2009). While the N protein seems to be not directly involved in pathogenesis it is essential for virion capsid formation and plays other roles in the virus life cycle. These include interaction with RVFV surface glycoproteins and involvement in transcription and replication (Liu et al., 2008). The N protein is strongly immunogenic but does not elicit neutralizing antibodies. However, immunization with recombinant N protein induces a partial immune protection in mice and RVFV-immunized sheep and mice exhibit lymphocytic proliferation *in vitro* in the presence of recombinant N protein (Lorenzo et al., 2010). The expression of type I IFN is upregulated in the liver and spleen of mice immunized with recombinant N shortly after RVFV challenge, compared to a delayed upregulation of the same gene in nonimmunized mice. In the acute phase of liver infection, however, there is a massive upregulation of type I and II IFN in the presence of high viral titers in nonimmunized mice associated with downregulation of several genes involved in the activation of B- and T-cells in the spleen, compared to normal expression in immunized mice. Furthermore, various genes with pro-apoptotic and pro-inflammatory effects are strongly upregulated, and anti-apoptotic genes downregulated in the liver of nonimmunized mice (Jansen van Vuren et al., 2011).

The postulated role of the M-segment encoded NSm1 and NSm2 in triggering apoptosis through the caspase 3, 8 and 9 pathways (Won et al., 2007) is inconclusive (Gerrard et al., 2007) .The S-segment encoded NSs protein has been identified as a major factor of virulence inhibiting host innate viral defenses (Billecocq et al., 2004), primarily by counteracting the antiviral IFN system by four major mechanisms: 1) blocking host mRNA transcription by sequestering the host cell protein p44 and XPB to inhibit maturation of transcription factor II H (TFIIH); 2) promoting degradation of the TFIIH subunit p62 in the nucleus of infected cells (Kalveram et al., 2011); 3) binding of the transcription factor Yin Yang 1 (YY1) on the interferon-β promoter to suppress its activation through the SAP30 complex (Le May et al., 2008), and 4) promoting the downregulation of double-stranded RNA-dependent protein kinase PKR early in infection, which prevents phosphorylation of eukaryotic initiation factor 2α (Ikegami et al., 2009; Habjan et al., 2009). SAP30 belongs to the Sin3A/NCoR/HDAC repressor complexes intervening in gene transcription regulation. Moreover, it has been shown that SAP30 interacts directly with YY1, a transcription factor involved in the regulation of expression of numerous genes, including IFN-β (Le May et al., 2008). The NSs therefore has multiple functions to counteract the IFN system, either at the transcriptional level or at translational level by degrading PKR. Removal of the NSs gene results in attenuation of RVFV, which has been demonstrated in a naturally occurring attenuated strain of RVFV and in recombinant RVFV lacking a fully functional NSs (Muller et al., 1995; Bird et al., 2011). Interestingly, a recent study showed that serial passages of RVFV in baby hamster kidney (BHK) 21 cells or *Aedes aegypti* cells (Aag2) result in large deletions of the NSs gene, while serial alternating passages between BHK21 cells and Aag2 cells did not induce its deletion. These findings suggest that host alternation is important to maintain stability of the NSs gene, thereby promoting RVFV capacity in evasion of the innate immune response (Moutailler et al., 2011). Mutants of RVFV that lack the NSs and/or NSm genes were recently assessed for the potential to infect and be transmitted by *Aedes* mosquitoes, which are the principal vectors for maintenance of the virus in nature and emergence of virus initiating disease outbreaks, and by *Culex* mosquitoes which are important amplification vectors. In *Ae. aegypti*, infection and transmission rates of the NSs deletion virus were similar to wild type virus while dissemination rates were significantly reduced. Infection and dissemination rates for the NSm deletion virus were lower compared to wild type. Virus lacking both NSs and NSm failed to infect *Ae. aegypti*. In *Cx. quinquefasciatus*, infection rates for viruses lacking NSm or both NSs and NSm were lower than for wild type virus. The double deleted viruses represent an ideal safe vaccine due to lack of ability to efficiently infect and be transmitted by mosquitoes (Crabtree et al., 2012).

MOLECULAR EPIDEMIOLOGY AND GENETIC DIVERSITY

Genetic variants of RVFV descendent from multiple lineages persist in Africa through vertical transmission in mosquitoes and infection of vertebrate hosts and large outbreaks emerge in years of heavy rainfall and floods (Bird et al., 2007; LaBeaud et al., 2007; Grobbelaar et al., 2011; Murithi et al., 2011; Swanepoel and Paweska, 2011), Figure 8.3. On a number of occasions, viruses from these lineages have been transported outside enzootic regions through the movement of infected animals and/or mosquitoes, causing large outbreaks in countries where the disease had not previously been noted, as exemplified by the outbreaks in Egypt in 1977, western Africa in 1988, and the Arabian Peninsula. Such viruses could potentially become established in their new territories through infection of wild and domestic ruminants and other animals and vertical transmission in local competent mosquito species. Dramatic environmental changes, a growing human/domestic animals/wildlife interface, increased trade and movement of animals throughout and outside Africa likely contributes to geographic dispersal of the virus.

Genomic reassortment is a potentially potent mechanism to generate genetic diversity and eventually drive the emergence of novel RVFV variants. The reassortment of RNA genome segments among viruses of the family *Bunyaviridae* has been reported in both *in vitro* and *in vivo* studies, including formation of reassortants in dually infected mosquitoes (Borucki et al., 1999; Gerrard et al., 2004; Briese et al., 2006). Reassortment events among RVFV strains have also been documented (Sall et al., 1999; Bird et al., 2007; Grobbelaar et al., 2011). However, the impact of reassortment on RVFV replication, fitness, and host virulence remains to be investigated. A recent study by Morrill et al. (2010) demonstrated that a single nucleotide heterogeneity at nucleotide 847

FIGURE 8.3   Flooded semi-arid areas in Garissa district, Northeastern Province of Kenya—an epicenter of the RVF outbreak in Kenya, 2006–2007 (Nguku et al., 2010).

of M-segment (M847) might substantially affect the RVFV virulence in mice. A recombinant RVFV ZH501 carrying adenine residue at M847 encoding glutamic acid compared to genetically identical virus but carrying a guanine residue encoding glycine amino acid at the corresponding site replicated more rapidly and to significantly higher titers in mice, of which most died within 8 days post inoculation, whereas the latter genetic virus variant had attenuated virulence to these animals. However, the effect of this single substitution in the Gn protein on RVFV virulence in other species still needs to be determined.

As a consequence of the expanding sequence database, the number of identified viral lineages of the virus has increased from 3 in an early analysis by Sall et al. (1997) to 15 lineages (designated A-O) in the recent report by Grobbelaar et al. (2011). Virus isolates from one area tend to cluster together within each lineage, but virus genetic variants with distant origins are found within different lineages, providing evidence of widespread dispersal and movement of RVFV throughout Africa. The strong phylogenetic linkage of virus strains from distant geographic locations suggests that the movement of infected livestock and the natural dispersal of mosquitoes allow the spread of RVFV throughout continental Africa, Madagascar, and the Arabian Peninsula. Irrespective of the genome segments analyzed, the genetic diversity of RVFV is low, approximately 4% and 1% at the nucleotide and protein coding levels, respectively. The low genetic diversity of RVFV may reflect the evolutionary constraint imposed on arboviruses by their altering replication in mammalian and arthropod hosts (Bird et al., 2007).

Recent Bayesian analysis suggests that the time of divergence of RVFV isolates from a most recent common ancestor, dated 1880–1890, coincides with the colonial period when the introduction of large concentration of susceptible sheep and cattle imported from Europe would have facilitated emergence of an unknown progenitor virus (Bird et al., 2007). The evolutionary rate of the virus is similar to that of other RNA viruses, suggesting that the principal factor for low genetic diversity among RVFV isolates is recent derivation from an ancestral virus in Africa (Grobbelaar et al., 2011).

EMERGENCE OF ENDEMIC LINEAGES, SINGLE AND MULTIPLE INTRODUCTIONS

In countries where RVFV activity was not previously detected, outbreaks of the disease in animal and human populations result from the spread of a single lineage of the virus characterized by minimal genetic diversity. For example, comparison of RVFV isolates from the 1977 Egyptian outbreak identified a single lineage of the virus with <0.33% nucleotide (nt) and <0.1% amino acid (aa) sequence differences. RVFV isolates from the 1987 Mauritanian outbreak had similar low genetic diversity (Bird et al., 2007). Until recently, due to limited data on the genetic diversity of RVFV, mechanisms associated with propagation of outbreaks in endemic regions were not well understood. However, early detection of the large 2006–2007 RVF outbreaks that occurred in East Africa provided an opportunity to conduct more detailed molecular epidemiology studies. Genome segment sequences obtained from virus specimens collected from domesticated livestock and wildlife, and representing all affected regions of Kenya, were all monophyletic with a virus isolate from the previous 1997–1998 outbreak in East Africa. However, among the 2006–2007 viruses analyzed, two separate sublineages (Kenya-1 and Kenya-2) were identified with increased genomic diversity relative to that observed among RVFV isolates from the Egyptian 1977–1979, and Mauritanian 1987 outbreaks (Bird et al., 2008). Results from a similar study analyzing RVFV isolates from humans and mosquitoes in Kenya and Tanzania indicate that the sequential RVF epidemics in the region were caused by three distinct lineages of the RVFV (Kenya-1, Kenya-2, and Tanzania-1), sometimes independently activated or introduced in distinct outbreak foci (Nderitu et al., 2011). While the shared evolutionary history of the 1997–1998 and 2006–2007 outbreak viruses was apparent based on phylogeny, further population genetics study on Kenya-1 and Kenya-2 lineages revealed that the two lineages were more closely related to the 1997–1998 RVFV prototype than to each other, indicating ongoing and separate evolutionary patterns since the previous outbreak. Moreover, it appears that the Kenya-1 lineage viruses had recently undergone demographic or spatial expansion, whereas the Kenya-2 lineage viruses had likely not. Interestingly, the timing of the Kenya-1 expansion event was calculated to have occurred approximately 2–4 years prior to the detection of the 2006–2007 outbreaks. These results indicate ongoing RVFV activity and evolution during the interepizootic period (IEP) and highlight the importance of a cryptic enzootic transmission cycle that allows for the establishment of RVFV endemicity and to precipitate explosive outbreaks. In Kenya, where the highest number of epizootics has been reported to date, the average IEP period is 3.6 years, a period likely representing the time required for the immunity of livestock populations to decrease to levels that are permissive for virus spread (Murithi et al., 2011).

Results of a molecular epidemiology study by Soumaré et al. (2012) suggest that Barkedji was a hub associated with three distinct introductions of the virus in Senegal from where it was then spread to other localities in West Africa. Barkedji, situated in the semi-arid region of central Senegal, was previously postulated to play a role as an important gateway for RVFV into Senegal and Mauritania based on serologic and entomologic surveys (Zeller et al., 1997; Traoré-Lamizana et al., 2001). Barkedji is an important crossroad of migration movements of wildlife between the southern and northern regions of Senegal, and to a larger extent, to southern Mauritania. Distinct introduction of RVFV in Barkedji appears to explain the maintenance of the endemic cycle at the regional scale. Endemicity without a permanent virus reservoir would be impossible in a single site except when there is a strictly periodic rainfall pattern, but it would be possible when there are herd movements and sufficient inter-site variability in rainfall, which drives mosquito emergence (Favier et al.,

2006). Active dispersal for most RVFV vectors is short, and it is estimated to be about 1 km, varying from less than 150 m for *Aedes* to approximately 2 km for *Culex theileri* (Gad et al., 1986; Service, 1997). The role in long-range dissemination of the virus across Africa could be fulfilled by infected members of roaming herds (LaBeaud et al., 2011a). Recent findings during an isolated RVF outbreak on a cattle dairy farm in Bela-Bela, Limpopo Province, South Africa in 2008 indicate that the intensity of RVFV transmission might be dependent on ruminant breed, individual susceptibility to infection, and the host preference of competent vectors at a given location and at a given time (Mapaco et al., 2012).

Although mosquito bite is the principal infection mechanism of RVF in animals, the different dynamics of the spread of contagious process during large epidemics suggests that other transmission mechanisms may also exist. Results of space-time analyses of RVF outbreaks in South Africa in 2008–2011 confirm the presence of an intense, short, initial transmission mechanism which could be attributed to active vector dispersal, and highlight the presence of another transmission mechanism of a lower intensity and over further distances up to 40 to 90 km, within about 2 weeks. The appearance of long-distance spread could be explained by emergence of the virus at several foci as a result of hatching of infected *Aedes* eggs or multiple re-introductions of infected vectors. However, detection of spatio-temporal clusters up to 20 km within 1 day, as observed in early 2009 in KwaZulu-Natal province, likely rules out active vector dispersal in favor of movements of infectious animals, passive vector dispersal or multiple local emergences (Métras et al., 2012).

The RVFV transmission during IEP without noticeable outbreak or clinical cases has been recently reported in African wildlife (Evans et al., 2008; LaBeaud et al., 2011a; Miller et al., 2011), in cattle in Mayotte (Cêtre-Sossah et al., 2012), in sheep and goats in Mozambique (Fafetine et al., 2013), in humans in Tanzania (Heinrich et al., 2012), Kenya (LaBeaud et al., 2009; LaBeaud et al., 2011b), and Gabon (Pourrut et al., 2010). It has been postulated that during an IEP in endemic regions, the virus persists in eggs of floodwater *Aedes* mosquito species or via low-level cycling between mosquitoes and vertebrates, but further studies are required to determine whether domestic and wildlife animals play any specific role in the virus maintenance between outbreaks and virus amplification prior to noticeable outbreaks in livestock and human populations. It remains intriguing why a low level of virus circulation during IEP does not result in clinical manifestations in livestock and humans. One explanation could be that since IEP transmission events occur usually in remote locations, sporadic clinical cases are either underreported or misdiagnosed. At present, it is unclear to what extent RVFV has become established in countries outside its historic enzootic areas in Africa. A number of outbreaks have been reported in Madagascar, but evidence to date indicates that they have resulted from repeated introductions of the virus from mainland Africa rather than from enzootic maintenance in the local environment (Carroll et al., 2011).

THE IMPACT OF LARGE SCALE VACCINATION ON THE EVOLUTION AND VIRULENCE OF RVFV

Results of recent genetic analysis indicate that the natural history of RVFV might be influenced by massive vaccination of ruminants in Africa with the live attenuated Smithburn neurotropic strain (SNS) of the virus (Grobbelaar et al., 2011). The SNS vaccine strain is only partially attenuated and, though approved for veterinary use in Africa, it has been shown to be abortogenic and teratogenic (Botros et al., 2006; Kamal, 2009). Accurate estimates of those conditions were never determined, but there has been a tendency to regard risks associated with animal use of the SNS vaccine as acceptable in the face of an outbreak (Swanepoel and Paweska, 2011). Vaccination of livestock might be a dangerous practice if carried out when RVF epizootics have already begun. During large outbreaks of RVF in Africa millions of doses of SNS vaccine were administered to livestock, particularly to sheep, at times when the virulent RVFV was actively circulating in the animal population. Livestock vaccines are sold in multidose vials and commonly administered with automatic syringes while needles are not changed often enough. This common practice results in further spread of the virus among animals and increased risk to human populations (Davies, 2010). Isolate SA 184/10 from a patient in South Africa potentially exposed to co-infection with the SNS vaccine strain and wild virus from a needle injury while vaccinating sheep was a reassortant grouping with the parent vaccine strain in lineage K. Moreover, lineages K, L and M contained neurotropic and hepatotropic laboratory strains, and isolates from Kenya, Zimbabwe, South Africa and Egypt, countries that used the SNS vaccine on a large scale during major outbreak of RVF (Grobbelaar et al., 2011). Although analysis of isolates from the first recorded human deaths caused by RVFV in 1975 in South Africa and isolates from Egypt did not show clear-cut evidence of reassortment, they fell into the group that manifested convergence of lineages D and E in the study by Bird at al. (2007), corresponding to lineages L and K (the vaccine lineage) in the study by Grobbelaar et al. (2011). It should be noted that the SNS vaccine virus was intensively passaged by intracranial inoculation of mice. In this context it could be relevant that viruses attenuated through intracranial passage in mice may acquire new tissue tropism and pathogenic properties (Swanepoel et al., 1992; Hayes, 2010).

RIFT VALLEY FEVER VIRUS VECTORS, THEIR ECOLOGY AND ROLE IN TRANSMISSION

RVFV has been isolated in the field from a wide range of mosquito species of which many were shown to be also susceptible and capable of virus transmission in the laboratory. Cumulative results from field and laboratory investigations indicate that the epidemiologically most important vectors of RVFV involve members of the subgenera *Neomelaniconion* of *Aedes* and *Culex* (Pepin et al., 2010).

Biting flies such as midges, phlebotomids, stomoxids and simulids might serve as mechanical transmitters of infection, however, non-vectorial transmission is not considered to be important in livestock, as opposed to humans (Swanepoel and Paweska, 2011).

Outbreaks of RVF in eastern and southern regions of Africa occur at irregular intervals of up to 15 years. The fate of the virus during IEP periods still constitutes a central enigma in the epidemiology of the disease. Cryptic maintenance and transmission cycles have been postulated but the exact mechanism is not well understood. Early observations made in Uganda, Kenya and South Africa suggested that the virus was circulating in *Eretmapodites* spp. mosquitoes and unknown vertebrates in indigenous African forest, and transmitted to livestock during exceptionally heavy rainfall seasons. However, isolation of RVFV from unfed *Ae. mcintoshi* mosquitoes in Kenya during IEP periods in 1982 and 1984 confirmed that the virus is enzootic in livestock rearing areas and indicated that it is maintained by transovarial transmission in aedines (Linthicum et al., 1983). Floodwaterbreeding aedine mosquitoes of the subgenera *Aedimorphus* and *Neomelaniconion* are associated with a shallow depression in the general topography, with water-saturated soil overlaying a poorly porous stratum that allows standing water after heavy rainfall. Such habitats are commonly present throughout the bushveld-savanna and higher-altitude grasslands of sub-Saharan Africa, and are colloquially described as dambos or pans (Swanepoel & Coetzer, 2004). Suitable habitats for aedine mosquitoes are also found in shallow depressions in the floodplains of rivers when floodwaters overflow riverbanks, especially in the coastal plains of eastern and southern Africa, and also in the headwaters to rivers (Linthicum et al., 1983). The aedine mosquitoes overwinter as eggs which can survive for long periods in dried mud, possibly for several seasons if the area remains dry. The distinct biology and ecological requirements of mosquito vectors have a potentially significant impact on RVFV epidemiology. *Aedes* spp. mosquitoes are associated with freshly flooded temporary or semipermanent fresh-water bodies, while *Culex* spp. utilize more permanent fresh-water bodies for breeding. The floodwater-breeding aedine mosquitoes oviposit on the soils surrounding the standing water (Pepin et al., 2010). It is obligatory that the eggs of aedine mosquitoes are exposed to a period of drying before they will hatch when the breeding niches flood. This biological requirement might provide a single opportunity but an ideal mechanism for survival of RVFV over long periods of time (Linthicum et al., 1985; Logan et al., 1991). When dambo habitats are flooded after heavy rainfall, biological transmission occurs via infected mosquito saliva to animals attracted to the water pools. Viraemia in vertebrate hosts is of rather short duration, typically 2-7 days (McIntosh et al., 1973; Davies & Karstad, 1981), implying that if vertical transmission in floodwater-breeding aedine mosquitoes is effective and regularly occurring, the invertebrate vector can function as major reservoir and the most important mechanism for survival of RVFV from season to season. However, of potential epidemiological importance is the demonstration that the larvae of *Cx. pipiens*, *Ae. mcintoshi* and *Ae. circumluteolus* become infected after feeding on liver homogenates prepared from an experimentally inoculated hamster (Turell et al., 1990).

When dambos remain flooded for more than 2 or 3 weeks, aedine mosquitoes are succeeded by *Culex* spp., which oviposit in small egg rafts on the surface of the water. These egg rafts lead to a population explosion of *Culex* spp. mosquitoes, which if infected upon feeding on viremic vertebrate hosts significantly contribute to further virus dispersal. While the floodwater *Aedes* spp. tend to remain in the immediate vicinity of the dambos and only feed at dusk and dawn, the more nocturnal *Culex* spp. are more likely to disperse to find vertebrate hosts for blood feeding (Pepin et al., 2010). This eventually leads to extensive dissemination of virus, and the resulting large outbreaks of the disease in livestock and human populations. Flooding of large areas also contributes to concentration of animals and humans on areas of dry land, thus further increasing the potential for virus transmission.

The known outbreaks in North and West Africa differ from the epidemiology of disease in sub-Saharan Africa, which were reported independently of rainfall and most likely in association with vectors which breed in rivers and dams. The large epizootic in Egypt in 1977–78 was probably facilitated by an increase in mosquito breeding sites after the building of the Aswan dam. In contrast to the Egyptian epizootic of 1977–78, the principal mosquito vectors of RVFV in sub-Saharan Africa tend to be zoophilic and sylvatic and therefore in this part of Africa humans become infected mainly from contact with infected animal tissues. The main vector in the Egyptian epizootic, *Cx. pipiens* is known to be peridomestic and anthropophilic, implying that infected humans might also serve as amplifying hosts for the infection of mosquitoes (Swanepoel & Paweska, 2011). In West Africa the *Aedes (Aedimorphus)* spp. typically breed in the small, temporary ground pools that occur after localized rains. This region had not experienced heavy rainfall or floods and the typical vectors, *Ae. mcintoshi*, *Ae. circumluteolus* and *Cx.* spp., were not present in high numbers during the 1987 outbreak (Fontenille et al., 1998). In this region breeding conditions are more suited for floodbreeding aedine mosquitoes of the subgenera *Aedimorphus,* including *Ae. vexans*, *Ae. ochraceus*, *Ae. dalzieli,* and *Ae. cumminsii* (Chevalier et al., 2010). The construction of the Manantali dam on the Senegal River and the Diama dam downstream on the border between Mauritania and Senegal increased potential mosquito breeding sites in areas where the virus was already known to be active, and also led to the concentration of people and livestock in the proximity of the dams during severe drought conditions. With the advent of a second large outbreak of RVF in this region in 1998–1999, there was apparently a shift in the dominant species towards *Mansonia uniformis* and *Cx. poicilipes* (Diallo et al., 2005). Irrigation for agriculture in the Tihama regions of Yemen and Saudi Arabia and the proximity of the Jizan Dam provide suitable breeding grounds for *Ae. vexans* and *Cx. tritaeniorhynchus* (Jupp et al., 2002). There was speculation that RVFV may have been imported into Saudi Arabia and Yemen from Africa with slaughter animals, or carried from Africa by wind-borne mosquitoes in 2000, but there were no known epidemics in the Horn of Africa at the time. It is more likely that infected animals were imported during the 1997–98 epidemic in East Africa, and that infection smoldered on the Arabian Peninsula until ideal circumstances for an epidemic occurred following heavy rains in 2000 (Swanepoel & Paweska, 2011). Competent mosquito vectors are present in RVF-free regions (Moutailler et al., 2008; Turell et al., 2008; Konrad and Miller, 2012), implicating the potential for the virus spread into these areas.

HOSTS AND THEIR ROLE IN MAINTENANCE OF RIFT VALLEY FEVER VIRUS

Results from natural cases, experimental studies and seroepidemiological surveys show that a wide range of animal species are susceptible to RVFV infection (Table 8.1). Domesticated ruminants are the primary species affected and likely the major amplifiers of the virus but serological evidence suggests that a large number of African herbivorous wildlife and other species might also play a role in the RVF epidemiology (Table 8.2). Newborn lambs and goat kids are extremely susceptible to RVFV infection with a very short incubation, rarely surviving longer than 2 days after the onset of illness. Mortality may exceed 90% in animals less than a week old. Lambs and kids older than 2 weeks and mature sheep and goats are less susceptible to the disease but some animals may die peracutely. Most develop an acute

TABLE 8.1 Rift Valley Fever Virus Host Range and Disease Severity

High (~100%) mortality	Severe illness, abortion, mortality	Severe illness, viremia, abortion	Infection, viremia	Refractory to infection
Lambs	Sheep	Monkeys	Horse	Guinea pigs
Calves	Cattle	Camels	Cats	Rabbits
Kids	Goats	Rats	Dogs	Pigs
Puppies	Water buffalo	Gray squirrels	Monkeys	Hedgehogs
Kittens	Humans			Frogs
White mice				Chickens
Hamster				Canaries
Field voles				Pigeons
Field mice				Parakeets
Door mice				

Easterday, 1965; Swanepoel and Coetzer, 2004, modified.

TABLE 8.2 African Wildlife Species Tested Positive for Antibody Against Rift Valley Fever Virus

Species	No. Positive/ No. tested (%positive)	Country	Reference
Topi (*Damaliscus korrigum*)	1/2(50)	Chad	Maurice, 1967
Red-fronted gazelle (*Eudorcas rufifrons*)	3/4 (75)	Chad	Maurice, 1967
Dama gazelle (*Nanger dama*)	2/7 (28.6)	Chad	Maurice, 1967
Scimitar-horned oryx (*Oryx dammah*)	2/3 (66.7)	Chad	Maurice, 1967
Common reedbuck (*Redunca redunca*)	1/2 (50)	Chad	Maurice, 1967
African buffalo (*Syncerus caffer*)	1/1 (100)	Chad	Maurice, 1967
Dorcas gazelle (*Gazella dorcas*)	6/12 (50)	Chad	Maurice, 1967
Impala (*A. melampus*)	5/8 (62.5)	Kenya	Evans et al., 2008
Thomson's gazelle (*Gazella thomsonii*)	7/8 (87.5)	Kenya	Evans et al., 2008
Waterbuck (*Kobus ellipsiprymnus*)	2/10 (20)	Kenya	Evans et al., 2008
Gerenuk (*Litocranius walleri*)	5/5 (100)	Kenya	Evans et al., 2008
African buffalo (*S. caffer*)	37/237 (15.6)	Kenya	Evans et al., 2008
Lesser kudu (*Tragelaphus strepsiceros*)	5/10 (50)	Kenya	Evans et al., 2008
African bush elephant (*Loxodonta africana*)	5/83 (6%)	Kenya	Evans et al., 2008
Warthog (*Phacochoerus aethipicus*)	2/81 (2.5)	Kenya	Evans et al., 2008
Giraffe (*Giraffa camelopardalis*)	1/34 (2.9)	Kenya	Evans et al., 2008
Burchell's zebra (*Equus burchellii*)	1/102 (1)	Kenya	Evans et al., 2008
Black rhinoceros (*Diceros bicornis*)	14/43 (32.6)	Kenya	Evans et al., 2008
Impala (*Aepyceros melampus*)	8/801(1)	Zimbabwe	Anderson & Rowe, 1998
Sable antelope (*Hippotragus niger*)	9/289 (3)	Zimbabwe	Anderson & Rowe, 1998
Waterbuck (*K. ellipsiprymnus*)	8/179 (4.5)	Zimbabwe	Anderson & Rowe, 1998

TABLE 8.2 African Wildlife Species Tested Positive for Antibody Against Rift Valley Fever Virus—cont'd

Species	No. Positive/No. tested (%positive)	Country	Reference
African buffalo (*S. caffer*)	34/541 (6.3)	Zimbabwe	Anderson & Rowe, 1998
African buffalo (*S. caffer*)	115/550 (21)	South Africa	LaBeaud et al., 2011a
White rhinoceros (*Ceratotherium simum*)	49/100 (49)	South Africa	Miller et al., 2011

disease with high fever, anorexia, weakness, listlessness and hyperpnea. Some animals may develop melena or fetid diarrhea and a bloodtinged, mucopurulent nasal discharge (Easterday et al., 1962; Swanepoel and Coetzer, 2004). Many sheep and goats undergo subclinical infection. Death rates vary from 5 to 60% for sheep, with highest mortality occurring in pregnant animals. The disease in calves resembles that in lambs and sheep, but a higher proportion of calves may develop icterus. Death generally occurs 2 to 8 days after infection, and mortality is usually less than 10%. Infection is frequently inapparent in adult cattle, but some animals develop acute disease. The death rate in cattle does not generally exceed 10%. Abortion appears to be an almost inevitable outcome of infection in pregnant sheep, goats and cattle. Animals may abort at any stage of gestation. However, abortion rates vary with epidemiological circumstances, and have ranged from 15 to 100% in different outbreaks, or in separate herds and flocks in a single outbreak. Frequently, abortion may be the only overt manifestation of disease in a herd or flock. High viremia is generally demonstrable in domestic ruminants at the onset of fever and may persist for up to a week (Swanepoel and Paweska, 2011).

Humans are highly susceptible to RVFV infection and develop a sufficient viremia to be a source of infection for mosquitoes and introduction of disease into uninfected areas (Figure 8.4). While RVF epidemics can involve hundreds if not thousands of individuals, the majority of infections in humans are inapparent or associated with moderate to severe, nonfatal, febrile illness. Less than 1% of human patients develop the hemorrhagic and/or encephalitic forms of the disease. The overall case fatality ratio is estimated to range from 0.5% to 2%, but it appears to be higher in recent outbreaks of the disease in East Africa and South Africa (Mohamed et al., 2010; Nguku et al., 2010; Swanepoel and Paweska, 2011). Human cases

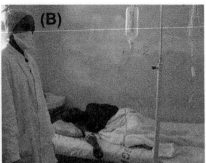

FIGURE 8.4 (A) Settlement of nomadic pastoralists in Kenya: RVFV infections are often diagnosed in pastoralists who had handled infected animals, their products or tissues. (B) RVF patient in Garissa Provincial Hospital, North-eastern province of Kenya, 2006. Of 340 laboratory confirmed cases during the RVF outbreak in Kenya in 2006–2007, 90 (26.4%) were fatal; serosurvey data suggest that up to 185, 000 people may have been infected (Nguku et al., 2010). A similar high case-fatality rate was reported during the disease outbreak in humans in Tanzania in 2007 (Mohamed et al., 2010).

with jaundice, neurological disease, or hemorrhagic complications are at increased risk of fatality (Madani et al., 2003). In a minority of patients the disease is complicated by the development of ocular lesions at the time of the initial illness or up to four weeks later. Estimates for the incidence of ocular complications range from less than 1% to 20% of human infections. The ocular disease usually presents as a loss of acuity of central vision, sometimes with development of scotomas. The lesions and the loss of visual acuity generally resolve over a period of months with variable residual scarring of the retina, but in instances of severe hemorrhage and detachment of the retina, there may be permanent uni or bilateral blindness (Swanepoel and Paweska, 2011). Economic effects of RVF can be catastrophic for meat and dairy producers. A high illness and mortality rate among affected livestock herds prompts the placement of strict international embargoes of livestock exports. These epidemics are particularly devastating for pastoral nomads and local herders. A high death rate among pregnant ruminants affects the next crop of newborns and the survival of locals who are economically and physically dependent on milk and meat. During large outbreaks, extensive numbers of human infections leads to formidable challenges in resource-limited health care settings (Rich and Wanyoike, 2010).

Results of the first study describing an experimental infection of nonhuman primates (NHPs) with RVFV was published by Findlay and Daubney (1931) who reported that infection of rhesus macaques (*Macaca mulata*) induced febrile responses and leukopenia but was not fatal. Later studies in rhesus macaque and cynomolgus macaque (*Macaca fascicularis*) seem to indicate that these species are more susceptible to RVFV infection by the aerosol than by peripheral exposure routes (Easterday, 1965). Until recently, rhesus macaques

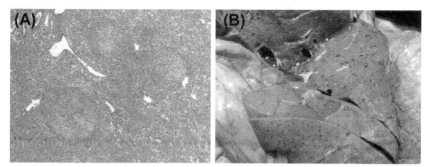

FIGURE 8.5   Histo- and gross pathology in bovine liver infected with RVFV. (A) Multifocal hepatic necrosis. (B) Diffuse hepatic necrosis with cholestasis which imparts yellow gold appearance of the liver; widespread capsular and parenchymal hemorrhages. *Courtesy of Dr Rick Last, Veterinary Pathology Services, Cascades, South Africa.*

infected with RVFV strain ZH501 provided the most realistic model of human infection; however, severe disease in these animals is infrequent, and large cohorts are needed to observe significant morbidity and mortality. A recent experimental study in the common marmoset (*Callithrix jacchus*) demonstrated that these animals are more susceptible to RVFV than rhesus macaques, and display signs of severe hemorrhagic manifestations and neurological impairment, thus presenting an ideal model for the evaluation of potential vaccines and therapeutics (Smith et al., 2012). Antibodies to RVFV (Olive et al., 2012), seroconversion after experimental infection and viremia were demonstrated in several species of NHPs and it appears that some South American NHP species are more susceptible to infection than African NHP species, including the green guenon (*Cercopithecus callitrichus*), the sooty mangabey (*Cercocebus fuliginosus*), and the Patas guenon (*Erythrocebus patas*) (Findlay, 1932). Spider monkeys (*Ateles ater*) seem to be refractory (Easterday, 1965), while baboons (*Papio anubis*) appears to be as susceptible as rhesus macaque (Davies at al., 1972). Available data are inconclusive regarding the possible role of wild primates in the maintenance of RVFV but it seems likely that NHPs might play the same role as humans in amplifying RVFV.

In both animals and humans, the primary site of RVFV tissue pathology is the liver (Figure 8.5). This finding is consistent among severe cases and has been clearly demonstrated by histopathological examination in natural cases and in experimentally infected animals (Coetzer, 1982; Shieh et al., 2010; Smith et al., 2012). The rapid onset of severe hepatic damage, particularly in ruminants, may explain many of the early clinical signs associated with severe RVF disease. Although RVF virus is primarily hepatotropic, during severe infections the virus can be found in virtually all tissues and cell types, indicating that the as-yet undiscovered pantropic cellular receptor is likely to be ubiquitous. Results of a recent experimental study in C57BL/6 mouse model suggest that RVFV pathogenesis is associated with a loss of liver function due to liver necrosis and hepatitis yet the long-term course of disease in surviving animals is neurologic disease, however associated with little pathology in the brain. The pathogenesis of the virus in the liver and brain was shown to be mostly driven by chemokine and pro-inflammatory cytokine responses (Gray et al., 2012).

RVFV antibodies have been detected in camels from Egypt (Hoogstraal et al., 1979), Kenya (Davies et al., 1985), Mauritania (Nabeth et al., 2001), Niger (Mariner et al., 1995), Nigeria (Ezeifeka et al., 1982), and Sudan (Eisa, 1984). Circumstantial evidence during the 1977–78 Egyptian outbreak suggests that the disease in camels could be fatal and induce abortion. Clinical signs, including fever and abortion in approximately 10% of pregnant females, were observed in free-ranging camel herds during the 2006–2007 outbreaks in Kenya (Munyua et al., 2010). High antibody prevalence was found in domesticated Asian water buffaloes during the 1977–78 epizootic in Egypt, with abortion and mortality rates of 7 to 12% recorded on some farms (Hoogstral et al., 1979).

Equines develop only low grade viremia following experimental infection. Domestic horses are resistant to RVFV; however, during the Egyptian epizootic there was a single isolation of virus from a horse and four abortions in donkeys were likely due to RVF (Swanepoel and Paweska, 2011). Negative serologic results obtained in zebras in Kenya seem to indicate that these animals do not support RVFV replication (Evans et al., 2008). Earlier experimental studies in domestic pigs (*Sus scrofa*) indicated that they are clinically resistant to RVFV infection (Easterday et al., 1962), but Scott (1963) demonstrated that the resistance in pigs might be dose-dependent. In a recent study in Egypt 37 (15.1%) of 245 pigs were serologically positive (Youssef, 2009). However, data currently available for wild and domestic pigs are too limited to understand their potential role in RVFV maintenance during IEP.

Experimental inoculation studies in domestic dogs and cats demonstrated that they are resistant to RVFV infection (Walker et al., 1970a, b; Keefer at al., 1972). Wild species of Carnivora were not extensively tested, but RVFV antibodies were detected in lions (*Panthera leo*) in southern (House at al., 1996) and East Africa (House et al., 1996; Evans et al., 2008). Domestic rabbits (Findlay and Daubney, 1931) and birds (Findley, 1931; Davies and Addy, 1979) are refractory to RVFV infection.

During the 1951 RVF epizootic in South Africa, abortions occurring among captive wild ruminants were presumably associated with RVFV infection. One farmed springbok (*Antidorcas marsupialis*) and one blesbok (*Damaliscus dorcas phillipsi*) aborted where RVF domestic ruminant and human cases were observed (Gear et al., 1955). These early observations led to increased interest in wild ruminants as reservoirs of the virus. Consequently, specific antibodies were demonstrated in many wild herbivorous and other wildlife species in Africa (Maurice, 1967, Davies, 1975; Davies and Karstad, 1981; Anderson and Rowe, 1998; Evans et al., 2008; LaBeuaud et al., 2011a; Table 8.2). Experimental RVF infection of African buffaloes (*Syncerus caffer*) in Kenya resulted in transient fever and viremia, and one of two pregnant females aborted. Viraemia lasted for at least 48h post intradermal inoculation, ranging from 3.8 to 5.5 Tissue Culture Infectious Dose $(TCID)_{50}$ / ml of blood in four of the five inoculated buffalos (Davies and Karstad, 1981). RVF was confirmed as a cause of abortion in captive-bred buffaloes in northeastern South Africa in 1999, with six buffalos and one waterbuck (*Kobus ellipsiprymnus*) aborting fetuses. A small outbreak among captive-bred buffaloes occurred again in South Africa in 2008 (Swanepoel and Paweska 2011).

A study by Evans et al. (2008) suggested circulation in RVFV among African buffaloes during the 1999–2006 IEP in Kenya. Similar results were recently reported by LaBeaud at al. (2011a) from South Africa where during an IEP of 2000–2006 seroconversion was detected among 7% of resampled African buffaloes residing in the Kruger National Park. However, this low level of seroconversion suggests that during IEPs, circulation of RVFV among African buffaloes leads to dead-end infection. Very high seroprevalence of RVFV-antibody were recently reported from Kenya in lesser kudu (*Tragelaphus strepsiceros* – 50%), impala (*Aepyceros melampus* – 62.5%), Thomson's gazelle (*Gazella thomsonii* - 87.5%), and Gerenuk, also known as a Wallers gazelle (*Litocranius walleri* - 100%), but a limited number of serum samples were tested (Table 8.2). Sera from lesser kudu, impala and Thomson's gazelle were taken during the IEP of 1999–2005, while gerenuk were bled during the large 2006–07 RVFV outbreak in Kenya (Evans et al., 2008). High population density of some of these animals in nature might contribute to their significant role in RVFV maintenance. On the other hand detection of RVFV-antibody in wildlife does not prove their role in a vector-vertebrate maintenance cycle. It might instead be explained by livestock-to-wildlife transmission during epizootics or even during IEPs. Although the possibility of asymptomatic and chronic carriers among wild herbivorous needs to be investigated, the currently available data seem to indicate that RVFV causes similar disease symptoms in wild and domestic ruminants. Therefore, there is no convincing evidence that wild ruminants act as reservoirs of RVFV. It rather appears that they may play a similar role as domestic ruminants in the virus cycle during IEP periods. Intensity and duration of viremia as well as modes of RVFV shedding in wild ruminants is largely unknown. Host preference of RVFV mosquito vectors to wild ruminants has not been investigated. It is notable, however, that the genetic lineage of RVFV found in aborted buffalo fetuses in the Kruger National Park (KNP) and in the dead waterbuck in Klaserie Game Reserve in 1999 was the same as that which infected captive buffalo, farm animals and humans along the Crocodile River outside the KNP in 2008, and also in captive buffalo outside the KNP to the north of Klaserie in the same year. The same virus spread to farming areas in the northeast of South Africa in 2008, and in 2009 it emerged in KwaZulu-Natal Province (Grobbelaar et al, 2011). These findings not only indicate protracted circulation of RVFV in a major wildlife conservation area of South Africa, but also provide molecular evidence of the virus spill over from wildlife to adjacent farming areas.

Of 83 sera collected in Kenya from African elephants (*Loxodonta africana*), five (6%) tested positive for RVFV-specific antibodies (Evans et al., 2008). In the study by Evans et al. (2008) of 43 sera collected from black rhinoceros (*Diceros bicornis*) in Kenya, 14 (32.6%) tested positive for RVFV-specific anti-bodies. High RVFV-antibody seroprevalence was also recently reported among white rhinoceros (*Ceratotherium simum*) from the Kruger National Park in South Africa, of 100 animals tested, 49% were positive (Miller et al., 2011). Due to low population densities of African elephants, black and white rhinoc-eros, it is however unlikely that these wildlife species play an important role in RVFV transmission cycle in nature.

RVFV has been isolated in suckling mice from pooled organs of Peter's lesser epauletted fruit bat (*Micropterus pusillus*), Aba leaf-nosed bat (*Hipposideros aba*) and Sundevall's leaf-nosed bat (*H. caffer*) wild-caught in the Republic of Guinea (Boiro et al., 1987). In contrast, samples of brain, liver, salivary glands and brown fat from 150 bats comprising seven spe-cies trapped in South Africa and Lesotho tested negative for RVFV antigen by an enzyme linked immunosorbent assay (ELISA), but this assay proved to have limited sensitivity. One Schreiber's long-fingered bat (*Miniopterus schreibersii*) and two Cape serotines (*Eptesicus capensis*) bats inoculated by oral or intramuscular route, respectively, did not develop any clinical signs. A low concentration of RVFV antigen was found in the liver and urine of *M. schreibersii* 4 days post inoculation, and low level of antigen was detected in the brown fat 18 days post inoculation in one of the *E. capensis* (Oelofsen and Van der Ryst, 1999). These results demonstrate that bats can be infected with RVFV, and further studies are required to determine the potential of different bat species in the maintenance of RVFV.

The role of rodents in the epidemiologic cycle of RVFV has long consti-tuted a subject of many investigations. Antibodies to RVFV have been detected in several rodent species in endemic RVF areas (Pretorius et al., 1997; Zeller et al., 1997; Diop et al., 2000; Youssef and Hadia, 2001) and the virus has been isolated from rats (*Rattus rattus*) during the 1977–78 epizootic in Egypt (Imam et al., 1979). Despite some reports favoring the rodents in the maintenance of RVFV, this role is incompletely demonstrated and published data are conflict-ing. Transmission of the virus between rodents via direct transmission through excreta or body fluids in the absence of vector mosquitoes has not yet been demonstrated.

GAPS IN KNOWLEDGE AND FUTURE RESEARCH

Longitudinal research aimed at uncovering exact mechanisms of RVFV main-tenance and persistence during IEP is essential for better understanding of the complex interactions between the viral, vectorial, host, ecologic and climatic factors which accumulate to drive large-scale emergence and transmission of the virus to susceptible hosts. Much still remains to be learned about the biology and ecology of the known and the potential RVFV vectors, including the role of vertical transmission of the virus. Neither vertical transmission rate, replication level in quiescent embryos nor the rate of horizontal transmission of RVFV from the progeny *Aedes* mosquitoes are known. Field studies have shown infection with RVFV of both male and female *Aedes* mosquitoes reared only from field-collected larvae in Kenya. These original findings could not be subsequently confirmed, and demonstration of transovarial transmission is hampered by the difficulty in establishing laboratory colonies of *Aedes* mosquitoes. Demonstration that the larvae of *Cx. pipiens*, *Ae. mcintoshi* and *Ae. circumluteolus* become infected after feeding on liver homogenates from an experimentally inoculated hamster is of epidemiological importance, and thus these laboratory findings should be confirmed in the field. RVFV has been shown to be relatively stable in protein-rich environment but it is sensitive to heat inactivation. In this context, the postulated persistence of the virus in mosquito-infected eggs deposited in mud for extended periods of drought and hot weather conditions needs to be investigated.

The role of domestic and wild vertebrate animals in the maintenance of the virus during IEP is not fully understood. Also the host preference of known RVFV mosquito species to wild vertebrates is largely unknown. Further studies are required to determine whether wild mammal species play any specific role in the virus maintenance between outbreaks and virus amplification prior to noticeable outbreaks in livestock and human populations. There is a need to enhance precision and confidence of existing RVF risk predication models. Higher specificity of forecast models could be achieved by increased entomologic and animal surveillance programs in order to determine potential for disease spread and herd immunity. High levels of herd immunity should limit the potential of disease outbreaks even when climatic risks exist. In this context availability of safe and efficacious vaccines and getting vaccines into livestock in low resource areas remains a challenge. Diagnostic validation of serological assays in wild mammals is very limited and assays for detection of specific IgM antibodies in wildlife are not available. Serologic testing of wild vertebrates would provide important clues to the host involved in transmission. The development of validated serologic assays and immunoreagents for surveying the diverse array of potential vertebrates in RVF endemic areas is a priority for research. Challenges associated with RVFV outbreaks and gaps in our current knowledge of this emerging pathogen should be addressed and coordinated by combined veterinary–human health control, prevention, surveillance and research programs under a "One Health" approach. The One Health concept, although still at the assessment and feasibility stage worldwide, has the potential to more cost-effectively and efficiently meet challenges posed by RVFV, especially in resource-poor areas.

CONCLUSIONS

The recent large RVF outbreaks in historically endemic areas, sudden emergence of the virus outside its traditional geographic boundaries, the presence of competent vectors in RVF-free regions, the intensification of international livestock and wildlife trade, effects of global climate change on the spread and establishment of arboviruses in new areas, the potentially dramatic health and socioeconomic consequences of RVFV introduction into RVF-free countries, and potential for deliberate release of this zoonotic agent, are some of the reasons for great international concern. Of particular interest is the mechanism leading to RVFV large scale re-emergence in endemic regions after long periods of silence and its capacity to spread into new territories by crossing significant natural geographic barriers, as exemplified by the virus spreading over Indian Ocean, Sahara desert, and Red Sea in the past three decades. Recent progress in studies on virus pathogenesis, development of new diagnostic tools and vaccines has contributed greatly to our understanding of this viral pathogen. Despite these efforts, safe and efficient commercial vaccines for animal use in RVF-free countries are still unavailable. There is also a general lack of safe vaccines and effective antiviral drugs for human use. Although recent advances in molecular epidemiology assists in understanding the natural history of the virus, we should remain humble about our ability to fully understand the complexity of dynamic interactions between the virus, vectorial, host, ecologic, climatic and anthropogenic factors governing the cryptic transmission of the virus during long interepizootic/interepidemic periods and the emergence of massive RVF outbreaks. These challenges make RVFV one of the most significant emerging viral threats to public and veterinary health in the 21st century.

REFERENCES

Andriamandimby, S., Randrianarivo-Solofoniaina, A., Jeanmaire, E., Ravolomanana, L., Razafimanantsoa, L., Rakotojoelinandrasana, T., Razainirina, J., Hoffman, J., Ravalohery, J. -P., Rafisandratantsoa, J. -T., Rollin, P. E., & Reynes, J. -M. (2010). Rift Valley fever during rainy seasons, Madagascar, 2008 and 2009. *Emerg. Infect. Dis.*, *16*, 963–970.
Anderson, G. W., Jr., & Rowe, L. W. (1998). The prevalence of antibody to the viruses of bovine virus diarrhoea, bovine herpers virus 1, Rift Valley fever, ephemeral fever and bluetongue and to *Leptospira* spp. in free-ranging wildlife in Zimbabwe. *Epid. Infect.*, *121*, 441–449.
Archer, B. N., Weyer, J., Paweska, J., Nkosi, D., Leman, P., San Tint, K., & Blumberg, L. (2011). Outbreak of Rift Valley fever affecting veterinarians and farmers in South Africa, 2008. *S. Afr. Med. J.*, *101*, 263–266.
Billecocq, A., Spiegel, M., Vialat, P., Kohl, A., Weber, F., Bouloy, M., & Haller, O. (2004). NSs protein of Rift Valley fever virus blocks interferon production by inhibiting host gene transcription. *J. Virol.*, *78*, 9798–9806.
Bird, B. H., & Nichol, S. T. (2012). Breaking the chain: Rift Valley fever virus control via livestock vaccination. *Curr. Opin. Virol.*, *2*, 315–323.

Bird, B. H., Khristova, M. L., Rollin, P. E., Ksiazek, T. G., & Nichol, S. T. (2007). Complete genome analysis of 33 ecologically and biologically diverse Rift Valley Fever virus strains reveals widespread virus movement and low genetic diversity due to recent common ancestry. *J. Virol., 81*, 2805–2816.

Bird, B. H., Githinji, J. W. K., Macharia, J. M., Kasiiti, J. L., Muriithi, R. M., Gacheru, S. G., Musaa, J. O., Towner, J. S., Reeder, S. A., Oliver, J. B., Stevens, T. L., Erickson, B. R., Morgan, L. T., Khristova, M. L., Hartman, A. L., Comer, J. A., Rollin, P. E., Ksiazek, T. G., & Nichol, S. T. (2008). Multiple virus lineages sharing recent common ancestry were associated with a large Rift Valley fever outbreak among livestock in Kenya during 2006–2007. *J. Virol., 82*, 11152–11166.

Bird, B. H., Maartens, L. H., Campbell, S., Erasmus, B. J., Erickson, B. R., Dodd, K. A., Spiropoulou, C. F., Cannon, D., Drew, C. P., Knust, B., McElroy, A. K., Khristova, M. L., Albarino, C. G., & Nichol, S. T. (2011). Rift Valley fever virus lacking the NSs and NSm genes is safe, nonteratogenic, and confers protection from viremia, pyrexia, and abortion following challenge in adult and pregnant sheep. *J. Virol., 85*, 12901–12909.

Boiro, I. O., Konstaninov, K., & Numerov, A. D. (1987). Isolation of Rift Valley fever virus from bats in Republic of Guinea. *Bull. Soc. Pathol. Exot., 80*, 62–67.

Borucki, M. K., Chandler, L. J., Parker, B. M., Blair, C. D., & Beaty, B. J. (1999). Bunyavirus superinfection and segment reassortment in transovarially infected mosquitoes. *J. Gen. Virol., 80*, 3173–3179.

Boshra, H., Lorenzo, G., Busquets, N., & Brun, A. (2011). Rift Valley Fever: Recent Insights into Pathogenesis and Prevention. *J. Virol., 85*, 60–6105.

Botros, B., Omar, A., Elian, K., Mohamed, G., Soliman, A., Salib, A., Salman, D., Saad, M., & Earhart, K. (2006). Adverse response of non-indigenous cattle of European breeds to live attenuated Smithburn Rift Valley fever vaccine. *J. Med. Virol., 78*, 787–791.

Bouloy, M., & Flick, R. (2009). Reverse genetics technology for Rift Valley fever virus: Current and future applications for the development of therapeutics and vaccines. *Antiviral Res., 84*, 101–118.

Bouloy, M., & Weber, F. (2010). Molecular biology of Rift Valley fever virus. *Open Virol. J., 4*, 8–14.

Brennan, B., Welch, S. R., McLees, A., & Elliott, R. M. (2011). Creation of a recombinant Rift Valley fever virus with a two-segmented genome. *J. Virol., 85*, 10310–10318.

Briese, T., Bird, B. H., Kapoor, V., Nichol, S. T., & Lipkin, W. I. (2006). Batai and Ngari viruses: M segment reassortment and association with severe febrile disease outbreaks in East Africa. *J. Virol., 80*, 5627–5630.

Carroll, S. A., Reynes, J. M., Khristova, M. L., Andriamandimby, S. F., Rollin, P. E., & Nichol, S. T. (2011). Genetic evidence for Rift Valley fever outbreaks in Madagascar resulting from virus introductions from the East African mainland rather than enzootic maintenance. *J. Virol., 85*, 6162–6267.

Cêtre-Sossah, C., Pédarrieu, A., Guis, H., Deferenz, C., Bouloy, M., Favre, J., Girard, S., Cardinale, E., & Albina, E. (2012). Prevalence of Rift Valley fever among ruminants, Mayotte. *Emerg. Infect. Dis., 18*, 972–975.

Chevalier, V., Pepin, M., Plee, L., & Lancelot, R. (2010). Rift Valley fever—a threat for Europe? *EuroSurveillance, 15*, 1950–1956.

Clements, A. C., & Pfeiffer, D. U. (2009). Emerging viral zoonoses: frameworks for spatial and spatiotemporal risk assessment and resource planning. *Vet. J., 182*, 21–30.

Coetzer, J. A. (1982). The pathology of Rift Valley fever. II. Lesions occurring in field cases in adult cattle, calves and aborted foetuses, Onderstepoort J. *Vet. Res., 49*, 11–17.

Crabtree, M. B., Kent, Crockett, R. J., Bird, B. H., Nichol, S. T., Erickson, B. R., Biggerstaff, B. J., Horiuchi, K., & Miller, B. R. (2012). Infection and transmission of Rift Valley fever viruses lacking the NSs and/or NSm genes in mosquitoes: potential role for NSm in mosquito infection. *PLoS Negl. Trop. Dis.*, *6*, e1639.

Daubney, R., Hudson, J. R., & Garnham, P. C. (1931). Enzootic hepatitis or Rift Valley fever. An undescribed virus disease of sheep, cattle and man from East Africa. *J. Pathol. Bacteriol.*, *34*, 545–579.

Davies, F. G. (1975). Observations on the epidemiology of Rift Valley fever in Kenya. *J. Hyg.*, *75*, 219–230.

Davies, F. G. (2010). The historical and recent impact of Rift Valley fever in Africa. *Am. J. Trop. Med. Hyg.*, *83*(Suppl. 2), 73–74.

Davies, F. G., & Addy, P. A. (1979). Rift Valley fever. A survey for antibody to the virus in bird species commonly found in situations considered to be enzootic. *Trans. R. Soc. Trop. Med. Hyg.*, *73*, 584–585.

Davies, F. G., Clausen, B., & Lund, L. J. (1972). The pathogenicity of Rift Valley fever virus for the baboon. *Trans. R. Soc. Trop. Med. Hyg.*, *66*, 363–365.

Davies, F. G., & Karstad, L. (1981). Experimental infection of the African buffalo with the virus of Rift Valley fever. *Trop. Anim. Health Prod.*, *13*, 185–188.

Davies, F. G., Koros, J., & Mbugua, H. (1985). Rift Valley fever in Kenya: The presence of antibody to the virus in camels (*Camelus dromedarius*). *J. Hyg.*, *94*, 241–244.

Diallo, M., Nabeth, P., Ba, K., Sall, A. A., Ba, Y., Mondo, M., Girault, L., Abdalahi, M. O., & Mathiot, C. (2005). Mosquito vectors of the 1998–1999 outbreak of Rift Valley Fever and other arboviruses (Bagaza, Sanar, Wesselsbron and West Nile) in Mauritania and Senegal. *Med. Vet. Entemol.*, *19*, 119–126.

Diop, G., Thiongane, Y., Thonnon, J., Fontenille, D., Diallo, M., Sall, S., Ruel, T. D., & Gonzalez, J. P. (2000). The potential role of rodents in the enzootic cycle of Rift Valley fever virus in Senegal. *Microb. Infect.*, *2*, 343–346.

Dodd, K. A., Bird, B. H., Metcalfe, M. G., Nichol, S. T., & Albarino, C. G. (2012). Single-dose immunization with virus replicon particles confers rapid robust protection against Rift Valley fever virus challenge. *J. Virol.*, *86*, 4204–4212.

Easterday, B. C. (1965). Rift Valley fever. *Adv. Vet. Sci.*, *10*, 65–127.

Easterday, B. C., Murphy, L. C., & Bennet, D. G. (1962). Experimental Rift Valley fever in calves, goats and pigs. *Am. J. Vet. Res.*, *23*, 1225–1230.

Eisa, M. (1984). Preliminary survey of domestic animals in the Sudan for precipitating antibodies to Rift Valley fever virus. *J. Hyg.*, *93*, 629–637.

Evans, A., Gakuya, F., Paweska, J. T., Rostal, M., Akoolo, L., Jansen Van Vuren, P., Manyibe, T., Macharia, J. M., Kziazek, T. G., Feiken, D. R., Reiman, R. F. K., & Njenga, M. K. (2008). Prevalence of antibodies against Rift Valley Fever virus in Kenyan wildlife. *Epidemiol. Infec.*, *136*, 1261–1269.

Ezeifeka, G. O., Umoh, J. U., Belino, E. D., & Ezeokoli, C. D. (1982). A serological survey for Rift Valley fever antibody in food animals in Kaduna and Sokoto States of Nigeria. *Int. J. Zoonoses*, *9*, 147–151.

Fafetine, J., Neves, L., Thompson, P. N., Paweska, J. T., Victor, P. M. G., Rutten, V. P. M.G., Coetzer, J. A. W., & J.A.W. (2013). Serological evidence of Rift Valley fever virus circulation in sheep and goats in Zambézia Province, Mozambique. *PLoS Negl. Trop. Dis.*, *7*, e2065.

Favier, C., Chalvet-Monfray, K., Sabatier, P., Lancelot, R., Fontenille, E., & Dubois, M. A. (2006). Rift Valley fever in West Africa: the role of space in endemicity. *Trop. Med. Int. Health*, *11*, 1878–1888.

Filone, C. M., Heise, M., Doms, R. W., & Bertolotti-Ciarlet, A. (2006). Development and characterization of a Rift Valley fever virus cell–cell fusion assay using alphavirus replicon vectors. *Virol.*, *356*, 155–164.

Findlay, G. M. (1931). The virus of Rift Valley fever or enzootic hepatitis. *Trans. R. Soc. Trop. Med. Hyg.*, *25*, 229–262.

Findlay, G. M. (1932). The infectivity of Rift Valley fever for monkeys. *Trans. R. Soc. Trop. Med. Hyg.*, *26*, 161–168.

Findlay, G. M., & Daubney, R. (1931). The virus of Rift Valley fever or enzootic hepatitis. *Lancet*, *221*, 1350–1351.

Fontana, J., Lopez-Montero, N., Elliott, R. M., Fernandez, J. J., & Risco, C. (2008). The unique architecture of Bunyamwera virus factories around the Golgi complex. *Cell. Microbiol.*, *10*, 2012–2028.

Fontenille, D., Traore-Lamizana, M., Diallo, M., Thonnon, J., Digoutte, J. P., & Zeller, H. G. (1998). New Vectors of Rift Valley Fever in West Africa. *Emerg. Infect. Dis.*, *4*, 289–293.

Gad, A. M., Feinsod, F. M., Allam, I. H., Eisa, M., Hassan, A. N., Soliman, B. A., El Said, S., & Saah, A. J. (1986). A possible route for the introduction of Rift Valley fever virus into Egypt during 1977. *J. Trop. Med. Hyg.*, *89*, 233–236.

Gear, J., De Meillon, B., Le Roux, A. F., Kofsky, R., Innes, R. R., Steyn, J. J., Oliff, W. D., & Schulz, K. H. (1955). Rift valley fever in South Africa: a study of the 1953 outbreak in the Orange Free State, with special reference to the vectors and possible reservoir hosts. *S. Afr. Med. J.*, *29*, 514–518.

Gerrard, S. R., Li, L., Barrett, A. D., & Nichol, S. T. (2004). Ngari virus is a Bunyamwera virus reassortant that can be associated with large outbreaks of hemorrhagic fever in Africa. *J. Virol.*, *78*, 8922–8926.

Gerrard, S. R., Bird, B. H., Albarino, C. G., & Nichol, S. T. (2007). The NSm proteins of Rift Valley fever virus are dispensable for maturation, replication and infection. *Virol.*, *359*, 459–465.

Gerrard, S. R., & Nichol, S. T. (2007). Synthesis, proteolytic processing and complex formation of N-terminally nested precursor proteins of the Rift Valley fever virus glycoproteins. *Virol.*, *357*, 124–133.

Gray, K. K., Worthu, M. N., Juelich, T. L., Agar, S. L., Pousssard, A., Ragland, D., Freiberg, A. N., & Holbrook, M. R. (2012). Chemotactic and inflamatory responses in the liver and brain are associated with pathogenesis of Rift Valley fever virus infection in the mouse. *PLoS Negl. Trop. Dis.*, *6*, e1529.

Grobbelaar, A. A., Weyer, J., Leman, P. A., Kemp, A., Paweska, J. T., & Swanepoel, R. (2011). Molecular epidemiology of Rift Valley fever virus. *Emerg. Infect. Dis.*, *12*, 2270–2276.

Habjan, M., Pichlmair, A., Elliott, R. M., Overby, A. K., Glatter, T., Gstaiger, M., Superti-Furga, G., Unger, H., & Weber, F. (2009). NSs protein of Rift Valley Fever Virus induces the specific degradation of the double-stranded RNA-dependent protein kinase (PKR). *J. Virol.*, *83*, 4365–4375.

Hayes, E. B. (2010). Is it time for a new yellow fever vaccine? *Vaccine*, *28*, 8073–8076.

Heinrich, N., Saathoff, E., Weller, N., Clowes, P., Kroidl, I., Ntinginya, E., Machibya, H., Maboko, L., Lösche, R. T., Dobler, G., & Hoelscher, M. (2012). High seroprevalence of Rift Valley fever and evidence for endemic circulation in Mbeya region, Tanzania, in a cross-sectional study. *PLoS Negl. Trop. Dis.*, *6*, e1557.

Hoogstraal, H., Meegan, J. M., & Khalil, G. M. (1979). The Rift Valley fever epizootic in Egypt 1977–78. II. Ecological and entomological studies. *Trans. R. Soc. Trop. Med. Hyg.*, *73*, 624–629.

House, C., Alexander, K. A., Kat, P. W., O'Brien, S. J., & Mangiafico, J. (1996). Serum antibody to Rift Valley fever virus in African carnivores. *Ann. N.Y. Acad. Sci.*, *791*, 345–349.

Ikegami, T. (2012). Molecular biology and genetic diversity of Rift Valley fever virus. *Antiviral Res.*, *95*, 293–310.

Ikegami, T., & Makino, S. (2011). The pathogenesis of Rift Valley fever. *Viruses*, *3*, 493–519.

Ikegami, T., Narayanan, K., Won, S., Kamitani, W., Peters, C. J., & Makino, S. (2009). Rift Valley fever virus NSs protein promotes post-transcriptional downregulation of protein kinase PKR and inhibits eIF2alpha phosphorylation. *PLoS Pathog.*, *5*, e1000287.

Imam, Z. E., El-Karamany, R., & Darwish, M. A. (1979). An epidemic of Rift Valley fever in Egypt, 2. Isolation of the virus from animals. *Bull. WHO*, *57*, 441–443.

Jansen van Vuren, P., Tiemessen, C. T., & Paweska, J. T. (2011). Anti-nucleocapsid immune responses counteract pathogenic effects if Rift Valley fever virus infection in mice. *PLoS ONE*, *6*, e2507.

Jouan, A., Le Guenno, B., Digoutte, J. P., Philippe, B., Riou, O., & Adam, F. (1988). An RVF epidemic in southern Mauritania. *Ann. Inst. Pasteur Virol.*, *139*, 307–308.

Jupp, P. G., Kemp, A., Grobbelaar, A., Leman, P., Burt, F. J., Alahmed, A. M., Al Mujalli, D., Al Khamees, M., & Swanepoel, R. (2002). The 2000 epidemic of Rift Valley fever in Saudi Arabia: Mosquito vector studies. *Med. Vet. Entomol.*, *16*, 245–252.

Kamal, S. A. (2009). Pathological studies on postvaccinal reactions of Rift Valley fever in goats. *Virol. J.*, *6*, 94.

Kamal, S. A. (2011). Observations on Rift Valley fever virus and vaccines in Egypt. *Virol. J.*, *8*, 532.

Kalveram, B., Lihoradova, O., & Ikegami, T. (2011). NSs protein of Rift Valley fever virus promotes posttranslational downregulation of the TFIIH subunit p62. *J. Virol.*, *85*, 6234–6243.

Keefer, G. V., Zebarth, G. L., & Allen, W. P. (1972). Susceptibility of dogs and cats to Rift Valley fever by inhalation or ingestion of virus. *J. Infec. Dis.*, *125*, 307–309.

Konrad, S. K., & Miller, S. N. (2012). A temperature-limited assessment of the risk of Rift Valley fever transmission and establishment in the continental United States of America. *Geospat. Health*, *6*, 161–170.

Kortekaas, J., Oreshkova, N., Cobos-Jiméez, V., Vloet, R. P. M., Potgieter, C. A., & Moormann, R. J. M. (2011). Creation of a nonspreading Rift Valley fever virus. *J. Virol.*, *85*, 12622–12630.

Ksiazek, T. G., Jouan, A., Meegan, J. M., Le Guenno, B., Wilson, M. L., Peters, C. J., Digoutte, J. P., Guillaud, M., Merzoug, N. O., & Touray, E. M. (1989). Rift Valley fever among domestic animals in the recent West African outbreak. *Res. Virol.*, *140*, 67–77.

LaBeaud, A. D., Ochiai, Y., Peters, C. J., Muchiri, E. M., & King, C. H. (2007). Spectrum of Rift Valley fever virus transmission in Kenya: insight from three distinct regions. *J. Trop. Med. Hyg.*, *76*, 795–800.

LaBeaud, A. D., Muchiri, E. M., Ndzovu, M., Mwanje, M. T., Muiruri, S., Peters, C. J., & King, C. H. (2009). Interepidemic Rift Valley fever virus seropositivity, Northeastern Kenya. *Emerg. Infect. Dis.*, *14*, 1240–1246.

LaBeaud, A. D., Cross, P. C., Getz, W. M., Glinka, A., & King, C. H. (2011a). Rift Valley fever virus infection in African buffalo (*Syncerus caffer*) herds in rural South Africa: evidence of interepidemic transmission. *Am. J. Trop. Med. Hyg.*, *84*, 641–646.

LaBeaud, A. D., Muiruri, S., Sutherland, L. J., Dahir, S., Gildengorin, G., Morril, J., Muchiri, E. M., Peters, C. J., & King, C. H. (2011b). Postepidemic analysis of Rift Valley fever virus transmission in northeastern Kenya: a village cohort study. *PLoS Negl. Trop. Dis.*, *5*, e1265.

Lagerqvist, N., Moiane, B., Bucht, G., Fafetine, J., Paweska, J. T., Lundkvist, A., & Falk, K. I. (2012). Stability of a formalin-inactivated Rift Valley Fever vaccine: evaluation of a vaccination campaign for cattle in Mozambique. *Vaccine*, *30*, 6534–6540.

Le May, N., Mansuroglu, Z., Leger, P., Josse, T., Blot, G., Billecocq, A., Flick, R., Jacob, Y., Bonnefoy, E., & Bouloy, M. (2008). A SAP30 complex inhibits IFN-beta expression in Rift Valley fever virus infected cells. *PLoS Pathog.*, *4*, e13.

Lihoradowa, O., & Ikegami, T. (2012). Modifying the NSs gene to improve live-attenuated vaccine for Rift Valley fever. *Exp. Rev. Vac.*, *11*, 1283–2012.

Linthicum, K. G., Davies, F. G., Bailey, C. L., & Kairo, A. (1983). Mosquito species succession in a dambo in an East African forest. *Mosq. News*, *43*, 464–470.

Linthicum, K. J., Davies, F. G., Kairo, A., & Bailey, C. L. (1985). Rift Valley fever virus (family *Bunyaviridae*, genus *Phlebovirus*). Isolations from diptera collected during an inter-epizootic period in Kenya. *J. Hyg.*, *95*, 197–209.

Liu, L., Celma, C. C. P., & Roy, P. (2008). Rift Valley fever virus structural proteins: Expression, characterization and assembly of recombinant proteins. *Virol. J.*, *5*, 82.

Logan, T. M., Linthicum, K. J., Davies, F. G., Binepal, Y. S., & Roberts, C. R. (1991). Isolation of Rift Valley fever virus from mosquitoes (Diptera: Culicidae) collected during an outbreak in domestic animals in Kenya. *J. Med. Entomol.*, *28*, 293–295.

Lorenzo, G., Martin-Foglar, R., Hevia, E., Boshra, H., & Brun, A. (2010). Protection against lethal Rift Valley fever virus (RVFV) infection in transgenic INFAR−/−-mice induced by different DNA vaccination regimes. *Vaccine*, *28*, 2937–2944.

Lozach, P. Y., Kuhbacher, A., Meier, R., Mancini, R., Bitto, D., Bouloy, M., & Helenius, A. (2011). DC-SIGN as a receptor for phleboviruses. *Cell Host Microbe*, *10*, 75–88.

Madani, T. A., Al-Mazrou, Y. Y., Al-Jeffri, M. H., Mishkhas, A. A., Al-Rabeah, A. M., Turkistani, A. M., Al-Sayed, M. O., Abodahish, A. A., Khan, A. S., Ksiazek, T. G., & Shobokshi, O. (2003). Rift Valley Fever Epidemic in Saudi Arabia: Epidemiological, Clinical, and Laboratory Characteristics. *Clin. Infect. Dis.*, *37*, 1084–1092.

Mandel, R. B., Koukuntla, R., Mogler, L. J., Carzoli, A. K., Freiberg, A. N., Holbrook, M. R., Martin, B. K., Staplin, W. R., Vahanian, N. N., Link, C. J., & Flick, R. (2010). A replication-incompetent Rift Valley fever vaccine: chimeric virus-like particles protect mice and rats against lethal challenge. *Virol.*, *397*, 187–198.

Mapaco, L. P., Coetzer, J. A. W., Paweska, J. T., & Venter, E. H. (2012). An investigation into outbreak of Rift Valley fever on a cattle farm in Bela-Bela, South Africa, in 2008. *J. S. Afr. Vet. Ass.*, *83*, E1–E7.

Mariner, J. C., Morrill, J., & Ksiazek, T. G. (1995). Antibodies to hemorrhagic fever viruses in domestic livestock in Niger: Rift Valley fever and Crimean-Congo hemorrhagic fever. *Am. J. Trop. Med. Hyg.*, *53*, 217–221.

Martin, V., Chevalier, V., Ceccato, P., Anyamba, A., De Simone, L., Lubroth, J., de La Rocque, S., & Domenech, J. (2008). The impact of climate change on the epidemiology and control of Rift Valley fever. *Rev. Sci. Tech.*, *27*, 413–426.

Maurice, Y. (1967). First serologic verification of the incidence of Wesselsbronn's disease and Rift Valley fever in sheep and wild ruminants in Chad and Cameroon. *Rev. Elev. Med. Vet. Pays Trop.*, *20*, 395–404.

McIntosh, B. M., Dickinson, D. B., & Dos Santos, I. (1973). Rift Valley fever. 3. Viremia in cattle and sheep. 4. The susceptibility of mice and hamsters in relation to transmission of virus by mosquitoes. *J. S. Afr. Vet. Ass.*, *44*, 167–169.

Meegan, J. M. (1981). Rift valley fever in Egypt: an overview of the epizootics in 1977 and 1978. In T. A. Swartz, M. A. Klinberg, N. Goldblum & C. M. Papier (Eds.), *Contributions to Epidemiology and Biostatistics* (pp. 100–113). Basel: Rift Valley Fever, S. Karger AG.

Métras, R., Porphyre, T., Pfeiffer, D. U., Kemp, A., Thomson, P., Collins, L. M., & White, R. G. (2012). Exploratory space-time analyses of Rift Valley fever in South Africa in 2008–2011. *PLoS Negl. Trop. Dis.*, *6*, e1808.

Miller, M., Buss, P., Joubert, J., Maseko, N., Hofmeyer, M., & Gerdes, T. (2011). Serosurvey for selected viral agents in white rhinoceros (*Ceratotherium simum*) in Kruger National Park, 2007. *J. Zoo Wildl. Med.*, *42*, 29–32.

Mohamed, M., Mosha, F., Mghamba, J., Zaki, S. R., Shieh, W. -J., Paweska, J., Omulo, S., Gikundi, S., Mmbuji, P., Bloland, P., Zeidner, N., Kalinga, R., Breiman, R. F., & Njenga, M. K. (2010). Epidemiologic and clinical aspects of a Rift Valley fever outbreak in humans in Tanzania, 2007. *Am. J. Trop. Med. Hyg.*, *83*(Suppl. 2), 22–27.

Morrill, J. C., & Peters, C. J. (2011). Protection of MP-12-vaccinated rhesus macaques against parenteral and aerosol challenge with virulent Rift Valley fever virus. *J. Infect. Dis.*, *204*, 229–236.

Morrill, J. C., Ikegami, T., Yoshikawa-Iwata, N., Lokugamage, N., Won, S., Teresaki, K., Zamoto-Niikura, A., Peters, C. J., & Makino, S. (2010). Rapid accumulation of virulent Rift Valley fever virus in mice from an attenuated virus carrying a single nucleotide substitution in the M RNA. *PLoS ONE*, *5*, e9986.

Morvan, J., Saluzzo, J. F., Fontenille, D., Rollin, P. E., & Coulanges, P. (1991). Rift Valley fever on the east coast of Madagascar. *Res. Virol.*, *142*, 475–482.

Moutailler, S., Krida, G., Schaffner, F., Vazeille, M., & Failloux, A. B. (2008). Potential Vectors of Rift Valley Fever Virus in the Mediterranean Region. *Vector Borne Zoonotic Dis.*, *8*, 749–754.

Moutailler, S., Roche, B., Thiberge, J. M., Caro, V., Rougeon, F., & Failloux, A. B. (2011). Host alternation is necessary to maintain the genome stability of rift valley fever virus. *PLoS Negl. Trop. Dis.*, *5*, e1156.

Muller, R., Poch, O., Delarue, M., Bishop, D. H., & Bouloy, M. (1994). Rift Valley fever virus L segment: correction of the sequence and possible functional role of newly identified regions conserved in RNA-dependent polymerases. *J. Gen. Virol.*, *75*, 1345–1352.

Muller, R., Saluzzo, J. F., Lopez, N., Dreier, T., Turell, M., Smith, J., & Bouloy, M. (1995). Characterization of clone 13, a naturally attenuated avirulent isolate of Rift Valley fever virus, which is altered in the small segment. *Am. J. Trop. Med. Hyg.*, *53*, 405–411.

Murithi, R. M., Munyua, P., Ithondeka, P. M., Macharia, J. M., Hightower, A., Luman, E. T., Breiman, R. F., & Njenga, M. K. (2011). Rift Valley fever in Kenya: history of epizootics and identification of vulnerable districs. *Epidemiol. Infect.*, *139*, 372–380.

Munyua, P., Murithi, R. M., Weinwright, S., Githinji, J., Hightower, A., Mutonga, D., Macharia, J., Ithodeka, P. M., Musaa, J., Breiman, R. F., Bloland, P., & Njenga, M. K. (2010). Rift Valley fever outbreak in Kenya, 2006-2007. *Am. J. Trop. Med. Hyg.*, *83*, 58–64.

Nabeth, P. Y., Kane, M. O., Abdalahi, M., Diallo, K., Ndiaye, K. B., Schneegans, F., Sall, A. A., & Mathiot, C. (2001). Rift Valley fever outbreak, Mauritania, 1998: Seroepidemiologic, virologic entomologic, and zoologic investigations. *Emerg. Infect. Dis.*, *7*, 1052–1054.

Narayanan, A., Kehn-Hall, K., Senina, S., Hill, L., van Duyne, R., Guendel, I., Das, R., Baer, A., Bethel, L., Turell, M., Hartman, A. L., Das, B., Bailey, C., & Kashanchi, F. (2012). Curcumin inhibits Rift Valley fever replication in human cells. *J. Biol. Chem.*, *40*, 33198–33214.

Nderitu, L., Lee, J. S., Omolo, J., Omulo, S., O'Guinn, M. L., Hightower, A., Mosha, F., Mohamed, M., Munyua, P., Nganga, Z., Hiett, K., Seal, B., Feikin, D. R., Breiman, R. F., & Njenga, M. K. (2011). Sequential Rift Valley fever outbreaks in Eastern Africa caused by multiple lineages of the virus. *J. Infect. Dis.*, *203*, 655–665.

Nfon, C. K., Marszal, P., Zhang, S., & Weingartl, H. M. (2012). Innate immune response to Rift Valley fever virus in goats. *PLoS Negl. Trop. Dis.*, *6*, e1623.

Nguku, P., Sharif, S. K., Mutonga, D., Amwayi, S., Omolo, J., Mohamed, O., Farnon, E. C., Gould, L. H., Lederman, E., Rao, C., Sang, R., Schnabel, D., Feikin, D. R., Hightower, A., Njenga, M. K., & Breiman, R. F. (2010). An investigation of a major outbreak of Rift Valley fever in Kenya: 2006-2007. *Am. J. Trop. Med. Hyg.*, *83*, 5–13.

Olive, M. -M., Goodman, S. M., & Reynes, J. -M. (2012). The role of wild mamals in the maintenance of Rift Valley fever virus. *J. Wildl. Dis.*, *48*, 241–266.

Oelofsen, M. J., & Van der Ryst, E. (1999). Could bats act as reservoir hosts for Rift Valley fever virus? *Onderstepoort J. Vet. Res.*, *66*, 51–54.

Pepin, M., Bouloy, M., Bird, B. H., Kemp, A., & Paweska, J. (2010). Rift Valley fever virus (Bunyaviridae: Phlebovirus): an update on pathogenesis, molecular epidemiology, vectors, diagnostics and prevention. *Vet. Res.*, *41*, 61.

Piper, C. J., Sorenson, D. R., & Gerrard, S. R. (2011). Efficient cellular release of Rift Valley fever virus requires genomic RNA. *PLoS One*, *6*, e18070.

Pourrut, X., Nkoghe, D., Souris, M., Paupy, C., Paweska, J., Padilla, C., Moussavou, G., & Leroy, R. M. (2010). Rift Valley fever seroprevalence in human rural populations of Gabon. *PLoS Negl. Trop. Dis.*, *4*, e763.

Pretorius, A. M., Oelofsen, M. J., Smith, M. S., & van der Ryst, E. (1997). Rift Valley fever virus: a seroepidemiologic study of small terrestial vertebrates in South Africa. *Am. J. Trop. Med. Hyg.*, *57*, 693–698.

Reguera, J. F., Weber, F., & Cusak, S. (2010). Bunyaviridae RNA polymerases (L-protein) have an N-terminal, influenza-like endonuclease domain, essential for viral cap-dependent transcription. *PLoS Pathog*, *6*, e1001101.

Rich, K. M., & Wanyoike, F. (2010). An assessment of the regional and national socio-economic impacts of the 2007 Rift Valley fever outbreak in Kenya. *Am. J. Trop. Med. Hyg.*, *83*, 52–57.

Rusnak, J. M., Gibbs, P., Boudreau, E., Clizbe, D. P., & Pittman, P. (2011). Immunogenicity and safety of an inactivated Rift Valley fever vaccine in a 19-year study. *Vaccine*, *29*, 3222–3229.

Sall, A. A., Zanotto, P. M. D.A., Zeller, H. G., Digoutte, J. P., Thiongane, Y., & Bouloy, M. (1997). Variability of the NSs protein among Rift Valley fever virus isolates. *J. Gen. Virol.*, *78*, 2853–2858.

Sall, A. A., Zanotto, P. M. D.A., Sene, O. K., Zeller, H. G., Digoutte, J. P., Thiongane, Y., & Bouloy, M. (1999). Genetic reassortment of Rift Valley fever virus in nature. *J. Virol.*, *73*, 8196–8200.

Schmaljohn, C. S., & Nichol, S. T. (2007). Bunyaviridae. In D. M. Knipe, P. M. Howley, D. E. Griffin, R. A. Lamb, M. A. Martin, B. Roizman & S. E. Straus (Eds.), *Fields Virology* (5th ed., pp. 1741–1789). Philadelphia, PA, USA: Lippincott, Williams & Wilkins.

Scott, G. R. (1963). Pigs and Rift Valley fever. *Nature*, *200*, 919–920.

Scott, T., Paweska, J. T., Arbuthnot, P., & Weinberg, M. S. (2012). Pathogenic effects of Rift Valley fever virus NSs gene are alleviated in cultured cells by expressed antiviral short hairpin RNAs. *Antivir. Ther.*, *17*, 643–656.

Service, M. W. (1997). Mosquito (Diptera:Culicidae) dispersal—the long and short of it. *J. Med. Entomol.*, *34*, 579–588.

Sherman, M. B., Freiberg, A. N., Holbrook, M. R., & Watowich, S. J. (2009). Single-particle cryo-electron microscopy of Rift Valley fever virus. *Virol.*, *387*, 11–15.

Shieh, W. -J., Paddock, C. D., Lederman, E., Rao, C. Y., Gould, L. H., Mohamed, M., Mosha, F., Mghamba, J., Bloland, P., Njenga, M. K., Mutonga, D., Samuel, A. A., Guarner, J., Breiman, R. F., & Zaki, S. R. (2010). Pathologic studies on suspect animal and human cases of Rift Valley fever from an outbreak in Eastern Africa, 2006–2007. *Am. J. Trop. Med. Hyg.*, *83*, 38–42.

Smith, D. R., Bird, B. H., Lewis, B., Johnston, S. C., McCarthy, S., Keeney, A., Botto, M., Donnelly, G., Shamblin, J., Albarino, C., Nichol, S. T., & Hensley, J. E. (2012). Development of a novel nonhuman primate model for Rift Valley fever. *J. Virol.*, *86*, 2109–2110.

Soumaré, P. O. L., Freire, C. C. M., Faye, O., Diallo, M., de Oliveira, J. V. C., Zanatto, P. M. A., & Sall, A. A. (2012). Phylogeography of Rift Valley fever virus in Africa reveals multiple introductions in Senegal and Mauritania. *PLoS ONE*, *7*, e35216.

Struthers, J. K., & Swanepoel, R. (1982). Identification of a major non-structural protein in the nuclei of Rift Valley fever virus-infected cells. *J. Gen. Virol.*, *60*, 381–384.

Swanepoel, R., & Coetzer, J. A. (2004). Rift Valley fever. In J. A. Coetzer & R. C. Tustin (Eds.), *Infectious Diseases of Livestock* (pp. 1037–1070). Cape Town: Oxford University Press Southern Africa.

Swanepoel, R., & Paweska, J. T. (2011). Rift Valley fever. In S. R. Palmer, L. Soulsby, P. R. Torgerson & D. W. G. Brown (Eds.), *Oxford Textbook of Zoonoses: Biology, Clinical Practise, and Public Health Control* (pp. 421–431). Oxford University Press.

Swanepoel, R., Erasmus, B. J., Williams, R., & Taylor, M. B. (1992). Encephalitis and chorioretinitis associated with neurotropic African horsesickness virus infection in laboratory workers. Part III. Virological and serological investigations. *S. Afr. Med. J.*, *81*, 458–461.

Traoré-Lamizana, M., Fontenille, D., Diall, M., Bâ, Y., Zeller, H. R., Mondo, M., Adam, F., Thonon, J., & Maïga, A. (2001). Arbovirus surveillance from 1990 to 1995 in the Barkedji area (Ferlo) of Senegal a possible natural focus of Rift Valley fever virus. *J. Med. Entomol.*, *38*, 480–492.

Turell, M. J., Linthicum, K. J., & Beaman, J. R. (1990). Transmission of Rift Valley fever virus by adult mosquitoes after ingestion of virus as larvae. *Am. J. Trop. Med. Hyg.*, *43*, 677–680.

Turell, M. J., Dohm, D. J., Mores, C. N., Terracina, L., Wallette, D. L., Jr., Hribar, L. J., Pecor, J. E., & Blow, J. A. (2008). Potential for North American mosquitoes to transmit Rift Valley fever virus. *J. Am. Mosquito Control Ass.*, *24*, 502–507.

Von Teichman, B., Engelbrecht, A., Zulu, G., Dungu, B., Pardini, A., & Bouloy, M. (2011). Safety and efficacy of Rift Valley fever Smithburn and Clone 13 vaccines in calves. *Vaccine*, *29*, 5771–5777.

Walker, J. S., Remmele, N. S., Carter, R. C., Mitten, J. Q., Schuh, L. G., Stephen, E. L., & Klein, F. (1970a). The clinical aspects of Rift Valley Fever virus in household pets. I. Susceptibility of the dog. *J. Infect. Dis.*, *121*, 9–18.

Walker, J. S., Stephen, E. L., Remmele, N. S., Carter, R. C., Mitten, J. Q., Schuh, L. G., & Klein, F. (1970b). The clinical aspects of Rift Valley Fever virus in household pets. II. Susceptibility of the cat. *J. Infect. Dis.*, *121*, 19–24.

Won, S., Ikegami, T., Peters, C. J., & Makino, S. (2007). NSm protein of Rift Valley fever virus suppresses virus-induced apoptosis. *J. Virol.*, *81*, 13335–13345.

Youssef, B. Z. (2009). The potential role of pigs in the enzootic cycle of Rift Valley fever at Alexandria Governorate, Egypt. *J. Egypt. Public Health Assoc.*, *84*, 331–344.

Youssef, B. Z., & Hadia, H. A. (2001). The potential role of *Rattus rattus* in enzootic cycle of Rift Valley fever in Egypt. 1 – Detection of RVF antibodies in *R. rattus* blood samples by both enzyme-linked immuno sorbent assay (ELISA) and immuno-diffusion technique (ID). *J. Egypt. Public Health Assoc.*, *76*, 431–441.

Zeller, H., Fontenille, D., Traorelamizana, M., Thiongane, Y., & Digoutte, J. P. (1997). Enzootic activity of Rift Valley fever virus in Senegal. *Am. J. Trop. Med. Hyg.*, *56*, 265–272.

Chapter 9

From Simian to Human Immunodeficiency Viruses (SIV to HIV)

Emergence from Nonhuman Primates and Transmission to Humans

Denis M. Tebit[+] and Eric J. Arts

Division of Infectious Disease, Case Western Reserve University, Cleveland, Ohio, USA
[+]Present Affiliation: Myles H. Thaler Center for AIDS and Human Retrovirus Research,
Department of Microbiology, University of Virginia, Charlottesville, Virginia, USA

INTRODUCTION

Simian immunodeficiency viruses (SIV) are members of the *Lentivirus* genus of *Retroviruses* that infect nonhuman primates such as monkeys, chimpanzees, and gorillas. For the most part, these SIVs are either nonpathogenic or result in slow progression to a debilitating immunodeficiency or mortality. Transmission of SIV to humans occurred almost a century ago, resulting in the eventual adaptation and emergence of human immunodeficiency virus (HIV), the etiological agent of acquired immune deficiency syndrome (AIDS). The first cases of AIDS were reported in 1981 among young homosexual men who presented with symptoms of various malignancies and infections that had not been previously reported among this group of individuals (CDC, 1981). At that time, the origins of this disease were unknown and even after the initial characterization of LAV/HTLV-III (lymphadenopathy-associated virus or human T-lymphotropic virus type III) in 1983 (which would later be named HIV) (Barre-Sinoussi et al., 1983), it would take another two years to identify an HIV-like virus in a nonhuman primate (Daniel, et al., 1985) and at least another 15–20 years to sort out the true origins of HIV (Gao et al., 1999). The tropical equatorial forest of West Central Africa is rich in a wide variety of nonhuman primate species which have been living in co-existence and evolving for millions of

The Role of Animals in Emerging Viral Diseases. http://dx.doi.org/10.1016/B978-0-12-405191-1.00009-0

years. Retroviruses have been present in the germ lines of primates for at least the latter half of their existence such that some retroviruses are termed endogenous—i.e., viruses that have integrated into their host's genome and are passed as one of two alleles to offspring. The gray mouse lemur and the fat-tailed dwarf lemur appeared to have diverged from the primate lineage about >40 million years ago (mya) and, thus far, these are the earliest examples where the germline did not harbor two prosimian endogenous retroviruses (Gifford et al., 2008; Gilbert et al., 2009).

Molecular time clocks estimating the evolutionary timescale of lentiviruses are good predictive markers to track established zoonotic introductions of these viruses from nonprimates to humans. The "modern" exogenous pre-SIVs appear to have been introduced into the Old World monkeys about 6–10 mya (Fabre et al., 2009). The absence of these SIVs in New World monkeys suggests that the introduction of lentiviruses into primates occurred following the split between the Old and New World monkeys, which corresponds to the breakup of the southern Gondwana and northern Laurasia continents about 60 mya. Dating of SIV's introduction into most Old World monkeys is difficult due to the reduced evolutionary rates observed with adaptation and attenuation of SIV infections. The timescale of SIV attenuation and adaptation in a new primate host following a zoonotic jump is largely unknown but could range between decades and millions of years. The Bioko Island of Equatorial Guinea appeared to have separated from West Africa about 10,000 years ago and yet SIV is found in its indigenous nonhuman primate population (Worobey et al., 2010). This chapter dwells on the historical path HIV took from its original hosts (nonhuman primates) to humans.

CLASSIFICATION, GENOME STRUCTURE, AND REPLICATION

Lentiviruses cause chronic and persistent infections notably in primates, bovines, felines, sheep, and horses. SIVs that infect nonhuman primates can be classified into two main groups: those that infect apes (chimpanzees and gorillas), and those that infect monkeys (sooty mangabey, red-capped mangabey, L'Hoests and others). Table 9.1 outlines the geographical distribution of the various species of chimpanzees, gorillas and monkeys that inhabit the forests of Africa, and that are also the natural reservoirs of SIV. Almost 30 different primate species representing at least seven lineages are infected by different lentiviruses. As shown in Figure 9.1, these primate lineages include members of the genus *Pan* (chimpanzees; (Keele et al., 2006) and *Gorilla* (gorillas)), *Cercocebus* (sooty mangabeys; (Hirsch et al., 1989)); *Cercopithecus* (The Guenons; (Courgnaud et al., 2002)), *Chlorocebus* ((African green monkeys) (Allen et al., 1991; Hirsch et al., 2004)), *Mandrillus* ((mandrills and drills), (Clewley et al., 1998)), *Colobus* and *Pilocolobus* ((colobus monkeys, (Courgnaud et al., 2001)). The *Macaca* (macaques) and other Asian great apes such as the orangutans are not natural carriers of SIV, suggesting again that SIVs most likely originated in the African monkey population long before the separation of the Old and New World monkeys. The closest relatives of HIV-1, SIVcpz and SIVgor were isolated from the African greater apes, specifically two members of the *Pan* chimpanzee subspecies and *Gorilla*, respectively. Until recently, there were some suggestions that chimpanzees might not be the natural hosts of SIVs. This xenotropic theory relates to the low SIVcpz seroprevalence among chimpanzees (~6%) compared to higher prevalence of SIVs in the natural monkey hosts. For example, at least 50% of the red-capped mangabey monkeys are infected with SIVrcm (Table 9.1; Aghokeng et al., 2010). Recent studies using fecal and urine samples from *P. troglodytes troglodytes* living in different regions of Africa suggest that prevalence of SIV among these apes was much higher than previously thought (Table 9.1).

The structural make-up of Lentiviral genome is complex with a set of unique accessory genes in addition to the three genes *gag*, *pol*, and *env* encoding the main structural and enzymatic genes found in all retroviruses. Most HIVs and SIVs are very similar in genomic structure with slight shifts in the positioning and type of some accessory genes (Figure 9.2). The various lineages of SIVs (Table 9.1) share about 40–50% identity in Gag and Pol proteins (Hirsch & Johnson, 1994). The accessory genes *vif*, *rev*, *vpr*, and *nef* are common to all lentiviruses. The role of *tat* (trans-activating transcriptional factor) and *rev* (regulator of virus gene expression) were initially defined as transcription and RNA regulatory factors in the mid-80s and soon after the discovery of this new human lentivirus (Haseltine et al., 1984; Rosen et al., 1985; Sodroski et al., 1984). As outlined below, the role(s) of the other accessory proteins *vif*, *vpr*, *vpx*, and *nef* were more poorly defined in susceptible cells until the discovery of host restriction factors that appear to restrict exontropic transfers of viruses between related species. This type of host restriction and counter-evolution in lentiviruses is described below. Two other accessory genes, *vpx* and *vpu*, are specific to different strains of immunodeficiency viruses. Specifically, *vpx* is found in HIV-2, SIVsmm, SIVmac, SIVstm, SIVrcm, SIVmnd-2 and SIVdrl while *vpu* is found in HIV-1, SIVcpz, SIVgor, SIVgsn, SIVmon, SIVmus and SIVden respectively (Figure 9.2a and b). Some SIVs, for example SIVagm and SIVmnd-1, possess neither *vpu* nor *vpx* (Figure 9.2c). Some of the functions of HIV-1 *vpr* are performed by *vpx* while *vpu* functions are performed by *nef* or *env* in viruses that lack *vpu* (Sauter et al., 2009).

SIV like all retroviruses encodes a reverse transcriptase enzyme for genome replication. This enzyme lacks proofreading activity and accounts for the extreme heterogeneity of the virus. The RT error rate is approximately 3×10^{-5} mutations per base pair per cycle (Mansky, 1998; Mansky & Temin, 1995; Preston et al., 1988) coupled with a high viral turnover rate of about 10^{10} viral particles/day in the infected human. Because HIV carries two copies of genomic RNA, strand switching during reverse transcription/genome replication could lead to recombination, another process to increase genetic diversity

TABLE 9.1 Nonhuman Primates Naturally Infected with SIV

Common name	Species	Virus strain	Geographic location	Seroprevalence (%)	Cross species transmission
Great Apes					
Central Chimpanzee[1]	*Pan troglodytes troglodytes*	SIVcpzPtt	Central Africa	5.9 (0–32)	Humans
Eastern Chimpanzee	*Pan troglodytes schweinfurthii*	SIVcpzPts	East Africa	13.5	Not reported
Gorillas	*Gorilla gorilla*	SIVgor	Central Africa	1.6 (0–4.6)	Humans (?)
Old World Monkeys					
Sootey mangabey[1]	*Cercocebus atys*	SIVsmm	West Africa	20–58	Humans, macaques
Red-capped mangabey	*Cercocebus torquatus*	SIVrcm	West central Africa	50 (25.3–74.6)	Experimental transmission
Mandrill[1]	*Mandrillus sphinx*	SIVmnd-1	West central Africa	14	Not reported
		SIVmnd-2	West central Africa	50	Rh macaque experimental transient infection
Drill	*Mandrillus leucophaeus*	SIVdrl	West central Africa	33.3 (21.5–55.7)	Not reported
Talapoin (Northern)	*Miophithecus ogouensis*	SIVtal	Central Africa	16 (9.3–28)	Rh macaque experimental transient infection

Mantled colobus	Colobus guereza	SIVcol	Central Africa	17.9 (11–28)	Not reported
Western Red colobus	Piliocolobus badius	SIVwrc	West Africa	40	Not reported
Olive colobus	Procolobus verus	SIVolc	West Africa	40	Not reported
Grivet	Chlorocebus aethiops	SIVagmGri	East Africa	>50	Not reported
Vervet[2]	Chlorocebus pygerythrus	SIVagmVer	East and South Africa	>50	Baboons (natural)
					White crowned mangabey (captive)
					Pig-tailed macaques (experimental)
Sabaeus	Chlorocebus sabaeus	SIVagmSab	West Africa	>60	Patas monkeys (natural)
					Rh macaques (experimental)
Tantalus	Chlorocebus tantalus	SIVagmTan	Central Africa	50 (18.7–81.2)	Not reported
Greater spot-nosed monkey	Cercopithecus nictitans	SIVgsn	Central Africa	1(0.5–1.9)	Chimpanzee
Blue monkey	Cercopithecus mitis	SIVblu	Central East Africa	>60	Not reported
Mona monkey	Cercopithecus mona	SIVmon	West Central Africa	Unknown	Not reported
Dent's mona	Cercopithecus denti	SIVden	West Africa	10	Not reported

Continued

TABLE 9.1 Nonhuman Primates Naturally Infected with SIV—cont'd

Common name	Species	Virus strain	Geographic location	Seroprevalence (%)	Cross species transmission
Old World Monkeys					
Moustached monkey	*Cercopithecus cephus*	SIVmus	Central Africa	1 (0.5–1.9)	Unknown
Red-eared monkey	*Cercopithecus erythrotis*	SIVery	Central Africa	unknown	Not reported
Red-tailed monkey	*Cercopithecus ascanius*	SIVasc	Central Africa	Unknown	Not reported
DeBrazza monkey	*Cercopithecus neglectus*	SIVdeb	Central Africa	40.5 (26.3–56.5)	Not reported
Sykes monkey	*Cercopithecus albogularis*	SIVsykes	East Africa	30–60	Rh macaques (experimental) transient infection
L'Hoests monkey[1]	*Cercopithecus l'hoesti*	SIVlho	East Africa	50	Pig tailed macaques (experimental)
Sun-tailed monkey	*Cercopithecus solatus*	SIVsun	Central Africa	unknown	Sun tailed macaques (experimental)
Preussis monkey	*Cercopithecus preussi*	SIVpre	Central Africa	unknown	Not reported

[1]AIDS or AIDS-like pathogenicity reported in Captivity.
[2]AIDS in STLV co-infected monkey

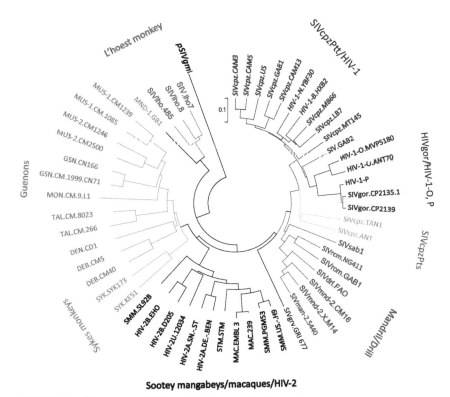

FIGURE 9.1 Illustration of HIV and SIV lineages. Phylogenetic tree representating the polymerase gene of dominant primate lentiviruses. HIV-1 group M clusters with SIVcpzPtt while groups O and P are more closely related to SIVgor; and HIV-2 is closely related to SIVsmm. The tree was rooted with sequences from SIVgml, the oldest known lentivirus *pol* sequence, which was obtained from the grey mouse lemur. The different colors represent the genus of various nonhuman primates.

(Preston et al., 1988; Temin, 1993). As described below, HIV-1 recombination has led to the generation of more than 40 "circulating recombinant forms" (CRF) and possibly millions of "unique recombinant forms" (URF) in the HIV pandemic.

Recombination has also played a very significant role in the evolution and cross species transmission of SIV. SIVcpz, the source and immediate progenitor of HIV-1, is a mosaic between SIVs from the red-capped mangabey (*Cercocebus torquatus*) and the greater spot-nosed monkey (*Cercopithecus* species) (Bailes et al., 2003). Close examination of the full length genomic structure from SIVcpz shows that its 3' end (*nef* gene and 3'LTR) are closely related to SIVrcm while the *vpu*, *tat*, *rev* and *env* genes cluster closely with SIVgsn (Bailes et al., 2003). It is also possible that SIVrcm, SIVgsn, or the primordial SIVcpz recombined via sequential infection of different primates that led to the stabilization of the current SIVcpz genome. Higher potential for this SIV

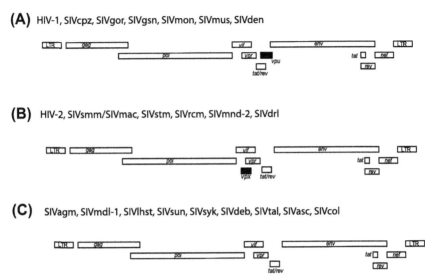

FIGURE 9.2   Genomic organization of simian and human immunodeficiency viruses.
(A) SIVs and HIV-1 have similar genomic organization but the latter encodes the *vpu* acces-
sory gene. (B) HIV-2 and some of its closely related lineages carry a *vpx*. (C) Some SIV strains
lack either *vpu* or *vpx*. See text for details.

recombination in chimpanzees relates to their hunting of *Cercopithecus* species (*C. nictitans*, *C. cephus*, and *C. mona*, respectively), known to harbor the
SIVgsn, SIVmus, and SIVmon lentiviruses.

GEOGRAPHICAL DISTRIBUTION AND ORIGIN OF SIVS

The geographical distribution of nonhuman primates has been a determining factor in the spread of SIV and HIV infections in Africa. Central Africa (specifically,
the Congo River basin) is the epicenter of the HIV-1 pandemic , while HIV-2 has
been largely restricted to the coast of west Africa (Guinea Bissau, Sierra Leone,
Liberia, Cote d'Ivoire). As mentioned above, humans are frequently exposed
to SIVs carried by most Old World nonhuman primates such as monkeys and
apes (chimpanzees and gorillas). Monkeys are comprised of numerous species
as shown in Figures 9.1 and 9.2; Table 9.1. Chimpanzees on the other hand
can be divided into two main species: *Pan troglodytes* (common chimpanzee)
and *Pan paniscus* (bonobos). Based on mitochondrial DNA sequences, four
subspecies of the *Pan troglodytes* have been defined, namely: *Pan troglodytes*
troglodytes (central), *Pan troglodytes verus* (western), *Pan troglodytes ellioti*
formerly called *Pan troglodytes vellerosus* (Nigeria-Cameroon) and *Pan troglodytes schweinfurthii* (eastern chimpanzee). *P.t.troglodytes* and *P.t.shweinfurthii*
are reported to have diverged more than 440,000 years ago. The *Pan paniscus*
are mostly located in the Democratic Republic of Congo (DRC). Two forms
of gorillas, namely *Gorilla gorilla* and *Gorilla beringei*, have been described

in central and east Africa, respectively. Except for the baboons, patas and the *P.t. vellerosus* chimpanzees, SIV infection has been detected by serological surveys in nearly all species of nonhuman primates in Africa (Aghokeng et al., 2006; Aghokeng et al., 2010).

Enzyme-linked immunosorbent assays (ELISAs) now contain antigens from diverse HIV and SIV strains which are also cross reactive to SIV-specific antibodies from nonhuman primates. Serological surveys using SIV specific antigen ELISAs have led to the identification of more than 40 nonhuman primate species (Aghokeng et al., 2006; Aghokeng et al., 2010). To complement this sensitive serological screen, recent technological advancements have facilitated the noninvasive collection of dry blood spots, fecal and urine wastes from wild-living nonhuman primates. Using blood derived from bush meat samples and pets in southern Cameroon, SIV prevalence was shown to vary between sampling sites as well as differ widely among monkey species (Aghokeng et al., 2010; Peeters, Courgnaud et al., 2002b). Interestingly, the most frequently hunted primate species had a low SIV seroprevalence at <8.5% (Peeters et al., 2002).

Several studies have now mapped the distribution of SIVs via analyses of nucleic acids derived from fecal and urine samples collected from 7,000 wild-living chimpanzees at 90 different sites stretching across west, central and east Africa (Santiago et al., 2003; Santiago et al., 2002; Worobey et al., 2004; Keele et al., 2006; Van & Peeters, 2007; Rudicell et al., 2010; Rudicell et al., 2011; Li et al., 2012). SIVcpzPtt is present in *P.t. troglodytes* at high endemic rates (>7%), which appear to be the reservoir for at least five independent zoonotic SIV introductions into humans and gorillas, leading to successful founder events. Although *P.t. schweinfurthii* was also found to carry SIV, there is no evidence for a successful transmission and founder event in humans (Li et al., 2012) (Tables 9.1 and 9.2). The other two chimpanzee species, *P.t. ellioti* and *P.t. verus*, do not harbor a "modern" SIVcpz suggesting that chimpanzees may have only acquired this current SIV after the divergence of this ape species approximately 440,000 years ago. Most large rivers in west and central Africa such as the Cross River (Nigeria-Cameroon Border), Sanaga River (running from Cameroon to Nigeria), the Congo River (DRC and Congo), Ubangui River (Gabon to DRC) have acted as natural boundaries between the habitats of chimpanzees and barriers for SIV transmission, due to the fact that chimpanzees are poor swimmers. For example, the Sanaga River, which runs between Eastern and Western Cameroon into Nigeria likely prevented contact between the Nigeria-Cameroon chimpanzee species (*P.t. ellioti*) with SIVcpz-infected *P.t. troglodytes* while *Pan paniscus* avoided SIVcpz from other ape species due to their isolation to the Congo basin found south of the Congo river/tributaries and north of the Kasai river. In general, the rates of SIVcpz infection vary from 30–50% between eastern and western chimpanzees in some communities. SIVcpz may be very rare or completely absent from some isolated troops of chimpanzees, especially in East Africa (Uganda and Kenya). Contrary to SIVcpz in chimpanzees, SIVsmm and SIVagm are more widely spread geographically and infect

TABLE 9.2 Important Milestones in SIV Nonhuman Primate to Human Cross-species Transmission

Year	Event / Finding	Reference
1983	Isolation of HIV-1	Barre-Sinoussi et al., 1983
1985	Sequence of HIV and genetic diversity reported	Wain-Hobson et al., 1985; Hahn et al., 1985
1986	Isolation of HIV-2	Clavel et al., 1986
1989	Partial characterization of SIVgab from Gabon	Peeters et al., 1989
1989	SIVsmm isolation and closeness to HIV-2	Hirsch et al., 1989
1990	Description of chimpanzee virus related to HIV-1	Huet et al., 1990
1992	Isolation of SIVgab from a wild captured chimpanzee	Peeters et al., 1992
1994	Isolation of first group O viruses	Gurtler et al., 1994
1996	Discovery of CXCR4 and CCR5 as HIV co-receptors	Feng et al., 1996; Dragic et al., 1996
1997	Ancient group O (1970s) from frozen Norwegian samples	Jonassen et al., 1997
1998	Sequence of an ancient (1959) HIV-1 sample from the DRC	Zhu et al., 1998
	Identification of HIV-1 group N	Simon et al., 1998
1999	First proof that HIV-1 originated from SIVcpzPtt	Gao et al., 1999
	Description of group M/O recombinant viruses in 2 patients	Peeters et al., 1999; Takehisa et al., 1999
2000	Description of SIVcpz env sequences similar to HIV-1 group N	Muller-Trutwin et al., 2000
2002	SIVcpz in wild chimpanzees	Santiago et al., 2002
	Discovery of APOBEC3G as an HIV host restriction factor	Sheehy et al., 2002
2003	SIVcpz described as a recombinant between SIVrcm and SIVgsn	Bailes et al., 2003
2004	Discovery of TRIM5-alpha, a factor that prevents HIV from infecting monkey cells	Stremlau et al., 2004

TABLE 9.2 Important Milestones in SIV Nonhuman Primate to Human Cross-species Transmission—cont'd

Year	Event / Finding	Reference
2005	SIV confirmed in chimpanzees living in the wild in Equatorial Africa	Nerrienet et al., 2005
2006	Confirmation that HIV-1 group N and M exists among chimpanzees in Equatorial forests of Africa	Keele et al., 2006
2006	HIV-1 group O-like sequences found among Gorillas in Eastern Cameroon	Van et al., 2006
2008	Sequencing of a 1960 DRC paraffin-embedded tissue HIV strain	Worobey et al., 2008
2009	Identification of gorilla-like group O sequences in humans and named "group P"	Plantier et al., 2009
	Tetherin as a deterrent in SIV cross species transmission	Sauter et al., 2009
2012	SIV is more prevalent among *P.t.schweinfurthii* than *P.t.troglodytes* in nature	Li et al., 2012

sooty mangabeys and African green monkeys at higher rates (Phillips-Conroy et al., 1994; Santiago et al., 2005). The jump of SIVs from monkeys (SIVrcm and SIVgsn) to chimpanzees most likely occurred through predatory activities. Nonhuman primates are reported to fight among each other as they struggle for food and survival in their communities. Chimpanzees hunt and eat smaller mammals including monkeys within their community. The increase of infection with age and lack of SIVcpz in most chimp infants confirms that blood-to-blood contact via aggression and heterosexual contacts are the primary means of transmission as opposed to vertical transmission. Coincidentally, the geographical region occupied by chimpanzees overlaps with that of the various *Cercopithecus* species carrying SIVgsn and SIVrcm. Again, the mixing and recombination of the primordial SIVcpz, SIVgsn, and SIVrcm within the *Cercopithecus* and *Pan troglodytes* species likely led to the modern SIVcpz, which in turn infected and adapted to the human species to give rise to HIV. This hypothesis is supported by the identical mosaic genomic structure observed for all the 30 sequenced SIV strains described to date.

These findings also suggest that a distinct series of cross species transmission and recombination events between several primate species were required to establish an SIV sub-species capable of establishing a successful zoonotic jump to humans. It is interesting to note that SIVcpz carried by *P.t.schweinfurthii* has not been observed in humans and may not have undergone the same evolutionary/cross species transmission process to infect humans. Recent studies have shown that the natural SIV prevalence in *P.t. schweinfurthii* is about 2.5 times higher compared to *P.t. troglodytes* (13.5 versus 5.9%; Table 9.1) (Li et al., 2012). The inability of SIVcpz from *P.t.schweinfurthi* to jump to humans may also be due to differential sensitivity to restriction factors in humans as described below (Li et al., 2012). In contrast to previous perceptions, SIVcpz is in fact pathogenic to chimpanzees reducing the health, reproduction, and life span of this natural host (Keele et al., 2009). Thus, in addition to representing a potential source for human infection, SIVcpz is a serious threat to the health of the chimpanzee populations living in the wild (Rudicell et al., 2010).

SIVs infecting gorillas (SIVgor) were only recently discovered in Southern Cameroon where about 200 fecal samples were tested (Van et al., 2006). SIV in gorillas came as a surprise considering that these apes are herbivores. To confirm this finding, a broader screening of about 2500 gorilla samples in central Africa indicated that highland gorillas at 4 of 30 different sites had a 5% SIV prevalence, i.e., lower than that observed in chimpanzees (Neel et al., 2010). Phylogenetic analyses of SIV sequences from gorillas showed a close but distinct clustering with SIVcpz sequences from chimpanzees. In particular, SIVgor clustered with SIVcpz from *P.t. troglodytes*, suggesting a likely jump from chimpanzees to gorillas approximately 100 to 200 years ago (Takehisa et al., 2009).

SIVsmm, the source of HIV-2, is highly prevalent (20–60%) in the sooty mangabeys of West Africa. In Liberia and Sierra Leone, SIVsmm prevalence among sooty mangabeys living in the wild is 22% but only 4% in those kept as household pets. Substantially lower SIV prevalence of those sooty mangabeys in captivity as compared to those in the wild suggest either sampling from specific wild mangabey populations or the possible resolution of SIVsmm infection in sooty mangabeys. There is no documented evidence for resolution of SIV or HIV infection in its natural host (Table 9.1; Silvestri, 2005). The high genetic diversity of HIV-2, especially in Sierra Leone, suggests frequent human contact with sooty mangabeys, which are also housed domestically as pets. While the prevalence of SIVsmm may be high in the Tai forests of Cote d'Ivoire, these monkeys are now rare in Liberia and Sierra Leone and reportedly extinct in Senegal, Guinea Bissau and some parts of Guinea Conakry (Chen et al., 1997; Chen et al., 1996). These findings support the possible transmission of SIVsmm to humans via the sooty mangabey pets. Interestingly, the initial SIVsmm adaptation and HIV-2 founder event is suspected to have occurred in Guinea Bissau, which is also the location of the highest HIV-2 prevalence (da Silva et al., 2008).

EMERGENCE, CROSS SPECIES TRANSMISSION AND CLASSIFICATION OF HIV

The most significant hint that HIV could have jumped from nonhuman primates came with the discovery in 1986 that HIV-2 from West Africa was antigenically distinct but morphologically similar to HIV-1, originating from East/Central Africa (Clavel et al., 1986). Shortly after this discovery, the origin of HIV-2 was linked to an SIV strain identified in a macaque that was captured from the wild (Chakrabarti et al., 1987). As described above, macaques are of Asian descent and, therefore, the report that they were carriers of SIV was surprising given that HIV-2 was localized to West Africa. It was later discovered that these captive macaques were unknowingly infected with blood from sooty mangabeys (Apetrei et al., 2005). The discovery that HIV-1 was transmitted from chimpanzees (Huet et al., 1990) and HIV-2 from sooty mangabeys (Hirsch et al., 1989) has been instrumental in tracing the evolutionary path of SIVs both in the human and nonhuman primate ancestry.

As described above, SIVcpz are the precursors of HIV-1 while HIV-2 likely originated from monkeys infected with SIVsmm. HIV-1 is subdivided into groups M, N, O, and P, while HIV-2 is divided into groups A to H. The various groups of HIV-1 and HIV-2 likely represent distinct transmission events from nonhuman primates to humans. HIV-1 group M was the first to be discovered and has been the most successful in establishing a human pandemic. One transmission event to gorillas led to the establishment of SIVgor, which is the closest SIV relative to HIV-1 groups O and P (Figure 9.1). Groups O, P, and N have been limited to west central Africa, all have their epicenter in Cameroon, and are responsible for less than 1% of HIV infections in Cameroon (<0.05% of all HIV-1 worldwide). As indicated by its mosaic genomic sequence, HIV-1 group N likely arose from a recombination event between a SIVcpz group N-like strain and an SIV progenitor of HIV-1 group M, which might have infected a chimpanzee species before being transmitted to humans (Simon et al., 1998).

With a 30–40% homology to HIV-1, HIV-2 was initially divided into subtypes A through H. However, only groups A and B were successfully transmitted from monkeys to humans. Groups C-H might have encountered a "dead end" as only single cases have been described and were never transmitted between humans or at least beyond the identified case (Gao et al., 1994; Chen et al., 1996; Chen et al., 1997; Santiago et al., 2005). HIV-2 group A dominates most of West Africa (Guinea Bissau and The Gambia) while B is found mostly in Cote d'Ivoire (Pieniazek et al., 1999). The geographical location of these HIV-2 strains also overlaps with the habitat of sooty mangabeys, the natural host of SIVsmm (Gao et al., 1992; Chen et al., 1996). The prevalence of HIV-2 has been declining steadily as the epidemic progresses. This is due in part to reduced transmission rates and poor adaptation of HIV-2 to human infection. HIV-2 infections support viral loads generally several folds lower than that in HIV-1 patients

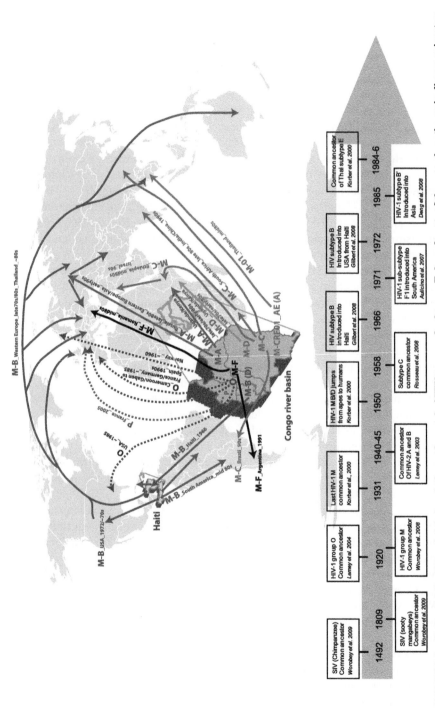

FIGURE 9.3 Estimated time line of global evolution and spread of HIV types and subtypes. Enlarged parts of the map show the main disease epicenters. The time line indicates the key events in SIV, HIV-1 groups M, N, O, P and HIV-2 evolution. *Figure was adopted with modifications from Tebit and Arts,* Lancet Infectious Diseases, *2011, volume 11, pages 45 to 55.*

(Popper et al., 1999). Lower viral loads correlate with very slow progression to AIDS as well as poor transmission efficiency (Rowland-Jones & Whittle, 2007).

The development of computer-based molecular clock models to estimate the time to the most recent common ancestors (tMRCA) for all forms of HIV (HIV-1 groups M, O, N and HIV-2) has been instrumental in understanding the patterns of HIV evolution. The tMRCA of HIV-1 groups M, O and N were estimated for the years 1908 (range 1884–1924), 1920 (1890–1940) and 1963 (1948–1977), respectively; while HIV-2 groups A and B are dated to originate in 1932 (1906–55) and 1935 (1907–61), respectively (Korber et al., 2000; Lemey et al., 2004; Wertheim & Worobey, 2009) (Figure 9.3). Based on these dates, it is clear that SIV transmission and establishment of the HIV lineages in humans occurred during the turn of the twentieth century, and notably HIV-1 group N is the youngest of the HIV lineages. However, the possibility of a deeper history of group N cannot be excluded given the sparse sampling. Dating of HIV-1 group P is the least defined (between 1845–1989) due to the presence of only a single reported case in humans (Sauter et al., 2011a; Figure 9.3). Analysis for SIV tMRCA is only possible with SIVsmm and SIVcpz, the precursors of HIV-2 and HIV-1 respectively. Assuming an unbiased molecular clock, the introduction of the primordial SIV in chimpanzees was estimated to be 1492 (1266–1685), while 1809 (1729–1875) was the estimated tMRCA for SIVsmm (Wertheim & Worobey, 2009) suggesting that this sooty mangabey virus was several centuries older than SIVcpz. The young age of SIV in these monkeys contrasts with the other ancient retroviruses found in primates. For example, the simian foamy virus appears to be at least 30 mya, based on congruence between the viral and primate host phylogenies (Switzer et al., 2005). Lentiviruses, to which SIV belongs, are likely 40–50 mya based on recent evidence of defective endogenous lentiviruses in rabbits and lemurs (Gifford et al., 2008; Katzourakis et al., 2009; van der Loo et al., 2009). One explanation of this young tMRCA for SIV is the loss of old lineages with extinction of some SIV-infected primate species to be later replaced by co-evolution of a new SIV infecting a new species or sub-species of primates. This process would involve the loss of deep SIV lineages and replacement by younger ones (Wertheim & Worobey, 2009).

SIV ADAPTATION TO THE HUMAN HOST

Transmission of a virus to a new host usually comes at a cost. As described later, viruses have to adapt to host conditions for infection. In the case of transmission between hosts of the same species, viruses such as HIV-1 must survive in the face of both innate and acquired immunity. However, HIV-1 has adapted to utilize a series of host factors to replicate in human cells. In the case of HIV-1, it first requires cellular receptors CD4 and CCR5 (or CXCR4) to gain entry into the host cell (Feng et al., 1996; Dragic et al., 1996). Replication of the viral genome then requires a supply of deoxynucleotide triphosphates that are synthesized by various host enzymes (Lori et al., 1992). Uncoating of the core and integration of the viral DNA genome into host chromosomes involve multiple factors (e.g., cyclophillin, LEDGF/p75), most of which are still unknown. Once integrated, efficient HIV-1 transcription of mRNA requires RNA pol II in combination with pTEFb. Every sequential step in the HIV lifecycle involves host factors, which makes this virus an obligate parasite (like all other viruses). A successful zoonotic transmission of SIVcpz from *P.t.troglodytes* to humans would require a series of adaptations, some of which may be present at low frequency in the intra-animal SIV quasispecies in the donor chimps. In many cases, chimpanzee host factor interacting with the SIV proteins for efficient replication may share high sequence/structural identity with the human homolog. Thus, adaptation of SIV protein/constituent may not be necessary. For example, human and chimpanzee Pol II enzymes are nearly identical (<0.7% nucleotide variation), so minimal adaptation may be necessary for initiation of transcription (Watanabe et al., 2004). Variation in other factors such as CCR5 is much greater, suggesting that chimpanzee CCR5 has been influenced by a selective sweep whereas human CCR5 has been subject to balancing selection (Wooding et al., 2005). As a consequence, only specific SIV env sequences may encode for glycoproteins with enhanced entry efficiency and survival in humans (Finzi et al., 2012).

SIV transmitted to humans may require multiple adaptations that are not readily found in the quasispecies due to low fitness or a replication defect in chimpanzee cells. Thus, a low level of SIV replication in human cells is required to generate a human-specific mutation that leads to adaptation. By scanning the genomes of SIVcpz and HIV-1, Wain et al. (2007) discovered a possible adaptation mutation at amino acid residue 30 in the SIV p17 matrix protein. A methionine at MAp17 position 30 is the main amino acid in SIVcpzPtt and SIVgor while Lys30 is found in SIVcpzPts. With the exception of HIV-1 subtype C which has a Met30, all other group M viruses have a Lys/Arg30. Interestingly, SIVs bearing the Met30Lys in MAp17 were found to replicate better in human than chimpanzee PBMCs suggesting the adaptation of SIVcpzPtt following the jump to humans (Wain et al., 2007). HIV-1 subtype C has the lowest replicative fitness of all isolates of the HIV-1 group M subtypes, which could be linked to poor adaptation of this HIV-1 lineage from SIVcpzPtt. It is unclear why SIVcpzPts did not establish a human epidemic but it is important to note that Lys/Arg30 is only one evolutionary change linked to many others necessary for SIV survival and adaptation in humans. One of the two HIV-1 group P strains (06CMU14788) harbors the "human-specific" Met30Lys in MAp17 (Vallari et al., 2011). The impact of this Met or Lys30 on replicative fitness of these two group P isolates is still unknown. In future, a study of this mutation on SIV and HIV-1 replication in human and chimpanzee CD4+ cells may provide clues on the possible adaptation of SIV in humans.

Entry of the virus into susceptible cells is the first "gatekeeper" to prevent new zoonotic introductions. Although both HIV and SIV use CD4 as the primary receptor for viral entry, SIV is generally less dependent on CD4 for infection than HIV-1 and can efficiently infect cells with low levels of surface CD4 expression (Edinger et al., 1997; Puffer et al., 2002). Keeping this in mind, Finzi et al. (2012) recently showed that there are lineage-specific differences on gp120 stoichmetry in SIV and HIV particles that accommodate for differences in CD4 avidity/affinity. SIV may harbor as many as 100 glycoprotein spikes per virus particle whereas less than a dozen gp120/gp41 trimers are present on an HIV-1 particle. CD4 binding by gp120 has also been linked to the size of the amino acid residue at position 375. Both SIVs and HIV groups (N, O, and P) have large amino acid residues (Trp, Met, Tyr, His, and Phe) while the pandemic form of HIV-1, group M, has a serine at this position (Finzi et al., 2012; Xiang et al., 2002). With a Ser at position 375, gp120 is less likely to assume a conformation conducive to human CD4 binding, which is necessary for host cell entry (Xiang, Kwong et al., 2002). This argument could be one reason why SIVcpz Ptt does not frequently jump into humans and why HIV-1 groups N and O never established global epidemics.

Strong and rapid adaptation of SIV for survival in a human body and replication in human cells is a major factor for the evolution of HIV and the establishment of this pandemic. Adaptations are difficult to track due to the complex evolutionary pathways converging on a common phenotype. SIV likely navigated through the initial fitness valley (cross species transmission) to emerge on the next ridge (human host) using many distinct evolutionary pathways through this Wright's fitness landscape. It is quite possible that several zoonotic introductions of SIVcpzptt into humans resulted in different evolutionary pathways, which radiated into the current HIV-1 group M diversity. Again, at least four zoonotic transmissions from four different SIV lineages in primates were responsible for establishing HIV-1 group M, N, O and P. The SIV transmission establishing group M was much more adapted to human replication and transmission, leading to rapid evolution into a "star" phylogeny. With divergence of group M, clades or subtype were established that share the same root/origin but ultimately "win" or "lose" in the epidemic based on relative virulence balanced with transmission efficiency. In the case of HIV-2, poor human adaptation of SIVsmm to humans likely led to establishment of an HIV with low pathogenicity. Prolonged infection may have aided in initial HIV-2 spread in humans but poor transmission efficiency in comparison to HIV-1 will lead to the eventual demise of this epidemic. A similar reduction in prevalence (albeit slower in time) is predicted for HIV-1 group M subtype D (M-D), which appears more virulent than the other HIV-1 isolates (Kaleebu et al., 2002; Baeten et al., 2007) while maintaining similar transmission efficiencies between humans. In contrast, the slow expansion of HIV-1 M-C in the global epidemic (compared to all group M subtypes) could be related to slow progression to disease in subtype C infected individuals coupled with a slight increase in transmission efficiency. In other words, HIV-1 M-C was either transmitted from chimpanzees (e.g., with a methionine at position 30 in the matrix p17) as a less virulent SIV or evolved to lower virulence in humans. As a consequence, HIV-1 M-C infected individuals may live longer and transmit the virus to more human hosts than other HIV-1 M strains.

A few studies have now compared the ex vivo models for pathogenic and transmission fitness of different group M subtypes. A fitness rank order has now been established for group M HIV-1 isolates based on thousands of pairwise competitions between primary HIV-1 isolates performed in human peripheral blood mononuclear cells, T cells, and macrophages. This potential surrogate for HIV-1 virulence suggests that HIV-1 subtypes B, D, and F (subtype BDF) have greater replicative fitness than subtype A, CRF01_AE, CRF02_AG (super-subtype A) and all are more fit than subtype C HIV-1 isolates. Interestingly, the trends in HIV subtype prevalence within the worldwide epidemic suggests a slow expansion of subtype BDF infections whereas subtype C HIV-1 has come to dominate the epidemic. In the early 1990s, HIV-1 subtype C was isolated to sub-Saharan Africa and was responsible for only 10–15% of all HIV-1 infection (<1 million subtype C infections). Currently, >15 million humans are infected with subtype C, i.e., >50% of all HIV-1 infections. Interestingly, this rapid spread of HIV-1 subtype C has been observed with several founder events in different regions around the world, namely, China, India, Brazil, and South Africa. Several studies now suggest a correlation between ex vivo "pathogenic" fitness of HIV-1 and virulence in patients. Higher virulence of HIV-1 is analogous to faster disease progression to AIDS. In contrast to *ex vivo* pathogenic fitness that can vary between the subtype BDF, super-subtype A, and subtype C, the *ex vivo* models for transmission fitness suggest that nearly all wild type HIV-1 isolates of different subtypes have an equal but low efficiency of establishing infection at sites of transmission. These transmission fitness studies involved infecting and competing different HIV-1 isolates in tissue explants found at the sites of initial virus exposure (e.g., human rectal, penile, cervical and vaginal tissue). Based on these *in vivo* and *ex vivo* observations on transmission efficiency and virulence, the classic models of $R°$ can be tested. Subtype C will continue to expand faster than any other HIV-1 subtype in the epidemic due to lower virulence while maintaining equal transmission efficiency. This of course assumes that the sexual habits of humans, i.e., the opportunity for transmission, are somewhat similar around the world. In other words, lower virulence of subtype C versus subtype D HIV-1 may lead to slower disease progression, longer asymptomatic periods of infection and, thus, more opportunity to transmit at equal efficiency.

RESTRICTION FACTORS PREVENTING CROSS-SPECIES TRANSMISSION OF RETROVIRUSES

Almost immediately following the discovery of HIV and its possible origins from nonhuman primates, studies have explored the susceptibility of different mammalian species to different primate lentiviruses. Part of this research attempted to define appropriate animal models for HIV-1 disease but mice, rabbits, and multiple other mammalian species were resistant for a multitude of reasons that were teased out during the next 25 years of research. Suffice it to say, HIV is restricted from infecting a large number of mammalian species at multiple steps in the virus lifecycle. The most common and obvious block is the inability to utilize the CD4 and CCR5 homologs aside from those in Old World monkeys. However, even when this block is overcome, new incompatibilities are found in most steps that involve a host factor. For example, if mouse cells are designed to express human CD4 and CXCR4 (or CCR5), a single amino acid difference between murine and human Cyclin T1 is sufficient to block HIV-1 Tat transactivation of HIV-1 mRNA transcription. However, most of these blocks are due to millions of years of divergent mammalian evolution and were not necessarily related to selection to prevent lentiviral infections.

Nearly every multiple and single cell organism has developed specific innate immune mechanisms to prevent or limit viral infections. In mammalian species, defense against pathogens is derived from both an innate and acquired immune system. This section in the chapter will only deal with specific arms of the intracellular innate immunity that primarily target lentiviruses. Interestingly, some of the most effective restriction mechanisms within host cells were only discovered via the counter-restriction mechanism found within lentiviruses. A detailed description of these restriction and counter-restriction mechanisms has been recently reviewed (Harris et al., 2012). Suffice to say that several of the accessory genes in primate lentiviruses express proteins to block a cellular restriction factor. Genes encoding the restriction factors in primates display one of the fastest evolutionary rates across the genome, suggesting ongoing positive selection for protective mechanisms to limit/prevent new zoonotic introductions. This host–pathogen "arms" race has been the subject of extensive genetic analyses and review (Kaiser et al., 2007; Malim & Emerman, 2008; Compton et al., 2012).

The known restriction and counter-restriction mechanisms observed with lentiviral infections could be briefly summarized as follows: (1) The HIV-1 Vif protein prevents the incorporation of a cytidine deaminase (APOBEC3F/G) into new virus particle. This enzyme will catalyze DNA cytosine-to-uracil deamination during reverse transcription, lead to G-to-A hypermutation, and error catastrophe/virus death. Nearly all lentiviruses (except EIAV) express a Vif protein that degrades the APOBEC3 enzyme of its host. However, xenotropic lentivirus infections (of the noncognate host species) are prevented or limited by the ability of APOBEC3 to hypermutate the proviral DNA and avoid the Vif-mediated degradation pathway. (2) Many new virus progeny cannot infect new cells, propagate, or transmit to a new host unless they are released from infected cell. Many viruses, including lentiviruses, downregulate their host receptors from the cell surface during the intracellular infection cycle. Stripping of CD4 and CCR5 from the cell surface and preventing new membrane expression is accomplished by both Nef and Vpu during HIV-1 infection, which ultimately promotes release of the new, budding virus particles from the cell surface. Cells may also express membrane proteins or glycans that actively trap virus on the cell surface. In particular, tetherin (BST-2 or CD317) is activated upon new virus infections. Both the N-terminal tail and the C-terminal glycosylaphosphatidylionsitol

(GPI) anchor can associate/embed into membranes, e.g., the cell and viral membranes, and thus, anchor a virus to cell plasma membrane to prevent virus release. HIV-1 counteracts this tetherin by expressing Vpu which binds to Tetherin via an interaction between the membrane spanning helices. This binding is followed by phosphorylation of the intracellular Vpu domain, interaction with E3 ubiquitin ligase complex, downregulation and degradation of Tetherin. (3) Many lentiviruses show limited replication in myeloid cell types such as macrophages and dendritic cells. Part of this restriction in related to low dNTP levels that often stalls lentivirus replication during the process of reverse transcription. Via the discovery of the anti-restriction mechanism of HIV-1/SIV Vpx, a new restriction mechanism has been described in myeloid cells involving SAMHD1. SAMHD1 appears to act as a dNTP phosphohydrolase which can reduce cellular dNTP levels (which are intrinsically low in macrophages and dendritic cells). Vpx counteracts restriction by binding the SAMHD1 dNTPase, acting as a scaffold for the E3 ubiquitin ligase complex, and ultimately leading to its degradation. (4) TRIM5α was the first restriction factor characterized as limiting new zoonotic introduction by new lentiviruses into primates. Following xenotropic entry of lentivirus into a host cell, TRIM5α binds to lentiviral capsid proteins with or without Cyclophilin A in the cytoplasm, modulates (and likely accelerates) the uncoating process, prevents successful reverse transcription, and leads to premature proteosomal degradation. HIV-1 is resistant to human TRIM5α but sensitive to TRIM5α from many Old and New World monkeys, e.g., owl monkey and rhesus macaques.

Viruses require a combination of their own and the hosts' cellular machinery to efficiently replicate. Due to species specificity, host cellular factors such as the latter restriction factors have been critical in preventing the spread of SIV across species barriers. The counter-restriction mechanism adopted by the viruses to evade these host restrictions generally involves the incorporation and evolution of the counter-measure host gene into the viral genome. The acquisition and function of these accessory genes are not always conserved across the lentiviruses, even the primate lentiviruses. For example, HIV-1 group M evades human tetherin by using the Vpu protein while HIV-2, because of its lack of Vpu, instead uses the Env protein for efficient release of infectious particles (Van et al., 2008; Van & Guatelli, 2008). Despite their acquisition of Vpu, the nonpandemic HIV-1 group O does not antagonize human tetherin while the rare Group N viruses gained only modest anti-tetherin Vpu activity during adaptation to humans. The Vpu, Nef, and Env of the recently discovered group P viruses do not antagonize tetherin, suggesting a lack of optimal adaptation in the human host (Sauter et al., 2011b). On the contrary, because vpu is absent in most SIVs (e.g., SIVcpz, SIVgor, SIVsmm), the Nef protein appears to have evolved to antagonize tetherin as well as to downregulate MHC-1 and CD4 (Jia et al., 2009). Differences in the antagonism of tetherin by HIV and these SIVs is mainly due to a five amino acid deletion in the N-terminal cytoplasmic tail of human tetherin which is critical to Nef sensitivity (Sauter et al., 2009). Like

HIV-2, some primate lentiviruses (e.g., SIVtantalus) have also been reported to engage tetherin with the Env protein, leading to its sequestration in intracellular compartments (Le & Neil, 2009; Gupta et al., 2009a; Gupta et al., 2009b). It is noteworthy to mention that the tetherin transmembrane and cytoplasmic tails of primate and nonprimate species are highly variable (Sauter et al., 2009).

Trim 5α is another restriction factor implicated as a barrier of SIV cross species transmission of primate lentiviruses (Hatziioannou et al., 2006). Trim 5α poorly inhibits the retroviruses found naturally in the same host species, but remain active against those that occur in other species.

THE ROLE OF HUMANS IN THE EMERGENCE AND GEOGRAPHICAL SPREAD OF IMMUNODEFICIENCY VIRUSES

The Equatorial forest of West Central Africa, especially the Congo River basin, has one of the highest populations and density of nonhuman primates. At the dawn of the 20th century, colonization by European nations and the need for space to accommodate the expansion of new cities, allowed the gradual deforestation of these regions. This resulted in increased movement and migration of both humans and nonhuman primates. With physical boundaries (e.g., rivers) in this geographical region, overlapping habitats between humans and nonhuman primates led to increased hunting of the latter for food, fur, and even for sport. It should be noted that transmission of SIV did not arise by consumption (eating) of these animal since most people of this area usually cook their food to temperatures that will kill and inactivate all viruses or at least SIV. Transmission most likely occurred by the direct contact between fresh primate blood/bodily fluids with exposed cuts and wounds of the hunters. Other zoonotic infections involving the simian foamy virus (SFV) have been reported in hunters who have come in contact with blood and bodily fluids of wild nonhuman primates. In fact Wolfe et al. (2004) found a prevalence rate of 1% SFV infections among hunters living in the central African forest (Wolfe et al., 2004). The sequences from these individuals clustered separately and suggested that hunters were infected by SFV from three separate nonhuman primates, namely: mandrill (*Mandrillus sphinx*), De Brazza's guenon (*Cercopithecus neglectus*), and gorilla (*Gorilla gorilla*) (Wolfe et al., 2004). With the exception of gorillas, both De Brazza's and mandrills are naturally infected with SIVs.

Deforestation and urbanization played a major role in cross species transmission of SIV to humans during the early postcolonial period. However, the use of unsterilized needles during vaccination campaigns might have led to an exponential HIV spread among humans. Among drug addicts who share needles, the transmission of HIV is about 10 times more efficient when compared with sexual routes of transmission (Mathers et al., 2008). Earlier suspicions that polio vaccines contaminated through cultivation of vaccine virus in infected nonprimate cells were unfounded after stored vials of polio vaccines were examined and found to be HIV negative (Plotkin & Koprowski, 2000). Overall, the incidence of HIV has been reduced in most countries over the past decade due in part to education and prevention campaigns. Also, HIV-related deaths have been reduced due to the introduction of effective antiviral drug treatments for HIV infected individuals. Interestingly, there has been little attention paid to "bush meat" hunters who remain at risk for new zoonotic infections and that may act as the "middle men" between the nonhuman primate forest habitats and human communities in villages and towns in West and Central Africa.

The jump of SIV to humans can be described as "opportunistic" as the transmitting host must come in contact with a susceptible recipient. It is feasible that SIV/HIVs had been jumping between monkey and human cohabitants of the tropical equatorial rain forests for hundreds to thousands of years. However, until the 20th century, neither primate population moved out of their habitats. SIVs that likely crossed species eventually died out in the human population alongside the infected individual. With deforestation and the availability of several means of transportation during the pre-independence and post-colonial times in Africa, most inhabitants of rural areas emigrated to towns and cities in search of better lifestyles. These authors subscribe to the "blame Emperor Leopold II" theory suggesting that the HIV epidemic was seeded between the 1870s and early 1900s based on the opening of the Congo basin through killing, mutilation, rape, and slavery imposed by Leopold's private army, Force Publique and in the name of economic opportunism. This horrific colonialism drove the human population from the Congo basin into urban centers. Kinshasa (formerly Leopoldville) in the DRC has one of the fastest growing populations over the past 80 years with a doubling rate of every 5 years since the turn of the century. HIV is thought to have originated from the Congo basin and then spread through humans in Kinshasa, a fact supported by the high genetic diversity in HIV subtypes in this region (Worobey et al., 2008).

The low prevalence of groups O, N, and P suggest that these viruses were less successful in crossing the species barrier than the SIV progenitor of group M. Introduction of the SIV progenitors of groups N and P in the new host met "dead ends" in transmission chains. A good example of an unsuccessful founder event was observed in Norway where a Norwegian sailor returned from West Africa infected with an HIV-1 group O strain in the 1960's. Despite this early introduction in Europe, the spread of this virus was limited to the immediate family of the index case, first his wife and then daughter via mother-to-infant transmission (Jonassen et al., 1997).

As described above, recombination and the generation of mosaic SIV/HIV strains played a significant role in the emergence of HIV-1. Group M of HIV-1 has nine subtypes (A-D, F-H, J-K) which differ in amino acid sequences ranging from 10% in pol, 20% in gag and 30% in env. These subtypes were isolated to specific geographical regions/human communities at the start of the epidemic but have rapidly expanded and intermixed (Tebit & Arts, 2011). Some HIV-1 group M subtypes have even recombined to form new strains which have spread as "new mosaics" termed Circulating Recombinant Forms (CRFs). Forty-eight CRFs have been described to date (HIV Sequence database, 2010), comprising about 20% of HIV strains in the epidemic. The continuous evolution of the epidemic has also seen the emergence of new sub-subtypes as is the case with A1-A4 and F1-F2 in group M. Genetic variation in HIV-1 group M remains the highest in Central Africa especially around the Congo River basin, a fact that is consistent with an old epidemic (Figure 9.3). The DRC is the origin of the two oldest HIV-1 strains described to date: A 1959 sample which clustered with subtype D and a 1960 sample that is closely related to subtype A (Zhu et al., 1998; Worobey et al., 2008). These two sequences have been used to determine the age of the global epidemic (Figure 9.3). The lower prevalence of HIV in the central African region argues for a stabilized epidemic. Compared to this region, the South African epidemic (Botswana, Zambia, Zimbabwe, Mozambique, Swaziland, Lesotho, and South Africa) which is dominated by subtype C has the highest rates of new HIV infections and now bears the greatest HIV burden in the world (UNAIDS, 2008). On average, about 20% of the human population in South Africa, Lesotho, Botswana and Zimbabwe are living with HIV contributing to almost 51% of all HIV-1 infections (UNAIDS, 2008). The highest prevalence during early times of the HIV epidemic (mid-1980s to late 1990s) was reported in East African countries (Uganda, Kenya and Tanzania). The high HIV prevalence in Southern Africa, which is dominated by subtype C, provides strong evidence of a founder effect, that is, the introduction of a single virus strain followed by a rapid spread (Rambaut et al., 2001; Figure 9.3). In Central Africa, genetic diversity is high in Cameroon with all groups and subtypes reported (Tebit & Arts, 2011). It is interesting to note that moving westwards from the Congo Basin to Cameroon, Nigeria, Cote d'Ivoire and Ghana, the number of subtypes in the human population is reduced and a mixture of nearly everything is filtered to a dominance of subtypes A, G, and CRF02_AG (Tebit et al., 2002). Crossing to Burkina Faso, pure subtypes A and G are almost absent and replaced by CRF06_cpx and second generation recombinants composing CRF02_AG and CRF06_cpx (Ouedraogo-Traore et al., 2003; Tebit et al., 2007, 2009), both of which have subtypes A and G as a major part of their genomes. Recent data from Angola also indicate a huge HIV-1 diversity similar to that in neighboring DRC suggesting heavy human mixing and migration within the DRC and Angola that was substantially greater than human mixing between this Congo basin region with West or East Africa. Again, the diversity of HIV-1 subtypes in West and East Africa (as observed with Southern Africa) is much less than in the Congo basin.

When HIV-1 was first introduced in Asia as subtype B (also known as Thai B′ because of its divergence from the American subtype B) in the mid-1980s, infection was mostly reported in China, India and Thailand (Deng et al., 2008). Fifteen years after this "Thai B" strain was first reported, CRF01_AE was introduced into Asia and has now taken over as the dominant strain in this region (Fontella et al., 2008; de Silva et al., 2010). Subtype C had a unique founder event in India leading to domination while recombinant forms, CRF07_B′C and CRF08_B′C are common in China. Northern Burma seems to be a fertile ground for human population mixing and high frequency of HIV transmission due to heroin trafficking and limited governance. As a consequence, Northern Burma has now seeded the largest expansion of HIV-1 genetic diversity in Asia (and likely, in the world). In general, CRFs and other unique recombinant forms comprise about 88% of the circulating viruses, which will have the potential to spread throughout south and south eastern Asia (Takebe et al., 2003).

In Eastern Europe, particularly among states of the former Soviet Union, the rapid spread of sub-subtype A1 is linked mainly to intravenous drug users whereas subtype B (mainly CRF03_AB) is spread through sexual transmission. Subtype B dominates in Western Europe although facilitation of travelling and immigration into this region has fueled a rise in non-B strains coming mostly from Africa (Perez-Alvarez et al., 2006; Figure 9.3). Portugal's ties with former Portuguese colonies of Cape Verde and Guinea Bissau in Africa could explain its high HIV diversity, especially the high prevalence of HIV-2, the highest observed outside Africa (Parreira et al., 2005; Hemelaar et al., 2006; Faria et al., 2012). HIV-2A has been reported in France within individuals of Senegalese origin (Faria et al., 2012).

The HIV epidemic in the Caribbean provided the clearest example of how recent emigration and immigration between continents has facilitated the spread of HIV. Angola and Cuba as well as DRC and Haiti have had long-standing relationships involving trade, travel, and support for wars. Recent studies applying a relaxed molecular clock on sequences from a 1982 sample from Haiti determined that HIV was likely introduced into Haiti from the DRC between 1962–1970 (Gilbert et al., 2007; Figure 9.3). This date coincides with reports of the oldest HIV samples dating from 1982 or 1983 in Haitian immigrants resident in the United States. The epidemic in Cuba is more diversified comprising CRF18_cpx, CRF19_cpx and subtype C but the most dominant are CRF20, CRF23 and CRF24, all of which have subtype C as the parental types (Cuevas et al., 2002).

The HIV-1 epidemic in the Americas is dominated by subtype B in the United States and Canada. In South America, Argentina and Brazil show the highest diversity with subtypes B, C, F as well as BC and BF recombinants reported. More than six BF mosaic strains have been described as CRFs (HIV Sequence database, 2010). However, recent studies suggest that most of the BFs in Argentina may actually be URFs, providing evidence for continued superinfection and recombination rather than transmission chains of a few CRFs (Tebit and Arts, Personal Communication). Subtype Cs of South America appear to cluster with those from Kenya, Ethiopia and Burundi suggesting an African origin (Fontella et al., 2008; Bello et al., 2008). The increase in pure C and BC recombinants in Rio do Sul in southern Brazil, from 35% to almost 70% from 1996 through to 2011 (Soares et al., 2005; Dias et al., 2009), supports the general fact that the subtype C epidemic is expanding faster than any other in South America.

Tracking the evolution of HIV-2 over time in some West African countries suggests that within the last two decades HIV-2 has shown a gradual but steady decrease while HIV-1 is gaining in prevalence. Surveys conducted in

Guinea Bissau during the early HIV-2 epidemic in 1987 revealed a drop in prevalence from 8.3% in 1987 to 2.4% in 2004 while that of HIV-1 rose from 0 to 4% during the same period (da Silva et al., 2008). This epidemiological data is backed by clinical findings which show that HIV-2 has a lower viral load, progresses more slowly to disease, and has a lower fitness than HIV-1 (Arien et al., 2005).

CONCLUSION AND PERSPECTIVE

The mid 1970s and early 1980s saw a silent but relentless dissemination of HIV throughout Africa, coinciding with the period when the virus crossed the Atlantic into Haiti before establishing a strong foothold in the United States. This period is critical because nonhuman primates, which are the original hosts of SIV, were able to establish several contacts with humans, who were then able to spread the virus through greater human-to-human contacts than prior to the turn of the 20th century. Even though we blame King Leopold II for this epidemic, the mixing of the human population and increased invasion into other primate habitats was inevitable with the exposure to Western culture, the rubber industry, and better means of transportation. After the first detection of AIDS cases and subsequent findings that HIV was the cause of AIDS, serological techniques were established facilitating the detection of HIV-1 antibodies, which in some cases cross-reacted with HIV-2 and SIVs. Due to close sequence relationship between HIV-2 and SIVsmm, the possible source of HIV-2 was determined to be sooty mangabeys. Later studies screening SIV-like sequences in fecal material of living chimpanzees and gorillas across Africa determine that the closest relative to HIV-1 group M and N was derived from *P.t. troglodytes*, whereas HIV-1 group O likely originated from *P.t. troglodytes* directly or via *G. gorilla*. Today, new sensitive and specific serological and molecular assays as well as complex computer-based programs have been used to track both HIV and SIV in human and nonprimate populations. Despite recent large-scale screening and description of many ape and monkey immunodeficiency viruses in the wild, there is a continuous risk that SIVs infecting different species of primates (humans or nonhumans) might recombine and generate new strains which could go undetected by commonly used screening assays. Because such new recombinant viruses could produce viruses whose virulence in humans could be severe, it will be prudent to develop even more assays that can continuously detect existing SIVs and even yet unknown SIVs. Until such a time, we should continue monitoring for known SIVs as well as human avoidance of direct contact with blood and bodily fluids from nonhuman primates.

ACKNOWLEDGMENTS

This chapter is dedicated to the memory of Dr. Elizabeth Bailes, one of the pioneers in the study of zoonotic transmission of lentiviruses between primates.

REFERENCES

Aghokeng, A. F., Ayouba, A., Mpoudi-Ngole, E., Loul, S., Liegeois, F., Delaporte, E., & Peeters, M. (2010). Extensive survey on the prevalence and genetic diversity of SIVs in primate bushmeat provides insights into risks for potential new cross-species transmissions 16. *Infect. Genet. Evol.*, *10*, 386–396.

Aghokeng, A. F., Liu, W., Bibollet-Ruche, F., Loul, S., Mpoudi-Ngole, E., Laurent, C., Mwenda, J. M., Langat, D. K., Chege, G. K., McClure, H. M., Delaporte, E., Shaw, G. M., Hahn, B. H., & Peeters, M. (2006). Widely varying SIV prevalence rates in naturally infected primate species from Cameroon. *Virology*, *345*, 174–189.

Allan, J. S., Short, M., Taylor, M. E., Su, S., Hirsch, V. M., Johnson, P. R., Shaw, G. M., & Hahn, B. H. (1991). Species-specific diversity among simian immunodeficiency viruses from African green monkeys. *J. Virol.*, *65*, 2816–2828.

Apetrei, C., Kaur, A., Lerche, N. W., Metzger, M., Pandrea, I., Hardcastle, J., Falkenstein, S., Bohm, R., Koehler, J., Traina-Dorge, V., Williams, T., Staprans, S., Plauche, G., Veazey, R. S., McClure, H., Lackner, A. A., Gormus, B., Robertson, D. L., & Marx, P. A. (2005). Molecular epidemiology of simian immunodeficiency virus SIVsm in U.S. primate centers unravels the origin of SIVmac and SIVstm. *J. Virol.*, *79*, 8991–9005.

Arien, K. K., Abraha, A., Quinones-Mateu, M. E., Kestens, L., Vanham, G., & Arts, E. J. (2005). The replicative fitness of primary human immunodeficiency virus type 1 (HIV-1) group M, HIV-1 group O, and HIV-2 isolates. *J. Virol.*, *79*, 8979–8990.

Baeten, J. M., Chohan, B., Lavreys, L., Chohan, V., McClelland, R. S., Certain, L., Mandaliya, K., Jaoko, W., & Overbaugh, J. (2007). HIV-1 subtype D infection is associated with faster disease progression than subtype A in spite of similar plasma HIV-1 loads. *J. Infect. Dis.*, *195*, 1177–1180.

Bailes, E., Gao, F., Bibollet-Ruche, F., Courgnaud, V., Peeters, M., Marx, P. A., Hahn, B. H., & Sharp, P. M. (2003). Hybrid origin of SIV in chimpanzees. *Science*, *300*, 1713.

Barre-Sinoussi, F., Chermann, J. C., Rey, F., Nugeyre, M. T., Chamaret, S., Gruest, J., Dauguet, C., Axler-Blin, C., Vezinet-Brun, F., Rouzioux, C., Rozenbaum, W., & Montagnier, L. (1983). Isolation of a T-lymphotropic retrovirus from a patient at risk for acquired immune deficiency syndrome (AIDS). *Science*, *220*, 868–871.

Bello, G., Passaes, C. P., Guimaraes, M. L., Lorete, R. S., Matos Almeida, S. E., Medeiros, R. M., Alencastro, P. R., & Morgado, M. G. (2008). Origin and evolutionary history of HIV-1 subtype C in Brazil. *AIDS*, *22*, 1993–2000.

CDC. (1981). Kaposi's sarcoma and Pneumocystis pneumonia among homosexual men—New York City and California. *MMWR CDC Surveill. Summ.*, *30*, 305–308.

Chakrabarti, L., Guyader, M., Alizon, M., Daniel, M. D., Desrosiers, R. C., Tiollais, P., & Sonigo, P. (1987). Sequence of simian immunodeficiency virus from macaque and its relationship to other human and simian retroviruses. *Nature*, *328*, 543–547.

Chen, Z., Luckay, A., Sodora, D. L., Telfer, P., Reed, P., Gettie, A., Kanu, J. M., Sadek, R. F., Yee, J., Ho, D. D., Zhang, L., & Marx, P. A. (1997). Human immunodeficiency virus type 2 (HIV-2) seroprevalence and characterization of a distinct HIV-2 genetic subtype from the natural range of simian immunodeficiency virus-infected sooty mangabeys. *J. Virol.*, *71*, 3953–3960.

Chen, Z., Telfier, P., Gettie, A., Reed, P., Zhang, L., Ho, D. D., & Marx, P. A. (1996). Genetic characterization of new West African simian immunodeficiency virus SIVsm: geographic clustering of household-derived SIV strains with human immunodeficiency virus type 2 subtypes and genetically diverse viruses from a single feral sooty mangabey troop. *J. Virol.*, *70*, 3617–3627.

Clavel, F., Guetard, D., Brun-Vezinet, F., Chamaret, S., Rey, M. A., Santos-Ferreira, M. O., Laurent, A. G., Dauguet, C., Katlama, C., & Rouzioux, C. (1986). Isolation of a new human retrovirus from West African patients with AIDS. *Science*, *233*, 343–346.

Clewley, J. P., Lewis, J. C., Brown, D. W., & Gadsby, E. L. (1998). A novel simian immunodeficiency virus (SIVdrl) pol sequence from the drill monkey, *Mandrillus leucophaeus*. *J. Virol.*, *72*, 10305–10309.

Compton, A. A., Hirsch, V. M., & Emerman, M. (2012). The host restriction factor APOBEC3G and retroviral Vif protein coevolve due to ongoing genetic conflict. *Cell Host Microbe.*, *11*, 91–98.

Courgnaud, V., Pourrut, X., Bibollet-Ruche, F., Mpoudi-Ngole, E., Bourgeois, A., Delaporte, E., & Peeters, M. (2001). Characterization of a novel simian immunodeficiency virus from guereza colobus monkeys (*Colobus guereza*) in Cameroon: a new lineage in the nonhuman primate lentivirus family. *J. Virol.*, *75*, 857–866.

Courgnaud, V., Salemi, M., Pourrut, X., Mpoudi-Ngole, E., Abela, B., Auzel, P., Bibollet-Ruche, F., Hahn, B., Vandamme, A. M., Delaporte, E., & Peeters, M. (2002). Characterization of a novel simian immunodeficiency virus with a vpu gene from greater spot-nosed monkeys (*Cercopithecus nictitans*) provides new insights into simian/human immunodeficiency virus phylogeny. *J. Virol.*, *76*, 8298–8309.

Cuevas, M. T., Ruibal, I., Villahermosa, M. L., Diaz, H., Delgado, E., Parga, E. V., Perez-Alvarez, L., de Armas, M. B., Cuevas, L., Medrano, L., Noa, E., Osmanov, S., Najera, R., & Thomson, M. M. (2002). High HIV-1 genetic diversity in Cuba. *AIDS*, *16*, 1643–1653.

da Silva, Z. J., Oliveira, I., Andersen, A., Dias, F., Rodrigues, A., Holmgren, B., Andersson, S., & Aaby, P. (2008). Changes in prevalence and incidence of HIV-1, HIV-2 and dual infections in urban areas of Bissau, Guinea-Bissau: is HIV-2 disappearing? *AIDS*, *22*, 1195–1202.

Daniel, M. D., Letvin, N. L., King, N. W., Kannagi, M., Sehgal, P. K., Hunt, R. D., Kanki, P. J., Essex, M., & Desrosiers, R. C. (1985). Isolation of T-cell tropic HTLV-III-like retrovirus from macaques. *Science*, *228*, 1201–1204.

de Silva, U. C., Warachit, J., Sattagowit, N., Jirapongwattana, C., Panthong, S., Utachee, P., Yasunaga, T., Ikuta, K., Kameoka, M., & Boonsathorn, N. (2010). Genotypic characterization of HIV type 1 env gp160 sequences from three regions in Thailand. *AIDS Res. Hum. Retroviruses*, *26*, 223–227.

Deng, X., Liu, H., Shao, Y., Rayner, S., & Yang, R. (2008). The epidemic origin and molecular properties of B': a founder strain of the HIV-1 transmission in Asia. *AIDS*, *22*, 1851–1858.

Dias, C. F., Nunes, C. C., Freitas, I. O., Lamego, I. S., de Oliveira, I. M., Gilli, S., Rodrigues, R., & Brigido, L. F. (2009). High prevalence and association of HIV-1 non-B subtype with specific sexual transmission risk among antiretroviral naive patients in Porto Alegre, RS, Brazil. *Rev. Inst. Med. Trop. Sao Paulo*, *51*, 191–196.

Dragic, T., Litwin, V., Allaway, G. P., Martin, S. R., Huang, Y., Nagashima, K. A., Cayanan, C., Maddon, P. J., Koup, R. A., Moore, J. P., & Paxton, W. A. (1996). HIV-1 entry into CD4+ cells is mediated by the chemokine receptor CC-CKR-5. *Nature*, *381*, 667–673.

Edinger, A. L., Mankowski, J. L., Doranz, B. J., Margulies, B. J., Lee, B., Rucker, J., Sharron, M., Hoffman, T. L., Berson, J. F., Zink, M. C., Hirsch, V. M., Clements, J. E., & Doms, R. W. (1997). CD4-independent, CCR5-dependent infection of brain capillary endothelial cells by a neurovirulent simian immunodeficiency virus strain. *Proc. Natl. Acad. Sci. USA.*, *94*, 14742–14747.

Fabre, P. H., Rodrigues, A., & Douzery, E. J. (2009). Patterns of macroevolution among Primates inferred from a supermatrix of mitochondrial and nuclear DNA. *Mol. Phylogenet. Evol.*, *53*, 808–825.

Faria, N. R., Hodges-Mameletzis, I., Silva, J. C., Rodes, B., Erasmus, S., Paolucci, S., Ruelle, J., Pieniazek, D., Taveira, N., Trevino, A., Goncalves, M. F., Jallow, S., Xu, L., Camacho, R. J., Soriano, V., Goubau, P., de Sousa, J. D., Vandamme, A. M., Suchard, M. A., & Lemey, P. (2012). Phylogeographical footprint of colonial history in the global dispersal of human immunodeficiency virus type 2 group A. *J. Gen. Virol.*, *93*, 889–899.

Feng, Y., Broder, C. C., Kennedy, P. E., & Berger, E. A. (1996). HIV-1 entry cofactor: functional cDNA cloning of a seven-transmembrane, G protein-coupled receptor. *Science*, *272*, 872–877.

Finzi, A., Pacheco, B., Xiang, S. H., Pancera, M., Herschhorn, A., Wang, L., Zeng, X., Desormeaux, A., Kwong, P. D., & Sodroski, J. (2012). Lineage-specific differences between human and simian immunodeficiency virus regulation of gp120 trimer association and CD4 binding 1. *J. Virol.*, *86*, 8974–8986.

Fontella, R., Soares, M. A., & Schrago, C. G. (2008). On the origin of HIV-1 subtype C in South America. *AIDS*, *22*, 2001–2011.

Gao, F., Bailes, E., Robertson, D. L., Chen, Y., Rodenburg, C. M., Michael, S. F., Cummins, L. B., Arthur, L. O., Peeters, M., Shaw, G. M., Sharp, P. M., & Hahn, B. H. (1999). Origin of HIV-1 in the chimpanzee *Pan troglodytes troglodytes*. *Nature*, *397*, 436–441.

Gao, F., Yue, L., Robertson, D. L., Hill, S. C., Hui, H., Biggar, R. J., Neequaye, A. E., Whelan, T. M., Ho, D. D., & Shaw, G. M. (1994). Genetic diversity of human immunodeficiency virus type 2: evidence for distinct sequence subtypes with differences in virus biology. *J. Virol.*, *68*, 7433–7447.

Gao, F., Yue, L., White, A. T., Pappas, P. G., Barchue, J., Hanson, A. P., Greene, B. M., Sharp, P. M., Shaw, G. M., & Hahn, B. H. (1992). Human infection by genetically diverse SIVSM-related HIV-2 in west Africa 54. *Nature*, *358*, 495–499.

Gifford, R. J., Katzourakis, A., Tristem, M., Pybus, O. G., Winters, M., & Shafer, R. W. (2008). A transitional endogenous lentivirus from the genome of a basal primate and implications for lentivirus evolution. *Proc. Natl. Acad. Sci. USA.*, *105*, 20362–20367.

Gilbert, C., Maxfield, D. G., Goodman, S. M., & Feschotte, C. (2009). Parallel germline infiltration of a lentivirus in two Malagasy lemurs. *PLoS. Genet.*, *5*, e1000425.

Gilbert, M. T., Rambaut, A., Wlasiuk, G., Spira, T. J., Pitchenik, A. E., & Worobey, M. (2007). The emergence of HIV/AIDS in the Americas and beyond. *Proc. Natl. Acad. Sci. USA. Nov 20*, *104*(47), 18566–18570. Epub 2007 Oct 31.

Gupta, R. K., Hue, S., Schaller, T., Verschoor, E., Pillay, D., & Towers, G. J. (2009a). Mutation of a single residue renders human tetherin resistant to HIV-1 Vpu-mediated depletion. *PLoS. Pathog.*, *5*, e1000443.

Gupta, R. K., Mlcochova, P., Pelchen-Matthews, A., Petit, S. J., Mattiuzzo, G., Pillay, D., Takeuchi, Y., Marsh, M., & Towers, G. J. (2009b). Simian immunodeficiency virus envelope glycoprotein counteracts tetherin/BST-2/CD317 by intracellular sequestration. *Proc. Natl. Acad. Sci. USA.*, *106*, 20889–20894.

Gurtler, L. G., Hauser, P. H., Eberle, J., Von, B. A., Knapp, S., Zekeng, L., Tsague, J. M., & Kaptue, L. (1994). A new subtype of human immunodeficiency virus type 1 (MVP-5180) from Cameroon. *J. Virol.*, *68*, 1581–1585.

Hahn, B. H., Gonda, M. A., Shaw, G. M., Popovic, M., Hoxie, J. A., Gallo, R. C., & Wong-Staal, F. (1985). Genomic diversity of the acquired immune deficiency syndrome virus HTLV-III: different viruses exhibit greatest divergence in their envelope genes. *Proc. Natl. Acad. Sci. USA.*, *82*, 4813–4817.

Harris, R. S., Hultquist, J. F., & Evans, D. T. (2012). The restriction factors of human immunodeficiency virus. *J. Biol. Chem.*

Haseltine, W. A., Sodroski, J., & Rosen, C. (1984). Structure and function of human leukemia and AIDS viruses. *Princess Takamatsu Symp.*, *15*, 187–196.

Hatziioannou, T., Princiotta, M., Piatak, M., Jr., Yuan, F., Zhang, F., Lifson, J. D., & Bieniasz, P. D. (2006). Generation of simian-tropic HIV-1 by restriction factor evasion. *Science*, *314*, 95.

Hemelaar, J., Gouws, E., Ghys, P. D., & Osmanov, S. (2006). Global and regional distribution of HIV-1 genetic subtypes and recombinants in 2004. *AIDS*, *20*, W13–W23.

Hirsch, V. M., Dapolito, G., McGann, C., Olmsted, R. A., Purcell, R. H., & Johnson, P. R. (1989). Molecular cloning of SIV from sooty mangabey monkeys. *J. Med. Primatol.*, *18*, 279–285.

Hirsch, V. M., & Johnson, P. R. (1994). Pathogenic diversity of simian immunodeficiency viruses. *Virus Res.*, *32*, 183–203.

Hirsch, V. M., Olmsted, R. A., Murphey-Corb, M., Purcell, R. H., & Johnson, P. R. (1989). An African primate lentivirus (SIVsm) closely related to HIV-2. *Nature*, *339*, 389–392.

Hirsch, V. M., Santra, S., Goldstein, S., Plishka, R., Buckler-White, A., Seth, A., Ourmanov, I., Brown, C. R., Engle, R., Montefiori, D., Glowczwskie, J., Kunstman, K., Wolinsky, S., & Letvin, N. L. (2004). Immune failure in the absence of profound CD4+ T-lymphocyte depletion in simian immunodeficiency virus-infected rapid progressor macaques. *J. Virol.*, *78*, 275–284.

HIV Sequence database (2010). Ref Type: Online Source, http://www.hiv.lanl.gov. Accessed August 10, 2010.

Huet, T., Cheynier, R., Meyerhans, A., Roelants, G., & Wain-Hobson, S. (1990). Genetic organization of a chimpanzee lentivirus related to HIV-1. *Nature*, *345*, 356–359.

Jia, B., Serra-Moreno, R., Neidermyer, W., Rahmberg, A., Mackey, J., Fofana, I. B., Johnson, W. E., Westmoreland, S., & Evans, D. T. (2009). Species-specific activity of SIV Nef and HIV-1 Vpu in overcoming restriction by tetherin/BST2. *PLoS. Pathog.*, *5*, e1000429.

Jonassen, T. O., Stene-Johansen, K., Berg, E. S., Hungnes, O., Lindboe, C. F., Froland, S. S., & Grinde, B. (1997). Sequence analysis of HIV-1 group O from Norwegian patients infected in the 1960s. *Virology*, *231*, 43–47.

Kaiser, S. M., Malik, H. S., & Emerman, M. (2007). Restriction of an extinct retrovirus by the human TRIM5alpha antiviral protein. *Science*, *316*, 1756–1758.

Kaleebu, P., French, N., Mahe, C., Yirrell, D., Watera, C., Lyagoba, F., Nakiyingi, J., Rutebemberwa, A., Morgan, D., Weber, J., Gilks, C., & Whitworth, J. (2002). Effect of human immunodeficiency virus (HIV) type 1 envelope subtypes A and D on disease progression in a large cohort of HIV-1-positive persons in Uganda. *J. Infect. Dis.*, *185*, 1244–1250.

Katzourakis, A., Gifford, R. J., Tristem, M., Gilbert, M. T., & Pybus, O. G. (2009). Macroevolution of complex retroviruses. *Science*, *325*, 1512.

Keele, B. F., Jones, J. H., Terio, K. A., Estes, J. D., Rudicell, R. S., Wilson, M. L., Li, Y., Learn, G. H., Beasley, T. M., Schumacher-Stankey, J., Wroblewski, E., Mosser, A., Raphael, J., Kamenya, S., Lonsdorf, E. V., Travis, D. A., Mlengeya, T., Kinsel, M. J., Else, J. G., Silvestri, G., Goodall, J., Sharp, P. M., Shaw, G. M., Pusey, A. E., & Hahn, B. H. (2009). Increased mortality and AIDS-like immunopathology in wild chimpanzees infected with SIVcpz. *Nature*, *460*, 515–519.

Keele, B. F., Van, H. F., Li, Y., Bailes, E., Takehisa, J., Santiago, M. L., Bibollet-Ruche, F., Chen, Y., Wain, L. V., Liegeois, F., Loul, S., Ngole, E. M., Bienvenue, Y., Delaporte, E., Brookfield, J. F., Sharp, P. M., Shaw, G. M., Peeters, M., & Hahn, B. H. (2006). Chimpanzee reservoirs of pandemic and nonpandemic HIV-1. *Science*, *313*, 523–526.

Korber, B., Muldoon, M., Theiler, J., Gao, F., Gupta, R., Lapedes, A., Hahn, B. H., Wolinsky, S., & Bhattacharya, T. (2000). Timing the ancestor of the HIV-1 pandemic strains. *Science*, *288*, 1789–1796.

Le, T. A., & Neil, S. J. (2009). Antagonism to and intracellular sequestration of human teth-
    erin by the human immunodeficiency virus type 2 envelope glycoprotein. *J. Virol.*, *83*,
    11966–11978.

Lemey, P., Pybus, O. G., Rambaut, A., Drummond, A. J., Robertson, D. L., Roques, P., Worobey, M.,
    & Vandamme, A. M. (2004). The molecular population genetics of HIV-1 group O. *Genetics*,
    167, 1059–1068.

Li, Y., Ndjango, J. B., Learn, G. H., Ramirez, M. A., Keele, B. F., Bibollet-Ruche, F., Liu, W.,
    Easlick, J. L., Decker, J. M., Rudicell, R. S., Inogwabini, B. I., Ahuka-Mundeke, S., Leen-
    dertz, F. H., Reynolds, V., Muller, M. N., Chancellor, R. L., Rundus, A. S., Simmons, N.,
    Worobey, M., Shaw, G. M., Peeters, M., Sharp, P. M., & Hahn, B. H. (2012). Eastern chim-
    panzees, but not bonobos, represent a simian immunodeficiency virus reservoir. *J. Virol.*, *86*,
    10776–10791.

Lori, F., di, M., V, de Vico, A. L., Lusso, P., Reitz, M. S., Jr., & Gallo, R. C. (1992). Viral DNA car-
    ried by human immunodeficiency virus type 1 virions. *J. Virol.*, *66*, 5067–5074.

Malim, M. H., & Emerman, M. (2008). HIV-1 accessory proteins—ensuring viral survival in a
    hostile environment. *Cell Host. Microbe.*, *3*, 388–398.

Mansky, L. M. (1998). Retrovirus mutation rates and their role in genetic variation. *J. Gen. Virol.*,
    79(Pt 6), 1337–1345.

Mansky, L. M., & Temin, H. M. (1995). Lower in vivo mutation rate of human immunodeficiency
    virus type 1 than that predicted from the fidelity of purified reverse transcriptase. *J. Virol.*, *69*,
    5087–5094.

Mathers, B. M., Degenhardt, L., Phillips, B., Wiessing, L., Hickman, M., Strathdee, S. A., Wodak,
    A., Panda, S., Tyndall, M., Toufik, A., & Mattick, R. P. (2008). Global epidemiology of inject-
    ing drug use and HIV among people who inject drugs: a systematic review. *Lancet*, *372*,
    1733–1745.

Muller-Trutwin, M. C., Corbet, S., Souquiere, S., Roques, P., Versmisse, P., Ayouba, A., Delarue,
    S., Nerrienet, E., Lewis, J., Martin, P., Simon, F., Barre-Sinoussi, F., & Mauclere, P. (2000).
    SIVcpz from a naturally infected Cameroonian chimpanzee: biological and genetic comparison
    with HIV-1 N. *J. Med. Primatol.*, *29*, 166–172.

Neel, C., Etienne, L., Li, Y., Takehisa, J., Rudicell, R. S., Bass, I. N., Moudindo, J., Mebenga, A.,
    Esteban, A., Van, H. F., Liegeois, F., Kranzusch, P. J., Walsh, P. D., Sanz, C. M., Morgan, D.
    B., Ndjango, J. B., Plantier, J. C., Locatelli, S., Gonder, M. K., Leendertz, F. H., Boesch, C.,
    Todd, A., Delaporte, E., Mpoudi-Ngole, E., Hahn, B. H., & Peeters, M. (2010). Molecular
    epidemiology of simian immunodeficiency virus infection in wild-living gorillas. *J. Virol.*, *84*,
    1464–1476.

Nerrienet, E., Santiago, M. L., Foupouapouognigni, Y., Bailes, E., Mundy, N. I., Njinku, B., Kfut-
    wah, A., Muller-Trutwin, M. C., Barre-Sinoussi, F., Shaw, G. M., Sharp, P. M., Hahn, B. H., &
    Ayouba, A. (2005). Simian immunodeficiency virus infection in wild-caught chimpanzees from
    cameroon. *J. Virol.*, *79*, 1312–1319.

Ouedraogo-Traore, R., Montavon, C., Sanou, T., Vidal, N., Sangare, L., Sanou, I., Soudre, R.,
    Mboup, S., Delaporte, E., & Peeters, M. (2003). CRF06-cpx is the predominant HIV-1 variant
    in AIDS patients from Ouagadougou, the capital city of Burkina Faso. *AIDS*, *17*, 441–442.

Parreira, R., Padua, E., Piedade, J., Venenno, T., Paixao, M. T., & Esteves, A. (2005). Genetic analy-
    sis of human immunodeficiency virus type 1 nef in Portugal: subtyping, identification of mosaic
    genes, and amino acid sequence variability. *J. Med. Virol.*, *77*, 8–16.

Peeters, M., Courgnaud, V., Abela, B., Auzel, P., Pourrut, X., Bibollet-Ruche, F., Loul, S., Liegeois,
    F., Butel, C., Koulagna, D., Mpoudi-Ngole, E., Shaw, G. M., Hahn, B. H., & Delaporte, E.
    (2002). Risk to human health from a plethora of simian immunodeficiency viruses in primate
    bushmeat. *Emerg. Infect. Dis.*, *8*, 451–457.

Peeters, M., Fransen, K., Delaporte, E., Van den Haesevelde, M., Gershy-Damet, G. M., Kestens, L., van der Groen, G., & Piot, P. (1992). Isolation and characterization of a new chimpanzee lentivirus (simian immunodeficiency virus isolate cpz-ant) from a wild-captured chimpanzee. *AIDS*, *6*, 447–451.

Peeters, M., Honore, C., Huet, T., Bedjabaga, L., Ossari, S., Bussi, P., Cooper, R. W., & Delaporte, E. (1989). Isolation and partial characterization of an HIV-related virus occurring naturally in chimpanzees in Gabon. *AIDS*, *3*, 625–630.

Peeters, M., Liegeois, F., Torimiro, N., Bourgeois, A., Mpoudi, E., Vergne, L., Saman, E., Delaporte, E., & Saragosti, S. (1999). Characterization of a highly replicative intergroup M/O human immunodeficiency virus type 1 recombinant isolated from a Cameroonian patient. *J. Virol.*, *73*, 7368–7375.

Perez-Alvarez, L., Munoz, M., Delgado, E., Miralles, C., Ocampo, A., Garcia, V., Thomson, M., Contreras, G., & Najera, R. (2006). Isolation and biological characterization of HIV-1 BG intersubtype recombinants and other genetic forms circulating in Galicia. *Spain. J. Med. Virol.*, *78*, 1520–1528.

Phillips-Conroy, J. E., Jolly, C. J., Petros, B., Allan, J. S., & Desrosiers, R. C. (1994). Sexual transmission of SIVagm in wild grivet monkeys. *J. Med. Primatol.*, *23*, 1–7.

Pieniazek, D., Ellenberger, D., Janini, L. M., Ramos, A. C., Nkengasong, J., Sassan-Morokro, M., Hu, D. J., Coulibally, I. M., Ekpini, E., Bandea, C., Tanuri, A., Greenberg, A. E., Wiktor, S. Z., & Rayfield, M. A. (1999). Predominance of human immunodeficiency virus type 2 subtype B in Abidjan, Ivory Coast. *AIDS Res. Hum. Retroviruses*, *15*, 603–608.

Plantier, J. C., Leoz, M., Dickerson, J. E., De, O. F., Cordonnier, F., Lemee, V., Damond, F., Robertson, D. L., & Simon, F. (2009). A new human immunodeficiency virus derived from gorillas. *Nat. Med.*, *15*, 871–872.

Plotkin, S. A., & Koprowski, H. (2000). No evidence to link polio vaccine with HIV.... *Nature*, *407*, 941.

Popper, S. J., Sarr, A. D., Travers, K. U., Gueye-Ndiaye, A., Mboup, S., Essex, M. E., & Kanki, P. J. (1999). Lower human immunodeficiency virus (HIV) type 2 viral load reflects the difference in pathogenicity of HIV-1 and HIV-2. *J. Infect. Dis.*, *180*, 1116–1121.

Preston, B. D., Poiesz, B. J., & Loeb, L. A. (1988). Fidelity of HIV-1 reverse transcriptase. *Science*, *242*, 1168–1171.

Puffer, B. A., Pohlmann, S., Edinger, A. L., Carlin, D., Sanchez, M. D., Reitter, J., Watry, D. D., Fox, H. S., Desrosiers, R. C., & Doms, R. W. (2002). CD4 independence of simian immunodeficiency virus Envs is associated with macrophage tropism, neutralization sensitivity, and attenuated pathogenicity. *J. Virol.*, *76*, 2595–2605.

Rambaut, A., Robertson, D. L., Pybus, O. G., Peeters, M., & Holmes, E. C. (2001). Human immunodeficiency virus. Phylogeny and the origin of HIV-1. *Nature*, *410*, 1047–1048.

Rosen, C. A., Sodroski, J. G., & Haseltine, W. A. (1985). The location of cis-acting regulatory sequences in the human T cell lymphotropic virus type III (HTLV-III/LAV) long terminal repeat. *Cell*, *41*, 813–823.

Rowland-Jones, S. L., & Whittle, H. C. (2007). Out of Africa: what can we learn from HIV-2 about protective immunity to HIV-1? *Nat. Immunol.*, *8*, 329–331.

Rudicell, R. S., Holland, J. J., Wroblewski, E. E., Learn, G. H., Li, Y., Robertson, J. D., Greengrass, E., Grossmann, F., Kamenya, S., Pintea, L., Mjungu, D. C., Lonsdorf, E. V., Mosser, A., Lehman, C., Collins, D. A., Keele, B. F., Goodall, J., Hahn, B. H., Pusey, A. E., & Wilson, M. L. (2010). Impact of simian immunodeficiency virus infection on chimpanzee population dynamics. *PLoS. Pathog.*, *6*, e1001116.

Rudicell, R. S., Piel, A. K., Stewart, F., Moore, D. L., Learn, G. H., Li, Y., Takehisa, J., Pintea, L., Shaw, G. M., Moore, J., Sharp, P. M., & Hahn, B. H. (2011). High prevalence of simian immunodeficiency virus infection in a community of savanna chimpanzees. *J. Virol.*, *85*, 9918–9928.

Santiago, M. L., Lukasik, M., Kamenya, S., Li, Y., Bibollet-Ruche, F., Bailes, E., Muller, M. N., Emery, M., Goldenberg, D. A., Lwanga, J. S., Ayouba, A., Nerrienet, E., McClure, H. M., Heeney, J. L., Watts, D. P., Pusey, A. E., Collins, D. A., Wrangham, R. W., Goodall, J., Brookfield, J. F., Sharp, P. M., Shaw, G. M., & Hahn, B. H. (2003). Foci of endemic simian immunodeficiency virus infection in wild-living eastern chimpanzees (*Pan troglodytes schweinfurthii*). *J. Virol.*, 77, 7545–7562.

Santiago, M. L., Range, F., Keele, B. F., Li, Y., Bailes, E., Bibollet-Ruche, F., Fruteau, C., Noe, R., Peeters, M., Brookfield, J. F., Shaw, G. M., Sharp, P. M., & Hahn, B. H. (2005). Simian immunodeficiency virus infection in free-ranging sooty mangabeys (*Cercocebus atys atys*) from the Tai Forest, Cote d'Ivoire: implications for the origin of epidemic human immunodeficiency virus type 2. *J. Virol.*, 79, 12515–12527.

Santiago, M. L., Rodenburg, C. M., Kamenya, S., Bibollet-Ruche, F., Gao, F., Bailes, E., Meleth, S., Soong, S. J., Kilby, J. M., Moldoveanu, Z., Fahey, B., Muller, M. N., Ayouba, A., Nerrienet, E., McClure, H. M., Heeney, J. L., Pusey, A. E., Collins, D. A., Boesch, C., Wrangham, R. W., Goodall, J., Sharp, P. M., Shaw, G. M., & Hahn, B. H. (2002). SIVcpz in wild chimpanzees. *Science*, 295, 465.

Sauter, D., Hue, S., Petit, S. J., Plantier, J. C., Towers, G. J., Kirchhoff, F., & Gupta, R. K. (2011a). HIV-1 Group P is unable to antagonize human tetherin by Vpu, Env or Nef. *Retrovirology*, 8, 103.

Sauter, D., Hue, S., Petit, S. J., Plantier, J. C., Towers, G. J., Kirchhoff, F., & Gupta, R. K. (2011b). HIV-1 Group P is unable to antagonize human tetherin by Vpu, Env or Nef. *Retrovirology*, 8, 103.

Sauter, D., Schindler, M., Specht, A., Landford, W. N., Munch, J., Kim, K. A., Votteler, J., Schubert, U., Bibollet-Ruche, F., Keele, B. F., Takehisa, J., Ogando, Y., Ochsenbauer, C., Kappes, J. C., Ayouba, A., Peeters, M., Learn, G. H., Shaw, G., Sharp, P. M., Bieniasz, P., Hahn, B. H., Hatziioannou, T., & Kirchhoff, F. (2009). Tetherin-driven adaptation of Vpu and Nef function and the evolution of pandemic and nonpandemic HIV-1 strains. *Cell Host. Microbe.*, 6, 409–421.

Sheehy, A. M., Gaddis, N. C., Choi, J. D., & Malim, M. H. (2002). Isolation of a human gene that inhibits HIV-1 infection and is suppressed by the viral Vif protein. *Nature*, 418, 646–650.

Silvestri, G. (2005). Naturally SIV-infected sooty mangabeys: are we closer to understanding why they do not develop AIDS? *J. Med. Primatol.*, 34, 243–252.

Simon, F., Mauclere, P., Roques, P., Loussert-Ajaka, I., Muller-Trutwin, M. C., Saragosti, S., Georges-Courbot, M. C., Barre-Sinoussi, F., & Brun-Vezinet, F. (1998). Identification of a new human immunodeficiency virus type 1 distinct from group M and group O. *Nat. Med.*, 4, 1032–1037.

Soares, E. A., Martinez, A. M., Souza, T. M., Santos, A. F., Da, H. V., Silveira, J., Bastos, F. I., Tanuri, A., & Soares, M. A. (2005). HIV-1 subtype C dissemination in southern Brazil. *AIDS*, 19(Suppl. 4), S81–S86.

Sodroski, J. G., Rosen, C. A., & Haseltine, W. A. (1984). Trans-acting transcriptional activation of the long terminal repeat of human T lymphotropic viruses in infected cells. *Science*, 225, 381–385.

Stremlau, M., Owens, C. M., Perron, M. J., Kiessling, M., Autissier, P., & Sodroski, J. (2004). The cytoplasmic body component TRIM5alpha restricts HIV-1 infection in Old World monkeys. *Nature*, 427, 848–853.

Switzer, W. M., Salemi, M., Shanmugam, V., Gao, F., Cong, M. E., Kuiken, C., Bhullar, V., Beer, B. E., Vallet, D., Gautier-Hion, A., Tooze, Z., Villinger, F., Holmes, E. C., & Heneine, W. (2005). Ancient co-speciation of simian foamy viruses and primates. *Nature*, 434, 376–380.

Takebe, Y., Motomura, K., Tatsumi, M., Lwin, H. H., Zaw, M., & Kusagawa, S. (2003). High prevalence of diverse forms of HIV-1 intersubtype recombinants in Central Myanmar: geographical hot spot of extensive recombination. *AIDS, 17,* 2077–2087.

Takehisa, J., Kraus, M. H., Ayouba, A., Bailes, E., Van, H. F., Decker, J. M., Li, Y., Rudicell, R. S., Learn, G. H., Neel, C., Ngole, E. M., Shaw, G. M., Peeters, M., Sharp, P. M., & Hahn, B. H. (2009). Origin and biology of simian immunodeficiency virus in wild-living western gorillas. *J. Virol., 83,* 1635–1648.

Takehisa, J., Zekeng, L., Ido, E., Yamaguchi-Kabata, Y., Mboudjeka, I., Harada, Y., Miura, T., Kaptu, L., & Hayami, M. (1999). Human immunodeficiency virus type 1 intergroup (M/O) recombination in Cameroon. *J. Virol., 73,* 6810–6820.

Tebit, D. M., & Arts, E. J. (2011). Tracking a century of global expansion and evolution of HIV to drive understanding and to combat disease. *Lancet Infect. Dis., 11,* 45–56.

Tebit, D. M., Nankya, I., Arts, E. J., & Gao, Y. (2007). HIV diversity, recombination and disease progression: how does fitness "fit" into the puzzle? *AIDS Rev., 9,* 75–87.

Tebit, D. M., Sangare, L., Tiba, F., Saydou, Y., Makamtse, A., Somlare, H., Bado, G., Kouldiaty, B. G., Zabsonre, I., Yameogo, S. L., Sathiandee, K., Drabo, J. Y., & Krausslich, H. G. (2009). Analysis of the diversity of the HIV-1 pol gene and drug resistance associated changes among drug-naive patients in Burkina Faso. *J. Med. Virol., 81,* 1691–1701.

Tebit, D. M., Zekeng, L., Kaptue, L., Salminen, M., Krausslich, H. G., & Herchenroder, O. (2002). Genotypic and phenotypic analysis of HIV type 1 primary isolates from western Cameroon. *AIDS Res. Hum. Retroviruses, 18,* 39–48.

Temin, H. M. (1993). Retrovirus variation and reverse transcription: abnormal strand transfers result in retrovirus genetic variation. *Proc. Natl. Acad. Sci. USA., 90,* 6900–6903.

UNAIDS. *HIV Epidemic Update 2008*; www.unaids.org/en/KnowledgeCentre/HIVData/GlobalReport (accessed on August 10, 2010). 8-12-0010. Ref Type: Online Source.

Vallari, A., Holzmayer, V., Harris, B., Yamaguchi, J., Ngansop, C., Makamche, F., Mbanya, D., Kaptue, L., Ndembi, N., Gurtler, L., Devare, S., & Brennan, C. A. (2011). Confirmation of putative HIV-1 group P in Cameroon. *J. Virol., 85,* 1403–1407.

van der Loo, W., Abrantes, J., & Esteves, P. J. (2009). Sharing of endogenous lentiviral gene fragments among leporid lineages separated for more than 12 million years. *J. Virol., 83,* 2386–2388.

Van, D. N., Goff, D., Katsura, C., Jorgenson, R. L., Mitchell, R., Johnson, M. C., Stephens, E. B., & Guatelli, J. (2008). The interferon-induced protein BST-2 restricts HIV-1 release and is downregulated from the cell surface by the viral Vpu protein. *Cell Host. Microbe., 3,* 245–252.

Van, D. N., & Guatelli, J. (2008). HIV-1 Vpu inhibits accumulation of the envelope glycoprotein within clathrin-coated, Gag-containing endosomes. *Cell Microbiol., 10,* 1040–1057.

Van, H. F., Li, Y., Neel, C., Bailes, E., Keele, B. F., Liu, W., Loul, S., Butel, C., Liegeois, F., Bienvenue, Y., Ngolle, E. M., Sharp, P. M., Shaw, G. M., Delaporte, E., Hahn, B. H., & Peeters, M. (2006). Human immunodeficiency viruses: SIV infection in wild gorillas. *Nature, 444,* 164.

Van, H. F., & Peeters, M. (2007). The Origins of HIV and Implications for the Global Epidemic. *Curr. Infect. Dis. Rep., 9,* 338–346.

Wain, L. V., Bailes, E., Bibollet-Ruche, F., Decker, J. M., Keele, B. F., Van, H. F., Li, Y., Takehisa, J., Ngole, E. M., Shaw, G. M., Peeters, M., Hahn, B. H., & Sharp, P. M. (2007). Adaptation of HIV-1 to its human host. *Mol. Biol. Evol., 24,* 1853–1860.

Wain-Hobson, S., Sonigo, P., Danos, O., Cole, S., & Alizon, M. (1985). Nucleotide sequence of the AIDS virus, LAV. *Cell, 40,* 9–17.

Watanabe, H., Fujiyama, A., Hattori, M., Taylor, T. D., Toyoda, A., Kuroki, Y., Noguchi, H., Ben-Kahla, A., Lehrach, H., Sudbrak, R., Kube, M., Taenzer, S., Galgoczy, P., Platzer, M., Scharfe, M., Nordsiek, G., Blocker, H., Hellmann, I., Khaitovich, P., Paabo, S., Reinhardt, R., Zheng, H. J., Zhang, X. L., Zhu, G. F., Wang, B. F., Fu, G., Ren, S. X., Zhao, G. P., Chen, Z., Lee, Y. S., Cheong, J. E., Choi, S. H., Wu, K. M., Liu, T. T., Hsiao, K. J., Tsai, S. F., Kim, C. G., Oota, S., Kitano, T., Kohara, Y., Saitou, N., Park, H. S., Wang, S. Y., Yaspo, M. L., & Sakaki, Y. (2004). DNA sequence and comparative analysis of chimpanzee chromosome 22. *Nature*, *429*, 382–388.

Wertheim, J. O., & Worobey, M. (2009). Dating the age of the SIV lineages that gave rise to HIV-1 and HIV-2. *PLoS. Comput. Biol.*, *5*, e1000377.

Wolfe, N. D., Switzer, W. M., Carr, J. K., Bhullar, V. B., Shanmugam, V., Tamoufe, U., Prosser, A. T., Torimiro, J. N., Wright, A., Mpoudi-Ngole, E., McCutchan, F. E., Birx, D. L., Folks, T. M., Burke, D. S., & Heneine, W. (2004). Naturally acquired simian retrovirus infections in central African hunters. *Lancet*, *363*, 932–937.

Wooding, S., Stone, A. C., Dunn, D. M., Mummidi, S., Jorde, L. B., Weiss, R. K., Ahuja, S., & Bamshad, M. J. (2005). Contrasting effects of natural selection on human and chimpanzee CC chemokine receptor 5. *Am. J. Hum. Genet.*, *76*, 291–301.

Worobey, M., Gemmel, M., Teuwen, D. E., Haselkorn, T., Kunstman, K., Bunce, M., Muyembe, J. J., Kabongo, J. M., Kalengayi, R. M., Van, M. E., Gilbert, M. T., & Wolinsky, S. M. (2008). Direct evidence of extensive diversity of HIV-1 in Kinshasa by 1960. *Nature*, *455*, 661–664.

Worobey, M., Santiago, M. L., Keele, B. F., Ndjango, J. B., Joy, J. B., Labama, B. L., Dhed'A, B. D., Rambaut, A., Sharp, P. M., Shaw, G. M., & Hahn, B. H. (2004). Origin of AIDS: contaminated polio vaccine theory refuted. *Nature*, *428*, 820.

Worobey, M., Telfer, P., Souquiere, S., Hunter, M., Coleman, C. A., Metzger, M. J., Reed, P., Makuwa, M., Hearn, G., Honarvar, S., Roques, P., Apetrei, C., Kazanji, M., & Marx, P. A. (2010). Island biogeography reveals the deep history of SIV. *Science*, *329*, 1487.

Xiang, S. H., Kwong, P. D., Gupta, R., Rizzuto, C. D., Casper, D. J., Wyatt, R., Wang, L., Hendrickson, W. A., Doyle, M. L., & Sodroski, J. (2002). Mutagenic stabilization and/or disruption of a CD4-bound state reveals distinct conformations of the human immunodeficiency virus type 1 gp120 envelope glycoprotein. *J. Virol.*, *76*, 9888–9899.

Zhu, T., Korber, B. T., Nahmias, A. J., Hooper, E., Sharp, P. M., & Ho, D. D. (1998). An African HIV-1 sequence from 1959 and implications for the origin of the epidemic. *Nature*, *391*, 594–597.

Chapter 10

Hantavirus Emergence in Rodents, Insectivores and Bats

What Comes Next?

Mathias Schlegel[1], Jens Jacob[2], Detlev H. Krüger[3], Andreas Rang[3] and Rainer G. Ulrich[1]

[1]*Friedrich-Loeffler-Institut, Federal Research Institute for Animal Health, Institute for Novel and Emerging Infectious Diseases, Greifswald - Insel Riems, Germany,* [2]*Julius Kühn-Institute, Federal Research Center for Cultivated Plants, Institute for Plant Protection in Horticulture and Forests, Vertebrate Research, Münster, Germany,* [3]*Institute of Medical Virology, Helmut-Ruska-Haus, Charité Medical School, Berlin, Germany*

INTRODUCTION

Hantaviruses are among the emerging pathogens with a high public interest. Public interest is focused on hantaviruses in the Americas causing human infections with a case fatality rate of about 35% (Krüger et al., 2011). The recent occurrence of human cases after exposure in the Yosemite National Park in 2012 also induced a strong public interest not only in the US, but also in Europe, as many tourists were potentially exposed to infection by a highly virulent hantavirus (Roehr, 2012). Further, the oscillations in the number of recorded human hantavirus disease cases in Europe is in the media spotlight, especially during years with a large number of cases, as in 2010 and 2012 in Germany (Ettinger et al., 2012; Heyman et al., 2011). This public as well as public health authority interest in hantaviruses also stimulates research efforts. Thus, the reasons for the emergence of hantaviruses, the driving forces for hantavirus outbreaks and the development of early warning tools have become major topics of investigation. The most recent detection of autochthonous hantavirus infections in wild and pet rats in the UK and their potential health impact on humans (Jameson et al., 2013a; Jameson et al., 2013b) also raises important questions about the presence of this rat-borne hantavirus, which is thought to be distributed world-wide, although molecular evidence is still scarce (Lin et al., 2012).

Currently the rapidly increasing knowledge about hantaviruses seems to be mainly driven by the discovery of novel hantaviruses. These findings increase

The Role of Animals in Emerging Viral Diseases. http://dx.doi.org/10.1016/B978-0-12-405191-1.00010-7

our knowledge of the host association of hantaviruses, but challenge our current picture of the origin and molecular evolution of hantaviruses. In contrast to the increasing number of hantaviruses described, the molecular characterization of hantaviruses, the definition of their virulence and identification of virulence-mediating gene products is still hampered by the lack of a reverse-genetics system to manipulate the genomic variation observed. Nevertheless, the molecular characterization of hantaviruses, their gene products, the steps of virus replication and immune evasion mechanisms have increasingly been characterized.

This review aims to evaluate the current knowledge of hantaviruses to identify gaps in knowledge and to develop potential strategies for future studies.

DISCOVERY OF HANTAVIRUSES

During the Korean conflict in 1951–1953 more than 3,000 United Nations soldiers were affected by an acute febrile illness, originally termed "Korean hemorrhagic fever" (KHF) and characterized by symptoms of high fever, chills, headache, generalized myalgia, abdominal and back pain, hemorrhagic manifestations and a case fatality rate of up to 10% (for review, see Johnson, 2001). Scientific approaches to identify the causative agent of this life-threatening disease were initiated and resulted at the end of the 1970s in the isolation of *Hantaan virus* (HTNV) and the identification of the striped field mouse *Apodemus agrarius* as its rodent reservoir (Lee et al., 1978; Lee et al., 1981).

In the following years it became clear that KHF in Korea and hemorrhagic fever disease occurring in Russia and China were similar to nephropathia epidemica (NE), a disease which had been described in Fennoscandia since the 1930s (Yanagihara and Gajdusek, 1987). The term "hemorrhagic fever with renal syndrome" (HFRS), as first suggested by Gajdusek (1962), was adopted by the World Health Organization in 1983 (WHO, 1983) for the designation of all clinical hantavirus infections throughout Eurasia including KHF and NE. The term *Hantavirus* (as a genus within the family *Bunyaviridae*) was introduced in 1985 for the group of HFRS-causing and related viruses by Schmaljohn et al. (1985).

The discovery of HTNV stimulated further research efforts to identify additional causative agents of HFRS. These efforts resulted in the isolation of a second HFRS-causing hantavirus in Asia, *Seoul virus* (SEOV) associated with different *Rattus* species (Lee et al., 1982). *Puumala virus* (PUUV) harbored by the bank vole *Myodes glareolus* (*Clethrionomys glareolus*) was identified as the causative agent of NE, a less severe form of HFRS (Brummer-Korvenkontio et al., 1980). During the same period the first indigenous American virus, the *Prospect Hill virus* (PHV), was isolated from the meadow vole *Microtus pennsylvanicus*, but found to be nonpathogenic to humans (Lee et al., 1985). Interestingly, PHV-related hantaviruses were identified in other *Microtus* species in North America, e.g., *Isla Vista virus* in *Microtus californicus*, and in Europe, for example, *Tula virus* in *M. arvalis* and *M. levis* (synonym *M. rossiaemeridionalis*; Plyusnin et al., 1994; Sibold et al., 1995; Song et al., 1995). All these *Microtus*-associated hantaviruses seem to have no or little pathogenic potential in humans (Vapalahti et al., 1996b; Schultze et al., 2002; Clement et al., 2003; Ulrich et al., 2004).

In May 1993 a cluster of acute respiratory distress syndrome (ARDS) deaths in the Four Corners region, located in the southwestern United States, were reported to represent a novel disease, Hantavirus Pulmonary Syndrome (CDC, 2008; Nichol et al., 1993). Within a very short period the causative agent of this disease was identified as a hantavirus initially designated Four Corners virus and later renamed *Sin Nombre virus* (SNV) with the common deer mouse (*Peromyscus maniculatus*) identified as the reservoir (Childs et al., 1994). Because of the inclusion of a myocardial depression, later the designation Hantavirus Cardiopulmonary Syndrome (HCPS) was suggested (Schmaljohn and Hjelle, 1997; Zavasky et al., 1999). Thereafter, a large number of novel hantaviruses were discovered in the Americas, all being associated with rodent reservoirs within the Cricetidae subfamilies Neotominae and Sigmodontinae. Importantly, for the South American hantavirus *Andes virus* (ANDV) a human-to-human transmission was reported (Chaparro et al., 1998; Padula et al., 1998; Pinna et al., 2004).

Although *Thottapalyam virus* (TPMV) was the first virus isolated from a non-rodent reservoir, the Asian house shrew *Suncus murinus* (Carey et al., 1971), it took more than 30 years for host range, genetics and molecular phylogeny investigations to confirm that TPMV was a shrew-borne hantavirus (Song et al., 2007a; Yadav et al., 2007). In the meantime, the introduction of broad-spectrum nested RT-PCR protocols resulted in the rapid detection of additional insectivore-borne hantaviruses (Klempa et al., 2007; Song et al., 2007b; Song et al., 2007c; Arai et al., 2007; Arai et al., 2008; Kang et al., 2009a; Kang et al., 2009b; Song et al., 2009). Recently, novel hantavirus sequences were detected in different bat species (Sumibcay et al., 2012; Weiss et al., 2012; Guo et al., 2013). To date it is still unclear whether these insectivore- and bat-associated viruses can cause infections and disease in humans.

Finally, it should be mentioned that retrospective investigations suggested the occurrence of hantavirus infections during military conflicts in the past. Multiple clinical reports about nephritis outbreaks during World War II (Rüdisser, 1942; Stuhlfault, 1943) might suggest hantavirus-caused HFRS. The oldest reference is a Chinese medical account of an HFRS-like disease dating to about A.D. 960 (Mertz et al., 1997)

VIRUS CLASSIFICATION

Hantaviruses represent the genus *Hantavirus* within the family *Bunyaviridae* (McCormick et al., 1982; Schmaljohn and Dalrymple, 1983; King et al., 2011). Although bunyaviruses seem to be phylogenetically related to arena- and orthomyxo-viruses, based on analysis of the RNA-dependent RNA polymerase (RdRp) sequences (see King et al., 2011), so far no higher taxon has been defined. The family *Bunyaviridae* comprises five genera containing human, animal and plant pathogens with different transmission routes (for review see Plyusin et al., 2011; Table 10.1). The genus *Tospovirus*

TABLE 10.1 Genera and Representatives of the Family *Bunyaviridae* and their Transmission Mode

Genus	Type species	Representatives	Transmission	Pathogenicity	
				Human	Animal
Orthobunyavirus	*Bunyamwera virus*	Bunyamwera virus	mosquitoes	+	–
		Akabane virus	mosquitoes, culicoid flies	–	+
		Bwamba virus	mosquitoes	+	–
		California encephalitis virus	mosquitoes	+	–
		Caraparu virus	mosquitoes	+	–
		Catu virus	mosquitoes	+	–
		Guama virus	mosquitoes	+	–
		Guaroa virus	mosquitoes	+	–
		Kairi virus	mosquitoes	–	+
		Madrid virus	mosquitoes	+	–
		Main Drain virus	mosquitoes, culicoid flies	–	+
		Manzanilla virus	mosquitoes, culicoid flies	–	+
		Marituba virus	mosquitoes	+	–
		Nyando virus	mosquitoes	+	–
		Oriboca virus	mosquitoes	+	–
		Oropouche virus	mosquitoes, culicoid flies	+	–
		Schmallenberg virus	mosquitoes, culicoid flies	–	+
		Wyeomyia virus	mosquitoes	+	–

Genus	Type species	Species	Host		
Hantavirus	Hantaan virus	Hantaan virus	rodents	–	+
		Andes virus	rodents	–	+
		Bayou virus	rodents	–	+
		Black Creek Canal virus	rodents	–	+
		Cano Delgadito virus	rodents	–	+
		Dobrava-Belgrade virus	rodents	–	+
		El Moro Canyon virus	rodents	–	–
		Isla Vista virus	rodents	–	–
		Khabarovsk virus	rodents	–	+
		Laguna Negra virus	rodents	–	+
		Muleshoe virus	rodents	–	–
		New York virus	rodents	–	+
		Prospect Hill virus	rodents	–	+
		Puumala virus	rodents	–	+
		Rio Mamore virus	rodents	–	+
		Rio Segundo virus	rodents	–	+
		Seoul virus	rodents	–	+
		Sin Nombre virus	rodents	–	+
		Thailand virus	rodents	–	–
		Thottapalayam virus	insectivores	–	+
		Topografov virus	rodents	–	–
		Tula virus	rodents	–	+
Phlebovirus	Rift Valley fever virus	Rift Valley fever virus	phlebotomines, mosquitoes	+	+
		Chandiru virus	phlebotomines	–	+
		Punta Toro virus	phlebotomines	–	+
		Sandfly fever Naples virus	phlebotomines	–	+
Nairovirus	Dugbe virus	Dugbe virus	ticks, mosquitoes, culicoid flies	–	+
		Crimean-Congo hemorrhagic fever virus	ticks	–	+
Tospovirus	Tomato spotted wilt virus	Tomato spotted wilt virus	thrips	–	–

Table modified from: *Index of Viruses – Bunyaviridae* (2006). In: ICTVdB – The Universal Virus Database, version 4. Büchen-Osmond, C (Ed), Columbia University, New York, USA
Taxonomy according to (Plyusnin et al., 2011).

contains only plant pathogens transmitted by arthropods whereas the genera *Orthobunyavirus*, *Nairovirus* and *Phlebovirus* comprise arthropod-borne bunyaviruses including zoonotic representatives causing disease in human and animals.

The species definition within the family *Bunyaviridae* is confronted with the same problems as other virus families (van Regenmortel et al., 2000). Thus, the criteria for the definition of a virus species within the different genera of the family differ (Plyusnin et al., 2011). According to the International Committee of Taxonomy of Viruses (ICTV) the definition of a hantavirus species should be based on the following criteria: A hantavirus species should (i) be found in a unique primary reservoir species, (ii) have an at least 7% difference in the amino acid (aa) sequences of the complete nucleocapsid (N) and glycoprotein precursor (GPC) proteins to all known species, (iii) show an at least four-fold difference in a two-way cross-neutralization test, and (iv) not have undergone natural reassortment events with other species (Plyusnin et al., 2011). To strictly fulfill these criteria, a comprehensive study of the virus to be defined as a novel species is needed. In fact, this would require the molecular identification of the reservoir under field conditions with a sympatric occurrence of related species, the determination of the complete coding sequences of N and GPC, cross-neutralization assays with defined sera, e.g., from a panel of wild animals with each being molecularly characterized to be infected with a defined hantavirus or from experimentally infected animals, and a broad investigation for potential reassortment events. In fact, reassortment has been reported for closely related hantaviruses *in vitro*, but also in nature (Li et al., 1995; Rodriguez et al., 1998; Kirsanovs et al., 2010; Klempa et al., 2003; Razzauti et al., 2008; Zou et al., 2008; Razzauti et al., 2009; Handke et al., 2010).

In the Ninth Report of the ICTV 23 hantavirus species were approved (King et al., 2011; Table 10.1). Not all of these species fulfill the above given ICTV criteria (Maes et al., 2009). Therefore, another scheme for the demarcation of hantavirus species (aa sequence distance >10% for N protein or >12% for GPC) and hantavirus groups (aa sequence distance >24% for N protein or >32% for GPC) has been proposed (Maes et al., 2009).

The problems associated with the hantavirus species definition are reflected in the previous discussion about the classification of *Apodemus*-associated hantaviruses in Europe (Klempa et al., 2003; Plyusnin et al., 2003). This scientific controversy resulted in the parallel use of a different nomenclature based on species or genetic lineage names, i.e., use of "Saaremaa virus" designation for all striped field mouse *A. agrarius*-derived sequences from Central and Eastern Europe and alternatively to use a different designation based on genetic lineages "DOBV-Aa" for Central and Eastern Europe strains from *A. agrarius* and "Saaremaa" for strains from Saaremaa island. As this parallel terminology brought confusion not only to the hantavirus scientific community but also to clinicians and public health authorities, recently the introduction of genotypes Dobrava, Kurkino, Sochi and Saaremaa within the species *Dobrava-Belgrade virus* was proposed (Klempa et al., 2013).

STRUCTURE AND GENOME ORGANIZATION OF HANTAVIRUSES

Hantaviruses are enveloped particles with a spherical shape and a broadly varying diameter of 70–210 nm (Goldsmith et al., 1995; Tao et al., 1987) (Figure 10.1A, B). This pleiomorphic size has been explained by the presence of additional copies of the genome segments within virions (Rodriguez et al., 1998). Electron microscopical investigations have revealed fuzzy projections of approximately 7 nm in length on the surface of the outer lipid envelope (Spiropoulou, 2011). These structures are formed by the virus-coded glycoproteins Gn and Gc. Inside the virion three ribonucleoprotein complexes formed by the single-stranded RNA genome segments of negative polarity associated with the viral N protein and the RdRp are located. The three genome segments are termed according to their size Small (S), Medium (M) and Large (L) (Schmaljohn and Dalrymple, 1983) (Figure 10.1C).

The S-segment with a varying length of 1530–2078 nucleotides (nt) encodes the N protein of 428–433 aa residues (Table 10.2). This size variation of the S-segment is caused mainly by a 3′-noncoding region (NCR) of variable length. The nt sequence identity level of the entire N-open reading frame (ORF) and the aa sequence of N protein is about 53–74% and 46–83%, respectively (Table 10.3). An additional putative ORF (NSs) of about 180–270 nt was predicted on

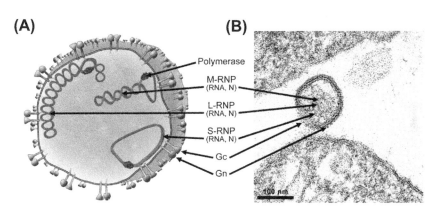

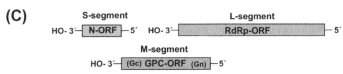

FIGURE 10.1   Schematic presentation (A) and electron-micrograph of a hantavirus particle (B), and hantavirus genome organization (C). The electron micrograph shows a negative staining of Tulavirus-infected Vero E 6 cells. The figures A and B were kindly provided by Dr. H. Granzow and the graphic design was prepared by M. Jörn (Friedrich-Loeffler-Institut, Greifswald-Insel Riems, Germany).

TABLE 10.2 Genome Organization and Coding Capacity of Hantaviruses

Hantavirus	Genome 5'-NCR, coding, 3'-NCR (nt)				NP	Protein (aa)		
	S	NSs	M	L		Gn	Gc	RdRp
Hantaan virus (HTNV)	36, 1290, 370	not	40, 3408, 168	37, 6456, 40	429	523	486	2151
Puumala virus (PUUV)	42, 1302, 486	+	40, 3447, 195	36, 6471, 43	433	528	485	2156
Dobrava Belgrade virus (DOBV)	35, 1290, 348	no	40, 3408, 187	37, 6456, 39	429	523	486	2151
Tula virus (TULV)	43, 1290, 499	+	55, 3426, 213	37, 6462, 43	429	528	484	2153
Seoul virus (SEOV)	42, 1290, 437	no	46, 3402, 203	36, 6455, 38	429	522	486	2151
Thailand virus (THAIV)	45, 1290, 549	no	-	-	429	-	-	-
Amur/Soochong virus	-, 1290, -	no	38, 3408, -	37, 6456, 40	429	523	486	2151
Da Bie Shan virus (DBSV)	-, 1290, -	no	46, 3405, 175	37, 6455, 40	429	524	485	2151
Sin Nombre virus (SNV)	42, 1287, 730	+	51, 3423, 222	35, 6462, 65	428	526	483	2153

Continued

New York virus (NYV)	42, 1287, 749	+	51, 3423, 194	-	428	526	486	-
El Moro Canyon virus (ELMCV)	42, 1287, 567	+	51, 3420, 330	-	428	526	484	-
Rio Segundo virus (RIOSV)	42, 1287, 420	+	-	-	428	-	-	-
Andes virus (ANDV)	42, 1287, 542	+	51, 3417, 203	35, 6462, 65	428	525	485	2153
Bayou virus (BAYV)	42, 1287, 629	+	48, 3426, 190	-	428	528	484	-
Black Creek Canal virus (BCCV)	42, 1287, 660	+	51, 3426, 191	-	428	-	-	-
Cano Delgadito virus (CADV)	39, 1287, 660	+	52, 3417, 206	43, 6462, 73	428	525	484	2153
Laguna Negra virus (LANV)	42, 1287, 575	+	51, 3417, 230	-	428	525	484	-
Muleshoe virus (MULV)	42, 1287, 660	+	-	-	428	-	-	-
Rio Mamore virus (RIOMV)	42, 1287, 597	+	51, 3417, 226	-	428	525	485	-
Choclo virus	42, 1287, 643	+	50, 3417, 218	-	428	525	485	-
Araraquara virus (ARAV)	42, 1287, 529	+	-	-	428	-	-	-

TABLE 10.2 Genome Organization and Coding Capacity of Hantaviruses—cont'd

Hantavirus	Genome 5'-NCR, coding, 3'-NCR (nt)				Protein (aa)			
	S	NSs	M	L	NP	Gn	Gc	RdRp
Prospect Hill virus (PHV)	42, 1302, 331	+	49, 3429, 229	36, 6467, 55	433	527	486	2155
Sangassou virus (SANGV)	45, 1290, 411	no	40, 3408, 203	37, 6456, 38	429	523	486	2151
Khabarovsk virus (KHAV)	42, 1302, 501	+	-, 3435, 222	-	433	527	486	-
Vladivostok virus (VLAV)	-	+	-	-	-	-	-	-
Yuanjiang virus (YUJV)	42, 1302, 517	+	-	-	433	-	-	-
Isla Vista virus (ISLAV)	42, 1302, 377	+	-	-	433	-	-	-
Muju virus (MUJV)	42, 1302, 513	+	39, 3429, 166	-	433	528	486	-
Hokkaido virus (HOKV)	-, 1302, -	+	40, 3447, 195	-	433	528	486	-
Topografov virus (TOPV)	42, 1302, 607	+	-, 3429, -	-	433	527	485	-
Thottapalayam virus (TPMV)	67, 1308, 155	no	39, 3366, 216	62, 6450, 63	436	503	482	2150

Virus								
Imjin virus (MJNV)	67, 1310, 199	no	40, 3363, 213	44, 6450, 75	437	503	482	2149
Nova virus (NVAV)	52, 1287, 500	no	-	33, 6474, 56	429	-	-	2157
Asama virus (ASAV)	-, 1302, 393	no	-, 3423, -	-	433	525	483	-
Seewis virus (SWSV)	46, 1290, 305	no	-	-	429	-	-	-
Kenkeme virus (KKMV)	47, 1290, 303	no	-	-	429	-	-	-
Jemez Springs virus (JMSV)	38, 1290, 282	no	-	-	429	-	-	-
Cao Bang virus (CBNV)	38, 1287, 508	no	40, 3420, 177	-	429	524	483	-
Rockport virus (RKPV)	32, 1287, 511	no	57, 3411, 179	42, 6464, 53	429	526	479	2154
Oxbow virus (OXBV)	38, 1287, 380	no	41, 3426, 76	-	429	526	483	-

-, no information or no complete sequence available in GenBank ; +, NSs-ORF present

TABLE 10.3 Pairwise Nucleotide Sequence Identities Between the Entire N-, GPC- and RdRp-encoding S-, M- and L-segment Sequences (above the diagonal) and Amino Acid Sequence Identities Between the Corresponding Proteins (below the diagonal) of Insectivore-borne Hantaviruses *Thottapalayam Virus* (TPMV), Imjin Virus (MJNV) and Rockport Virus (RKPV) and Different Rodent-borne Hantaviruses

Segment and virus species	% identity with virus species									
	HTNV	SEOV	DOBV	PUUV	TULV	SNV	ANDV	TPMV	MJNV	RKPV
S-segment ORF										
HTNV	-	74.5	73.9	60.9	63.1	62.4	63.0	52.6	53.5	62.0
SEOV	83.2	-	73.8	59.6	62.5	62.8	63.0	52.3	53.7	62.0
DOBV	83.0	81.6	-	61.6	61.7	62.9	61.2	52.5	52.7	61.8
PUUV	60.5	61.9	61.0	-	73.0	67.2	68.3	53.1	52.3	66.8
TULV	63.1	62.6	62.8	79.2	-	70.3	68.3	53.1	53.2	68.4
SNV	62.7	62.0	62.5	70.0	73.5	-	76.6	52.8	53.7	69.5
ANDV	65.1	64.8	64.4	72.1	74.7	86.0	-	55.1	53.6	71.0
TPMV	46.4	45.1	46.2	44.0	44.2	47.7	46.5	-	66.7	52.8

MJNV	45.8	44.6	45.3	44.2	43.5	46.3	46.8	69.7	–	52.5
RKPV	63.2	62.7	64.1	71.4	76.5	76.9	79.0	46.8	44.7	–
M-segment ORF										
HTNV	–	71.8	71.3	57.3	58.6	57.4	57.8	50.0	50.7	58.1
SEOV	77.0	–	70.8	57.7	58.6	56.6	57.5	50.6	51.0	57.2
DOBV	77.3	77.1	–	57.7	58.0	58.1	57.5	50.4	51.0	57.0
PUUV	52.6	52.6	52.1	–	71.5	65.4	64.9	50.5	50.3	63.1
TULV	54.5	53.8	54.3	78.3	–	67.1	65.5	51.3	51.5	65.0
SNV	54.2	52.3	53.2	66.0	69.0	–	71.3	50.7	50.4	62.9
ANDV	54.2	53.5	53.6	65.7	67.2	77.7	–	52.3	50.7	62.8
TPMV	41.7	41.2	42.5	40.9	41.7	41.8	42.6	–	68.7	50.8
MJNV	42.4	42.0	41.9	41.5	42.6	41.5	43.4	71.7	–	50.6
RKPV	52.9	52.8	52.4	61.7	62.7	62.1	62.2	41.8	41.3	–
L-segment ORF										
HTNV	–	74.3	74.7	65.9	65.2	65.8	65.9	62.3	62.3	65.8
SEOV	85.0	–	74.9	66.5	65.8	66.4	65.8	61.6	61.8	65.6
DOBV	85.2	85.4	–	66.5	66.0	66.7	66.0	61.9	62.4	66.0
PUUV	68.9	68.6	69.5	–	75.0	70.8	71.2	62.8	62.8	70.9

Continued

TABLE 10.3 Pairwise Nucleotide Sequence Identities Between the Entire N-, GPC- and RdRp-encoding S-, M- and L-segment Sequences (above the diagonal) and Amino Acid Sequence Identities Between the Corresponding Proteins (below the diagonal) of Insectivore-borne Hantaviruses *Thottapalayam Virus* (TPMV), Imjin Virus (MJNV) and Rockport Virus (RKPV) and Different Rodent-borne Hantaviruses—cont'd

Segment and virus species	% identity with virus species									
	HTNV	SEOV	DOBV	PUUV	TULV	SNV	ANDV	TPMV	MJNV	RKPV
TULV	68.4	68.8	68.8	85.0	-	71.3	71.4	62.5	62.1	70.4
SNV	69.2	69.0	69.5	77.8	78.5	-	75.4	62.0	62.3	71.4
ANDV	68.6	68.1	68.2	77.2	78.1	86.7	-	61.8	62.6	71.0
TPMV	62.3	61.9	61.8	61.6	61.4	61.9	62.1	-	74.2	61.4
MJNV	62.2	61.3	61.5	61.6	61.4	61.8	61.4	81.6	-	63.1
RKPV	67.8	67.8	68.5	75.7	76.2	77.9	76.5	61.2	61.5	-

Abbreviations and accession numbers: HTNV, *Hantaan virus* (NC_005218, NC_005219, NC_005222); SEOV, *Seoul virus* (NC_005236, NC_005237, NC_005238); DOBV, *Dobrava-Belgrade virus* (NC_005233, NC_005234, NC_005235); PUUV, *Puumala virus* (NC_005224, NC_005223, NC_005225); TULV, *Tula virus* (NC_005227, NC_005228, NC_005226); SNV, *Sin Nombre virus* (NC_005216, NC_005215, NC_005217); ANDV, *Andes virus* (NC_003466, NC_003467, NC_003468); TPMV, *Thottapalayam virus* (NC_010704, NC_010708, NC_010707); MJNV, Imjin virus (EF641805, EF641799, EF641807); RKPV, Rockport virus (HM015223, HM015219, HM015221).

the S-segment for certain hantaviruses, i.e., hantaviruses associated with rodents of the Cricetidae family (Plyusnin and Morzunov, 2001) (see Table 10.2). This ORF was found to be conserved in its position within the S-segment and its size, but the aa sequences of the putative NSs proteins are highly divergent, most likely due to the selection pressure of the overlapping N protein encoding ORF. Although the putative NSs ORF of PUUV is intact in rodent-derived strains, this ORF seems to be not necessary in type I interferon-deficient Vero E6 cells (Rang et al., 2006). The putative NSs is assumed to play a role in the adaptation of the virus to the rodent host (Ulrich et al., 2002). In addition, the NSs proteins of PUUV and TULV were shown to have moderate interferon antagonist activity (Jaaskelainen et al., 2007; Jaaskelainen et al., 2008) (see below and Table 10.4). Recently, in the S-segment of the shrew-borne Seewis virus in the opposite orientation to the N-ORF, a putative 612 nt-long additional ORF was predicted (Schlegel et al., 2012b).

The M-segment is 3543–3801 nt long and encodes the GPC which is cotranslationally cleaved into the amino-terminal Gn protein of 503–528 aa residues and the carboxy-terminal Gc protein of 479–486 aa residues (Table 10.2). The nt sequence identity level of the entire GPC-ORF and the aa sequence of GPC is about 50–72% and 42–77%, respectively (Table 10.3). Both glycoproteins are integral transmembrane proteins with C-terminal hydrophobic anchor domains (Elliott et al., 1991; Spiropoulou, 2001). The processing of the GPC is mediated by a cellular protease at a conserved WAASA motif (Lober et al., 2001). The long cytoplasmic tail of the Gn protein of about 150 aa contains a highly conserved YRTL-motif, which is proposed to be a glycoprotein trafficking signal and the tail of the Gc contains an endoplasmic reticulum retention signal (KKXX-motif) (Spiropoulou, 2001). In addition, both glycoproteins contain conserved N-glycosylation sites (NXS/T) (Spiropoulou, 2001).

The L-segment of 6529–6578 nt has the coding information for a 2147–2155 aa-residue long RdRp that functions as a replicase, transcriptase and endonuclease (Plyusnin et al., 1996). The nt sequence identity level of the entire RdRp-ORF and the aa sequence of RdRp is higher than those observed for N- and GPC-ORF and the corresponding proteins, and reaches about 62–75% and 62–85%, respectively (Table 10.3). This protein comprises typical conserved sequence motifs preA, A, B, C, D and E at aa positions 884–902, 964–980, 1050–1077, 1091–1101, 1152–1164 and 1171–1181 including an XDD motif that is thought to be essential for catalytic activity (Jonsson and Schmaljohn 2001; Kukkonen et al., 2005; Kang et al., 2009b). Molecular and biochemical characterization of the RdRp and its domains are hampered by the lack of its high-level heterologous expression and the lack of a suitable reverse-genetics system for hantaviruses.

The 5'-terminal noncoding region (NCR) of all three segments is 35–67 nt long, whereas the 3'-terminal NCR of the three genome segments differ strongly in the length. The L-segment contains the shortest 3'-NCR of 38–75 nts, whereas the 3'-NCR of the S-segment has a size of about 300–700 nt.

TABLE 10.4 Innate Immune Response Evasion Mechanisms of Hantaviruses

Virus	Protein	Effect	Reference
ANDV and PHV	Gn/Gc	Cytoplasmic retention of STAT1	Spiropoulou et al., 2007
NYV	cytoplasmic Gn-tail	Block of TBK1-mediated ISG-induction	Alff et al., 2006
PUUV	NSs	Reduction of poly-IC induced IFN-beta reporter gene activity	Jaaskelainen et al., 2007
NYV	cytoplasmic Gn-tail	Destabilisation of TBK1/TRAF3 complex	Alff et al., 2008
HTNV	NP	Interaction with Importin-alpha block of nuclear import of NFkB	Taylor et al., 2009
ANDV	NP + Gn/Gc	Reduction of Sendai virus induced IFN-beta reporter gene activity	Levine et al., 2010
SNV	Gn/Gc		
TULV	cytoplasmic Gn-tail	Reduction of IFN-beta reporter gene activity	Matthys et al., 2011

ANDV, *Andes virus*; PUUV, *Puumala virus*; PHV, *Prospect Hill virus*; NYV, *New York virus*; HTNV, *Hantaan virus*; SNV, *Sin Nombre virus*; TULV, *Tula virus*; IFN, Interferon; NP, nucleocapsid protein; Gn/Gc, glycoproteins; NSs, non-structural protein; STAT1, signal transducer and activator of transcription; TBK1, TRAF family member-associated NF-kappa-B activator (TANK) binding kinase 1; IC, polyinosine-polycytidylic acid; ISG, interferon stimulated gene; TRAF3, tumor necrosis factor receptor associated factor 3; NFkB, nuclear factor 'kappa-light-chain-enhancer' of activated B cells

Hantavirus-typical 5'- and 3' proximal sequences have been identified for all three segments which are highly conserved and form, due to their inverted complementary sequences, panhandle-like structures (Schmaljohn and Darymple, 1983; Elliot et al., 1991; Antic et al., 1992; Plyusnin et al., 1996).

REPLICATION

Hantaviruses enter the host cell by receptor-mediated endocytosis (Figure 10.2). Decay-accelerating factor (DAF, CD55), the complement receptor for the globular head domain of the complement C1q, and beta-integrins have been reported to be involved in hantavirus binding and uptake (Hepojoki et al., 2012).

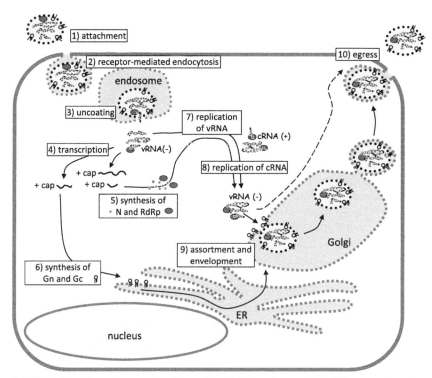

FIGURE 10.2   **Hantavirus replication cycle. (1) Hantaviruses bind different cell surface receptors. Pathogenic hantavirus species preferentially recruit beta3-integrins. PHV and TULV have been reported to interact with beta1-integrins. Furthermore, decay-accelerating factor (CD55) and the complement receptor gC1qR can be involved in hantavirus binding. (2) Most hantaviruses are internalized by clathrin-dependent endocytosis. Uptake of ANDV was reported to proceed via clathrin-independent pathways. (3) Uncoating is triggered at low pH by Gc-controlled membrane fusion, which leads to release of the genomic ribonu-cleoprotein complex. (4) Transcription is initiated at the panhandle structure of the genomic minus-stranded RNA (vRNA). Translation of viral transcripts requires hijacking of capped RNA 5´ends from the cellular mRNA pool. N protein has been reported to be a key player both for cap-stealing and directing mature viral transcripts to the translation machinery. (5) N protein and RdRp are translated by free ribosomes in the cytoplasm. (6) Gn and Gc are transmembrane glycoproteins which derive from a precursor polyprotein, which is cleaved during translation into the ER. (7) Replication of the genomic RNA (vRNA) to plus-stranded complementary replication intermediates (cRNA) most likely requires N-protein levels suf-ficient to decorate the synthesized RNA. (8) The cRNA panhandle structure controls initiation of the vRNA synthesis. This specific structure most likely prevents encapsidation of the cRNA. (9) The three vRNA complexes assort and bud into the Golgi. SNV and BCCV have been reported to assemble at the cell surface. (10) Mature virions are transported to the surface and egress without obvious cytopathic effects to the host cell.**

Interestingly, HFRS- and HCPS-causing hantaviruses bind integrin alpha IIa/beta3 and alpha v/beta3. In contrast nonpathogenic PHV or TULV enter the cells via integrin alpha5/beta1. Based on the selected recruitment of intergrin alpha IIa/beta3 and alpha v/beta3 by pathogenic hantaviruses *in vitro*, this virus/integrin interaction has been suggested to contribute to development of pathogenic symptoms in humans by modulation of endothelial function (Mackow and Gavrilovskaya, 2009).

Uptake of some hantaviruses is mediated by clathrin-coated vesicles, including HTNV, BCCV and SEOV (Jin et al., 2002). However, ANDV was reported to be internalized via clathrin-independent mechanisms. From early or late endosomes the genomic ribonucleoprotein complex is released after membrane fusion. Fusion of virus and cellular membrane can be triggered by reduction of the pH which activates the fusogenic loop of Gc (Hepojoki et al., 2012).

Upon release into the cytoplasm the replication cycle can be divided in three phases. The first phase is characterized by transcription and accumulation of viral proteins. Low levels of N protein favor production of viral mRNA transcripts, which are about 100 nucleotides shorter compared to the corresponding minus-stranded RNA templates (vRNA). Furthermore, the mRNA transcripts lack a 3'-poly A tail and contain capped heterologous 5' RNA extensions derived from the cellular mRNA pool. The N protein has been shown to bind capped mRNAs in processing bodies, which mediate storage and decay of cellular mRNA. Furthermore, the N protein was able to substitute the translation initiation factor eIF4F. Therefore, the N protein seems to be important both for primary transcription and translation of the viral mRNAs (Hussein et al., 2011).

Transition to the second phase of the replication cycle is driven by accumulation of high levels of N protein, which are required for production of minus-strand genomic RNA (vRNA) via the plus-stranded complementary replication intermediate (cRNA). Details of the transcription/replication mechanisms are only poorly understood. It seems clear, that the panhandle structures formed by the highly conserved complementary 3'- and 5'- terminal nucleotides serve as promoter for initiation of transcription and replication by the viral RdRp. RNA synthesis by the RdRp was reported to proceed via a so-called "prime and realign" mechanism (Garcin et al., 1995). Due to this mechanism the tri-phosphorylated 5'-terminus of the viral RNA is removed. RNAs with tri-phophorylated 5'-ends are prominent activators of the cellular pathogen recognition receptor RIG-I (Elliott and Weber, 2009). Thus, this transcription/replication mechanism can be considered as a strategy to escape activation of innate antiviral immune responses.

The third replication phase starts with the recruitment of the genomic vRNA associated with N protein and RdRp to the ER-Golgi intermediate compartment or the cis-Golgi. Recruitment is controlled by the cytoplasmic C-terminal tail of the viral glycoproteins inserted in the membrane of the Golgi. Binding of the glycoproteins triggers assembly and envelopment of the three genomic ribonucleoprotein complexes by budding into the lumen of the Golgi (Shi and Elliott, 2004). From the Golgi virions are transported out of the cell with recycling endosomes. Some New World hantaviruses, i.e., SNV and BCCV, have been reported to assemble and mature at the cell surface (Ravkov et al., 1997). Thus, hantaviruses seem to use different pathways for uptake and release of infectious virions.

The segmented genome organization allows generation of reassorted progeny viruses in host cells infected with two different hantaviruses. Between different hantavirus species reassortment is restricted *in vivo* due to the narrow host range, which prevents dual infection of the same host. In addition to this primary limitation, *in vitro* studies revealed that further viral factors repress reassortment between unrelated hantaviruses. At least for some species, this repression can be overcome, as documented by isolation of stable reassortants produced after dual infection in cell culture (McElroy et al., 2004; Kirsanovs et al., 2010; Handke et al., 2010).

The multiple steps of the replication cycle are only poorly understood. Development of a reverse genetics system for hantaviruses is urgently required to develop new approaches against this life-threatening pathogen.

GEOGRAPHICAL DISTRIBUTION OF HANTAVIRUSES

The current knowledge of the geographical distribution of hantaviruses is based on the recording of human cases and the detection of hantaviruses in reservoirs. The frequency of the detection of human infections is determined on one hand by the clinical outcome of the infection. Thus, for PUUV infections in Europe a high portion of human infections are not recorded due to unspecific or mild symptoms of the disease. On the other hand recording is influenced by the awareness of the clinicians and the presence of formal reporting systems. In several countries hantavirus cases are notifiable. Thus, the number of human infections for a large number of European countries has been recorded (Heyman et al., 2011). Studies in reservoirs are mainly driven by outbreak investigations, in order to determine molecular epidemiological evidence for the causative agent of a cluster or outbreak of human disease (Ettinger et al., 2012; Faber et al., 2013). In addition, rodent monitoring programs are running in several European countries. The current knowledge on the geographical distribution of shrew- and mole-borne hantaviruses is scarce as the data are usually based on low numbers of specimens investigated. It will be interesting to follow the geographical distribution of bat-associated hantaviruses in the future.

In general, the geographical distribution of hantaviruses follows the presence of their reservoir hosts. Thereby, in Europe and Asia, hantaviruses can be found which are exclusively harbored by "Old World" small mammal species. In Asia the majority of human cases are caused by infections with hantaviruses associated to rodents of the Murinae subfamily. Following the geographical distribution of the bank vole in almost all parts of Europe, in the majority of European countries PUUV infections have been found in the reservoir and in humans (Heyman et al., 2011; Vapalahti et al., 2003). However, in the United Kingdom PUUV infections have not been found in rodents or in humans (Vapalahti et al., 2003). In Germany, there appears to be a heterogenic distribution of the PUUV in bank voles (Ettinger et al., 2012; Faber et al., 2013; our unpublished data). In contrast, in the Americas hantaviruses are mainly associated with rodents of the New World, i.e., representatives of the subfamilies Neotominae, as the deer mouse-associated SNV in North America, and Sigmodontinae, such as the rice rat-associated ANDV in southern America. In some countries the presence of several hantaviruses is known. Thus, in Germany at least five different rodent- and shrew-borne hanta-viruses occur, challenging the diagnostic capacity of the responsible laboratories.

Microtus-associated hantaviruses are naturally distributed in parts of the Old and New World. Thus, in North America PHV and *Isla Vista virus* have been detected, whereas in Europe and Asia TULV, *Vladivostok virus* and *Khabarovsk virus* were found (Horling et al., 1996; Lee et al., 1985; Song et al., 1995; Kariwa et al., 1999).

SEOV transmitted by different *Rattus* species is thought to be distributed world-wide, as its reservoir host is abundant in almost all parts of the world. Indeed SEOV infections have been detected not only in rats in Asia but also in rats from some European and American countries (Heyman et al., 2004; Easterbrook et al., 2007; Cueto et al., 2008; Heyman et al., 2009; Woods et al., 2009; Lin et al., 2012). Recently, SEOV infections were reported for wild and pet Norway rats in the United Kingdom causing human infections (Jameson et al., 2013a; Jameson et al., 2013b). Except for a laboratory infection with SEOV, there are no additional reports on human SEOV infections in Europe (for review see Krüger et al., 2001).

Only very limited information is available about hantavirus-infected small mammals and human infections for Africa and Australia (Ulrich et al., 2002). Few reports illuminate the presence of hantavirus-specific antibodies in humans in Central and East African countries (Gonzalez et al., 1984; Coulaud et al., 1987; Dupont et al., 1987; Gonzalez et al., 1989; Rodier et al., 1993). First molecular evidence for the presence of hantaviruses in Africa was reported for a rodent (Klempa et al., 2006) and a shrew (Klempa et al., 2007). The rodent-associated hantavirus, named Sangassou virus, is the first African hantavirus which has been isolated in cell culture and characterized. Recently, serological evidence for human infections and disease with hantavirus-typical symptoms has been reported (Klempa et al., 2010). There is only one report about the presence of hantavirus-specific antibodies in wild-trapped rodents in Australia from Guinea (LeDuc et al., 1986).

Worldwide, hundreds of thousands of hantavirus disease cases are estimated to occur annually, with the highest numbers in China and Korea (Krüger et al., 2011). Increasing numbers of disease cases are also reported from Europe (Heyman et al., 2011). Oscillations of the number of recorded human cases have been observed in Finland, Sweden, Belgium and Germany (Heyman et al., 2011). In Germany alone, over 2800 clinical cases were notified in 2012—more than in any of the previous 12 years since implementation of the federal reporting system (http://www3.rki.de/SurvStat).

DISEASE IN HUMANS

Virus transmission to humans occurs via inhalation of aerosolized virus-contaminated rodent urine, saliva, and feces, rarely by rodent bites. Humans are usually considered as a dead-end host that does not transmit the virus further. In contrast to the asymptomatic course in the natural reservoir hosts, about 20% of immune-naïve persons develop clinical symptoms after infection. The severity of the hantavirus disease can range from mild to severe with case fatality rates of up to 35–50%. Infections by virus types circulating in Asia and Europe lead to hemorrhagic fever with renal syndrome, HFRS, whereas hantavirus disease caused by American virus types is mainly associated with cardiopulmonary failure and is therefore called hantavirus cardiopulmonary syndrome, HCPS (Peters et al., 1999; Krüger et al., 2011). It is highly probable that hantavirus disease occurs in Africa, too (Klempa et al., 2010).

Both HFRS and HCPS are associated with blood coagulation disturbances and changes in vascular permeability leading to edema and inflammations in affected organs (Krüger et al., 2011). Moreover, both syndromes can include renal as well as pulmonary symptoms. Therefore, the general use of the term "hantavirus disease" has been proposed (Clement et al., 2012).

The clinical course of hantavirus disease comprises different phases (Lee and van der Groen, 1989; Mertz et al., 2006). The incubation time before onset of symptoms is typically 2–3 weeks; however, periods up to 6 weeks have been documented (Kramski et al., 2009). In the early (prodromal) phase of about 3–5 days, HFRS and HCPS patients exhibit fever, myalgia, malaise, headache, backache, abdominal pain, and often nausea and diarrhea. For HFRS patients, also vision disorders have been reported.

In the next phases (about 2–7 days) following the prodrome, hypotension occurs and can result in cardiogenic shock and death. In HCPS patients lung edema and even lung failure develop and require, depending on severity, supplemental oxygen, intubation with mechanical ventilation, or extracorporal membrane oxygenation (ECMO). HFRS patients mainly show renal impairment or failure and require, in severe cases, hemodialysis during this oligouric/anuric phase. In addition to cardiogenic shock, lung or renal failure are also common reasons for fatal outcome.

The beginning of the diuretic phase is a positive prognostic sign for the patient. Clinical improvement is usually rapid. After a few days, the convalescence period starts which can last over several weeks. Hantavirus syndromes are regarded as acute diseases and it is a matter of discussion to what extent they can result in long-term sequela.

There are wide differences between the severities of clinical courses dependent on the hantavirus type causing infection and probably additional genetic properties of the human host. In some cases the different phases of disease are difficult to distinguish.

DIAGNOSIS OF HANTAVIRUS INFECTIONS

The laboratory diagnosis of human hantavirus infections is usually based on serological assays (for review see Krüger et al., 2001). Classically, the detection of hantavirus-specific IgM and IgG antibodies is based on indirect immuno-fluorescence assays (IFA) using hantavirus-infected Vero E6 cells. Currently, commercial IFA kits are available offering the parallel investigation of a single serum sample for different antibody specificities. Due to the necessity of a high-containment laboratory for virus antigen production and the lack of cell-culture isolates for the majority of novel hantaviruses, recombinant antigens have been used frequently as surrogates for the viral antigens. The N protein was found to be the antigen of choice for diagnostic applications as it induces an early and long-lasting antibody response (Hedman et al., 1991; Sjolander et al., 1997). Recombinant hantavirus antigens have been produced in large amounts in *Escherichia coli*, yeast *Saccharomyces cerevisiae* and insect cells (Yoshimatsu et al., 1993; Kallio-Kokko et al., 2000; Krüger et al., 2001; Razanskiene et al., 2004). These recombinant antigens are used in IgM and IgG indirect and capture enzyme-linked immunosorbent assay (ELISA) formats, Western blot and line immunoblot assays. The monoclonal antibody capture format is especially advantageous when using total lysates from hantavirus-protein expressing insect cells allowing a purification of the diagnostic antigen directly on the ELISA plate. A commercial line assay offers the parallel analysis of a single sample with different antigens in one test. In addition to the commercially available ELISA and immunoblot assays, rapid immunochromatographic assays are available which can be used as bedside (or field) tests (Hujakka et al., 2003; Sirola et al., 2004).

The cross-reactivity of the N proteins of closely related hantavirus species, e.g., within the groups HTNV/SEOV/DOBV, PUUV/TULV and SNV/ANDV, resulted in the initial assumption that the exclusive use of one antigen of each group might be sufficient for a highly sensitive diagnostic assay. However, several investigations underline the necessity to use the homologous antigen for a highly sensitive serological detection (Zoller et al., 1995; Vapalahti et al., 1996a; Meisel et al., 2006). Cross-reactivity resulted in the inability of all N-protein based serological assays to allow a differentiation of the causative hantavirus species (Krüger et al., 2001). Although it seems to be possible by endpoint titration, especially when using truncated N protein derivatives to evaluate the most likely source of infection (Araki et al., 2001), the focus reduction neutralization assay is the gold standard for differentiation (Lundkvist et al., 1997a). However, this assay is limited to a retrospective analysis in late convalescent sera and to the identification of the most related virus used in the assay. Investigations of the antibody response in rodents are usually based on indirect IFA and ELISA in-house assays or a rapid assay (Vaheri et al., 2008). Future serological investigations in reservoirs would profit from the development of competitive or double-antigen sandwich test formats that are independent from a species-specific secondary antibody conjugate.

In humans (and most likely spillover-infected nonreservoirs) the viremia and therefore the possibility to detect viral nucleic acid in blood is short-termed (Groen et al., 1995; Horling et al., 1995; Plyusnin et al., 1999). Therefore, nucleic acid detection as well as virus isolation from patients has been successful only rarely. In contrast, reservoirs are in general believed to be persistently infected, perhaps life-long. Therefore, investigations in reservoirs allow a molecular characterization of the infecting hantavirus species and strain. The initial screening of known hantaviruses is usually based on conventional One Step or nested RT-PCR or quantitative (real-time) RT-PCR (RT-qPCR) approaches (see, for example, Klempa et al., 2006; Kramski et al., 2007; Mohamed et al., 2013). The search for novel hantaviruses profits from the higher sequence conservation in the L-segment that allowed the development of a broad-spectrum nested RT PCR assay (Klempa et al., 2006). The future application of Next Generation Sequencing approaches will most likely open the possibility of finding and characterizing novel hantaviruses, but might be currently too expensive and insensitive for application in standard diagnostics. The nucleic acid array technology may offer a high throughput of diagnostic samples (Nordström et al., 2004), but might be especially interesting when combining hantavirus-specific oligonucleotide probes with those of other zoonotic pathogens.

HANTAVIRUS RESERVOIRS AND TRANSMISSION

Hantaviruses have been thought initially to be exclusively rodent-borne pathogens (Henttonen et al., 2008). Hantaviruses were detected in rodents of the families Cricetidae and Muridae, but so far not in other rodent families (Table 10.5). The family Muridae, subfamily Murinae, associated viruses form a clade in the phylogenetic tree and is represented by the prototype species HTNV, the *Rattus*-associated SEOV, the *Apodemus*-associated DOBV, the African Sangassou virus and related viruses (see Figure 10.3). Within the family Cricetidae hantaviruses were identified in representatives of the subfamilies Arvicolinae, Sigmodontinae and Neotominae. The Arvicolinae- and Sigmodontinae/ Neotominae-associated hantaviruses form two well-supported clades (Figure 10.3). Interestingly, *Microtus* species related hantaviruses were found in Europe, Asia and Northern America (see above). In contrast PUUV, the most important human pathogenic hantavirus in Europe, and related viruses, such as Muju virus and Hokkaido virus, were only found in Europe and Asia, but so far not in the Americas (Kariwa et al., 1999; Kariwa et al., 1995; Song et al., 2007d; Plyusnina et al., 2008), although *Myodes* species are present in northern America (see Figure 10.4). Large numbers of hantaviruses associated with rodents of the New World rodent subfamilies Sigmodontinae and Neotominae have been expectedly identified only in the New World.

TPMV, described originally as an arbovirus detected in a shrew species in India (Carey et al., 1971), was later confirmed to be a hantavirus, the first shrew-borne hantavirus ever detected (Song et al., 2007a). Thirty years after the first description of TPMV and mediated by the development of broad-spectrum RT-PCR assays rapidly a large number of Soricomorpha-associated hantaviruses have been discovered. Thus, novel hantaviruses have been detected in representatives of the subfamilies Soricinae and Crocidurinae, including Old World shrews in Africa, Asia, Europe and New World shrews (Figure 10.3).

TABLE 10.5 Taxonomy of the Orders Rodentia, Soricomorpha and Chiroptera with Taxa Where Hantaviruses Were Detected So Far

Order[*]	Suborder	Family	Hantavirus detection
Rodentia	Sciuromorpha	Aplodontiidae, Sciuridae, Gliridae	-
	Castorimorpha	Castoridae, Heteromyidae, Geomyidae	-
	Myomorpha	Dipodidae, Platacanthomyidae,	-
		Spalacidae, Calomyscidae, Nesomyidae	-
		Cricetidae	+
		Muridae	+
	Anomalomorpha	Anomaluridae, Pedetidae	-
	Hystricomorpha	Ctenodactylidae, Bathyergidae,	-
		Hystricidae, Petromuridae,	-
		Thryonomyidae, Erethizontidae,	-
		Chinchillidae, Dinomyidae[†*],	-
		Caviidae, Dasyproctidae, Cuniculidae,	-
		Ctenomyidae, Octodontidae,	-
		Abrocomidae, Echimyidae,	-
		Myocastoridae, Capromyidae,	-
		Heptaxodontidae[†]	-
Soricomorpha		Nesophontidae[†], Solenodontidae	-
		Soricidae	+
		Talpidae	+
Chiroptera		Pteropodidae	-
		Rhinolophidae	+
		Hipposideridae, Megadermatidae,	-
		Rhinopomatidae, Craseonycteridae,	-
		Emballonuridae	-
		Nycteridae	+
		Myzopodidae, Mystacinidae,	-
		Phyllostomidae, Mormoopidae,	-
		Noctilionidae, Furipteridae,	-
		Thyropteridae, Natalidae, Molossidae	-
		Vespertilionidae	+

*Taxonomy according to Wilson and Reeder, 2005a; Wilson and Reeder, 2005b
[†]extinct family
[†*]extinct family with one recent species

Interestingly, these shrew-borne hantaviruses are not monophyletic. In addition, four hantaviruses have been identified in different mole species, including the Rockport virus (RKPV) mostly related to Cricetidae-associated hantaviruses (Figure 10.3) (Kang et al., 2011).

Recently novel short L-segment hantavirus sequences have been detected in the bat species *Nycteris hispida* (family Nycteridae) and *Neoromicia nanus*

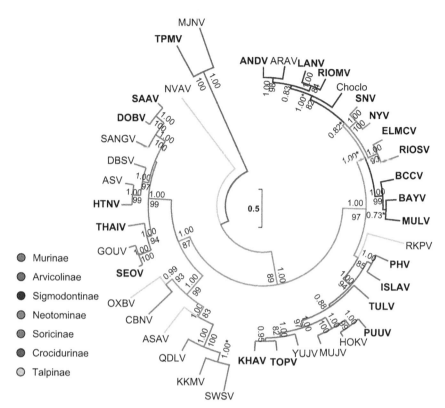

FIGURE 10.3   **Phylogenetic relationships of selected hantaviruses (taken from Schlegel, 2013). Bayesian phylogenetic tree of the complete N protein-encoding ORF nucleotide sequences of different rodent- and insectivore-borne hantaviruses and their associations to the Rodentia and Soricomorpha subfamilies. Posterior probabilities for Bayesian analysis are given above the branches and bootstrap values for the corresponding Maximum Likelihood (ML) tree under the branches. Only values ≥0.7 and ≥70% are shown. * indicates differences in the topology between ML- and Bayesian-tree at this node. Abbreviations of official hantavirus species are given in bold.** Virus abbreviations: ANDV, Andes virus; ARAV, Araraquara virus; ASAV, Asama virus; ASV, Amur/Soochong virus; BAYV, Bayou virus; BCCV, Black Creek Canal virus; DBSV, CBNV, Cao Bang virus; Da Bie Shan virus; DOBV, Dobrava-Belgrade virus; ELMCV, El Moro Canyon virus; GOUV, Gou virus; HOKV, Hokkaido virus; HTNV, Hantaan virus; ISLAV, Isla Vista virus; KHAV, Khabarovsk virus; KKMV, Kenkeme virus; LANV, Laguna Negra virus; MUJV, Muju virus; MULV, Muleshoe virus; MJNV, Imjin virus; NYV, New York virus; NVAV, Nova virus; OXBV, Oxbow virus; PHV, Prospect Hill virus; PUUV, Puumala virus; QDLV, Quiandao Lake virus; RIOMV, Rio Mamore virus; RIOSV, Rio Segundo virus; RKPV, Rockport virus; SAAV, Saarema virus; SANGV, Sangassou virus; SEOV, Seoul virus; SERV, Serang virus; SNV, Sin Nombre virus; SWSV, Seewis virus; THAIV, Thailand virus; TPMV, Thottapalayam virus; TOPV, Topografov virus; TULV, Tula virus; VLAV, Vladivostok virus; YUJV, Yuanjiang virus.

(family Vespertilionidae) (Sumibcay et al., 2012; Weiss et al., 2012). The most recent multiple detection of related sequences in several individuals of the bat species *Pipistrellus abramus, Rhinolophus affinis, Rhinolophus sinicus, and Rhinolophus monoceros* confirmed that bats, in addition to rodents, shrews and

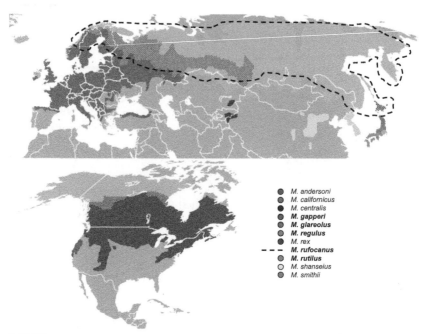

●	*M. andersoni*
●	*M. californicus*
●	*M. centralis*
●	***M. gapperi***
●	***M. glareolus***
○	***M. regulus***
●	*M. rex*
– –	***M. rufocanus***
○	***M. rutilus***
○	*M. shanseius*
●	*M. smithii*

FIGURE 10.4 Geographical distribution of the bank vole *Myodes glareolus* and other *Myodes* species where hantaviruses could be found are in bold and marked with a star, where hantavirus-specific antibodies are found with a circle or antigen with a filled circle. Hantavirus references: *M. gapperi*/Prospect Hill virus (Yanagihara et al., 1987); *M. glareolus*/Puumala virus (Brummer-Korvenkontio et al., 1980; *M. regulus*/Muju virus (Song et al., 2007d); *M. rufocanus*/Hokkaido virus (Daud et al., 2007; Kariwa et al., 1999; Kariwa et al., 1995); *M. rutilus*/hantavirus antigen (Chumakov et al., 1981; Traavik et al., 1984; Zhang et al., 2007a). Rodent data according to: IUCN 2012. *The IUCN Red List of Threatened Species. Version 2012.2.* <http://www.iucnredlist.org>. Downloaded on 13 April 2013.

moles, represent reservoirs for a novel type of hantaviruses (Guo et al., 2013). These recent findings of hantaviruses in different bat species within the families Vespertilionidae and Rhinolophidae may stimulate additional studies in further mammal orders to prove the presence of additional novel hantaviruses.

Usually a hantavirus is found in a single reservoir species, although there is increasing evidence that at least some hantaviruses might be capable of replication in several closely related rodent species. Thus, SEOV was detected in different species of the genus *Rattus* (Johansson et al., 2010). TULV has been initially found in the common vole *Microtus arvalis* and the sibling vole (East European vole) *M. levis* (synonym *M. rossiaemeridionalis*; Plyusnin et al., 1994; Sibold et al., 1995), but was later also molecularly detected in *M. agrestis*, *M. subterraneus*, and even *Arvicola amphibius* (Song et al., 2002; Schmidt-Chanasit et al., 2010; Scharninghausen et al., 2002; Schlegel et al., 2012a). These multiple findings underline the difficulties of the identification of the reservoir of a hantavirus. The "one hantavirus-one reservoir" association assumption is also challenged by the finding of different hantavirus species in the same (potential) reservoir. The finding of HTNV and a certain DOBV genotype in *Apodemus agrarius* with separated geographical distribution in Asia and central Europe might be explained by the occurrence of different genetic lineages of the striped field mouse *Apodemus agarius corea* and *A. agrarius agrarius* in Asia and Europe, respectively (Lee et al., 1978). Similarly, recently Seewis virus (SWSV) and Asikkala virus (ASIV) have been found in the pygmy shrew *Sorex minutus* which might be explained by a spillover of SWSV (Schlegel et al., 2012b; Radosa et al., 2013).

In contrast to all other animal bunyaviruses, the transmission of hantaviruses between rodent reservoirs is thought to be exclusively horizontal, i.e., indirect by inhalation of virus-contaminated dust or directly by biting during aggressive behavior (Douron et al., 1984; Glass et al., 1988; Hardestam et al., 2008; Lee et al., 1978; Mackow and Gavrilovskaya, 2001; Schonrich et al., 2008; Schultze et al., 2002). Hantaviruses were found to be quite stable outside the reservoir for a certain period of time (Kallio et al., 2006a). Currently, there are no hints for a vertical transmission in rodents or a role for arthropod transmission. The transmission route of insectivore- and bat-associated hantaviruses is still unknown.

The major route of transmission to humans seems to be also indirect by inhalation of virus-contaminated rodent excreta, but bites might represent an alternative rare mode of transmission (Douron et al., 1984; Schultze et al., 2002). A human-to-human transmission has been reported exclusively for ANDV (Padula et al., 1998). There are no indications for a perinatal transmission of hantaviruses from the mother to the fetus (Hofmann et al., 2012).

HANTAVIRUS EMERGENCE

Hantaviruses belong to the group of emerging and re-emerging hemorrhagic fever-causing viruses that includes other bunyaviruses but also representatives of the families *Arena-*, *Filo-*, and *Flaviviridae* (Peters, 2006; Vorou et al., 2007; Murphy, 2008).

The emergence of viruses is thought to be mediated by different scenarios. First, viruses might really occur for the first time in the human population. Such an emergence might be caused by ecological changes in reservoir species that allow contact with humans which used to be prevented, e.g., by geographical barriers. Alternatively, the virulence of an already existing pathogen might be increased by genetic modification due to mutation or genetic rearrangements and subsequent selection or population-driven bottleneck events in the reservoir population. Further, outbreaks of zoonotic viruses might be driven by a massive reproduction of the reservoir and/or the vector and thereby an increased probability of human exposure to the pathogen. Finally, novel viruses might have existed already in the human population or the reservoirs, but have not been detected previously due to insufficient diagnostic tools or a lack of analysis of the reservoirs.

The emergence of hantaviruses seems to be driven by several of these scenarios. Thus, the first description of SNV in 1993 clearly indicated the role of a cluster of a disease within an unexpected patients group, the awareness of the physicians and the availability of a diagnostic assay allowing the detection of a so-far unknown virus. A potential reason for the discovery of this novel disease at that time was an increased population density and ratio of SNV-infected deer mouse, that might be caused by increased food resources due to an El Niño Southern Oscillation (Hjelle and Glass, 2000). Interestingly, retrospective studies demonstrated that New World hantaviruses were already present in the rodent population a long time before the first description of HCPS. Clusters and outbreaks of human PUUV infections in Europe were also explained by a population outbreak of reservoirs (see below).

Hantavirus emergence is clearly linked to the occurrence of the reservoir. Therefore, changes in the geographic distribution of a certain potential or real reservoir species might cause the emergence of a hantavirus in a previously hantavirus-free region. Indeed, changes in the geographical distribution of small mammals have been documented in the past and present. During previous years the striped field mouse seemed to increase its geographical distribution in Austria (Sackl et al., 2007). Currently we know very little about these ongoing processes that might be caused by land use and habitat changes and might be influenced also by climate change. Ongoing barcoding of life (BoL) projects may help in documenting and understanding these processes. One has to take into consideration also neozoa as a potential source of hantavirus emergence. Indeed, one study has reported a hantavirus infection in the muskrat (*Ondatra zibethicus*) in Germany (Vahlenkamp et al., 1998), a species introduced to Europe for pelt production. In this respect the novel findings of hantaviruses in bats are highly important, as several bat species can migrate over long distances (Hutterer et al., 2005).

Variation of hantavirus virulence is still very difficult to define. A few studies reported on an increased virulence *in vivo*, e.g., by a single aa exchange in the GPC (Ebihara et al., 2000), or different replication kinetics reflected in different-sized foci *in vitro*, caused by two aa exchanges in the N protein and two aa exchanges in the RdRp, whereas the glycoproteins were identical (Sundström et al., 2011). Alternatively, cell-culture passaging resulted in the attenuation of a PUUV strain that thereby lost its ability to infect the bank vole reservoir (Lundkvist et al., 1997b). This attenuation was mapped to the coding region of the L-segment and to the noncoding region of the S-segment (Lundkvist et al., 1997b; Nemirov et al., 2003).

Spillover infections of nonreservoirs have been thought to be rare but very important processes that might result in a host switch event with dramatic consequences. A lacking co-divergence between certain hantavirus species and their reservoirs has been interpreted as a result of host switch events in the past (Vapalahti et al., 1999). Recent studies have demonstrated spillover infections might be more frequent than previously expected. Thus, the multiple finding of TULV sequences in different *Microtus* species or of *A. agrarius*-borne DOBV, genotype Kurkino, nucleic acid in *A. flavicollis* raises important questions on the host specificity of hantaviruses and factors determining it (see Table 10.6).

TABLE 10.6 Reported Nucleic Acid Detection of Insectivore-borne Seewis Virus and Different Rodent-borne Hantaviruses in Rodent and Insectivore Species, Which are not Postulated or Identified as the Predominant Host for this Particular Hantavirus

Virus species*	Reservoir host species	Natural spillover infections	
		Species	Reference
Hantaan virus	Apodemus agrarius	Apodemus peninsulae, Rattus norvegicus	Zhang et al., 2007b; Zou et al., 2008
Seoul virus	Rattus norvegicus	Rattus flavipectus, Rattus tanezumi, Rattus losea, Mus musculus	Zhang et al., 2007b, GenBank Acc. No. EF210133**; GenBank Acc. No. HQ992814; Shi et al., 2003; GenBank Acc. No. GU592939
Dobrava-Belgrade virus	Apodemus agrarius, Apodemus flavicollis	Apodemus flavicollis, Apodemus sylvaticus, Mus musculus, Sorex araneus, Microtus arvalis, Myodes glareolus, Meriones tamariscinus	Schlegel et al., 2009; Garanina et al., 2009; Weidmann et al., 2005; Zhuravlev et al., 2008; Garanina et al., 2009; Zhuravlev et al., 2008
Puumala virus	Myodes glareolus	Apodemus flavicollis, Microtus agrestis	Schlegel et al. unpublished data
Tula virus	Microtus arvalis	Microtus levis, Microtus agrestis, Microtus subterranus, Microtus gregalis, Lagurus lagurus, Arvicola amphibius, Myodes glareolus	Plyusnin et al., 1994; Scharninghausen et al., 2002; Schlegel et al., 2012a; Schmidt-Chanasit et al., 2010; Song et al., 2002; GenBank Acc. No. AF442620-1; GenBank Acc. No. AF 442618-9; Schlegel et al., 2012a; Schlegel et al. unpublished data

Continued

TABLE 10.6 Reported Nucleic Acid Detection of Insectivore-borne Seewis Virus and Different Rodent-borne Hantaviruses in Rodent and Insectivore Species, Which are not Postulated or Identified as the Predominant Host for this Particular Hantavirus—cont'd

Virus species*	Reservoir host species	Natural spillover infections	
		Species	Reference
Isla Vista virus	Microtus californicus	Peromyscus maniculatus, Peromyscus californicus	Song et al., 1995
Sin Nombre virus	Peromyscus maniculatus	Rheitrodontomys megalotis, Peromyscus leucopus	Hjelle et al., 1995; Rawlings et al., 1996
El Moro Canyon virus	Reithrodontomys megalotis	Peromyscus maniculatus	Rawlings et al., 1996
Andes virus	Oligoryzomys longicaudatus	Rattus rattus, Abrothrix longipilis, Loxodontomys micropus, Oligoryzomys flavescens, Oligoryzomys chacoensis	Medina et al., 2009; Gonzalez Della Valle et al., 2002
Bayou virus	Oryzomys palustris	Sigmodon hispidus	Torrez-Martinez et al., 1998
Laguna Negra virus	Calomys laucha	Calomys callosus, Akodon simulator	Levis et al., 2004
Rio Mamore virus	Oligoryzomys microtis	Holochilus sciureus, Oligoryzomys fornesi	Rosa et al., 2005
Seewis virus	Sorex araneus	Sorex daphaenodon, Sorex tundrensis, Sorex minutus, Neomys anomalus	Yashina et al., 2010; Schlegel et al., 2012b; GenBank Acc. No. EU418604

*Hantavirus taxonomy according to (Plyusnin et al., 2011).
**Additional GenBank entries of Rattus flavipectus: FJ803201-04, FJ803209, FJ803213, FJ884369, FJ884391, FJ884401, GU592943.

However, the finding of the putative spillover infections highlights also the problems for identification of a reservoir for a defined hantavirus species (Hjelle and Yates, 2001).

Spillover infections may represent also the basis for the genetic interaction of two hantaviruses. Thus, spillover infections of TULV into bank vole may result in an interaction of TULV with PUUV that is adapted to the bank vole. Indeed, *in vitro* investigations have demonstrated the reassortment between PUUV and a *Microtus*-associated hantavirus, i.e., PHV (Handke et al., 2010). Interestingly, this reassortment is based on the exchange of the M-segment. Similarly, *in vitro* reassortment studies with DOBV genotypes Kurkino and Dobrava resulted in M-segment reassortants (Kirsanovs et al., 2010). Similarly, *in vitro* studies resulted in the formation of reassortants between two related strains of SNV, but in decreased reassortment probabilities between members of different species (Rodriguez et al., 1998). In line, reassortment processes have been found to occur in nature between related strains of the same hantavirus species but not between different species. This observation in nature might be due to the isolation of different hantaviruses caused by their primary association to a single reservoir. Intramolecular RNA recombination has been demonstrated as an additional process in natural hantavirus evolution as evidenced for the TULV S-segment (Sibold et al. 1999).

Genetic changes due to mutation have been determined for naturally SNV-, TULV- or PUUV-infected reservoir rodents to be about $1–5 \times 10^{-3}$ (Plyusnin et al., 1996; Feuer et al., 1999). In addition, in these persistently infected reservoirs, a complex population of closely related virus variants (quasispecies) has been reported which is likely caused by point mutations. The previously reported increased virus titers of focus-assay purified hantaviruses might be due to the removal of defective interfering particles from the virus stock (Rang et al., 2006).

The emergence of hantaviruses might be also influenced by coinfections in the reservoir. Rodents have been identified to harbor a large number of zoonotic viruses other than hantaviruses. Thus, orthopox viruses, lymphocytic choriomeningitis virus and other arenaviruses, tick-borne encephalitis virus and Ljungan virus have been detected in rodents (Meerburg et al., 2009; Ulrich et al., 2009). In addition, rodents were found to be infected by their "own" beta and gamma herpesviruses that might modulate their immune system function (Ehlers et al., 2007). Additional rodent viruses, such as adenoviruses, papilloma and other DNA and RNA viruses may interfere with the immune competence of the reservoir as well (Klempa et al., 2009; Phan et al., 2011; Schulz et al., 2012). In addition, different *Leptospira* spp., *Rickettsia* spp., *Bartonella* spp. and other bacteria as well as blood parasites such as *Babesia* spp. have been detected in rodents (Houpikian and Raoult, 2001; Sinski et al., 2006; Mayer-Scholl et al., 2011; Schex et al., 2011). Further, certain intestine parasites are known for their immunosuppressive activity and were found in rodents. Currently little is known about the potential interactions of hantaviruses and other pathogens.

An initial study in field voles in the UK recently investigated the individual infection risks for a community of microparasites consisting of cowpox virus, *Babesia microti*, *Bartonella* spp. and *Anaplasma phagocytophilum* (Telfer et al., 2010).

The intimate interplay between hantaviruses and the reservoir might be also influenced by genetic factors of the reservoir. Thus, tumor necrosis factor alpha promoter polymorphisms or certain major histocompatibility factor alleles may influence the outcome of a hantavirus infection in the reservoir (Deter et al., 2008; Guivier et al., 2010).

The emergence of hantaviruses might be not only driven by virus and/or reservoir properties, but also by environmental factors influencing the exposure of humans. Interestingly, hantavirus seems to be quite stable outside the virus host, which is in line with an indirect transmission route via virus-contaminated aerosols (Kallio et al., 2006a; Hardestam et al., 2007). This route of transmission might also be reflected in the fact that outbreaks of hantavirus-like disease have been retrospectively suggested for World War II (Rüdisser, 1942; Stuhlfault, 1943) and the Korean conflict 1951–1953 representing the starting point for the identification of the hantaviruses (Lee et al., 1978).

ROLE OF RODENTS IN DRIVING OUTBREAKS

Rodent population dynamics

In the northern hemisphere, multiannual fluctuation of population size of small mammals, sometimes across species, has been observed in many species including hantavirus reservoir rodents (Krebs and Myers, 1974; Lindström et al., 2001; Korpimäki et al., 2004; Hörnfeldt, 2004; Jacob and Tkadlec, 2010; Cornulier et al., 2013). Despite many years of research since the first scientific publication about small mammal cycles in northern Europe and North America (Elton, 1924; Elton, 1942), there is still debate regarding the processes behind the regular rise and fall of populations.

Many hypotheses have been developed to explain multiannual fluctuations in rodent population abundance including extrinsic effects of predation, climate, food resources, landscape structure, disease as well as intrinsic effects of genetics, behavior and demography (Lindström et al., 2001; Smith et al., 2006) or effect combinations of intrinsic and extrinsic factors at different phases of the cycle (Andreassen et al., 2013). Predation matters for fluctuation in population size in Fennoscandian rodent species (Korpimäki et al., 2004) but causal relationships for other regions remain unclear (Aars et al., 2006).

Climate could affect rodent population dynamics directly or indirectly via several pathways and may explain the large-scale dampening of cyclic dynamics observed in the last 20 years in some European small herbivores (Cornulier et al., 2013). Climate seems to be related to population dynamics in a number of important hantavirus reservoir species. For instance, the dynamics of bank voles

(*Myodes glareolus*) is affected by climate (Palo, 2009), most likely through climate effects on food availability (Tersago et al., 2009). Similarly, population dynamics of deer mice (*Peromyscus maniculatus*) can be positively correlated to rainfall (Brown and Ernest, 2002; Yates et al., 2002; Orrock et al., 2011) and rainfall-driven food availability but there are also contrasting results in this species (Loehman et al., 2012) and in the *Akodon montensis* – Jaborá and Ape Aime hantavirus system (Owen et al., 2010).

Impact of rodent population dynamics on hantavirus seroprevalence in rodent hosts

It seems likely that the prevalence of hantavirus-specific antibodies in the rodent host is density dependent in one way or the other because for the spread of many pathogens contact among individuals is a prerequisite for transmission. Consequently, density dependent or delayed density dependent transmission of vector-borne pathogens has been suggested theoretically (Thrall et al., 1993). For some pathogens including hantaviruses that are carried by rodent species that fluctuate in density multiannually, delayed density dependent presence of antibodies was empirically confirmed to be more important than seasonal changes in rodent population size (Mills et al., 1999a; Mills et al., 1999b; Yates et al., 2002; Davis et al., 2005).

However, density dependence is only valid for IgG antibodies. Maternal antibodies are transferred *in utero* and do not indicate infection but immunity (Kallio et al., 2010). Prevalence of maternal antibodies correlates with earlier seroprevalence of IgG antibodies in bank vole populations and can delay transmission of PUUV because of transiently immune young individuals (Kallio et al., 2010; Kallio et al., 2006b).

Naturally, prevalence can only be density dependent if the pathogen in question is present in the population in the first place. The latter is not necessarily the case for hantavirus species such as PUUV in bank voles and DOBV in *Apodemus* (Heyman et al., 2011). For example, PUUV is not found in bank voles in large areas of eastern Germany and DOBV is virtually absent from western German *Apodemus* populations (Schlegel et al., 2009; Faber et al., 2013; our unpublished data) and it is absent from Great Britain (Bennett, 1997).

For most hantavirus reservoir rodent species a positive correlation of the abundance of the rodent reservoir and hantavirus prevalence in the rodent host has been reported from field studies. In *Oligoryzomys longicaudatus* seroprevalence of ANDV is positively correlated with the abundance of the rodent host (Andreo et al., 2011). Similarly, there is a positive correlation of bank vole population size and PUUV seroprevalence (Dobly et al., 2012; Olsson et al., 2002), at least in the initial phase of the virus spreading in the rodent host population before PUUV outbreak conditions are reached (Tersago et al., 2011). For *Peromyscus* there are reports that SNV is transmitted density dependently (Luis et al., 2012) but there are also contradictory findings for *P. maniculatus* (Dizney and Ruedas, 2009; Graham and Chomel, 1997) where the season proved to be more important for seroprevalence than population abundance (Bagamian et al., 2012). A spatially detailed trapping study indicated that delayed density dependence of SNV seroprevalence in *P. maniculatus* acts at regional but not at local scale with a time lag of about 6–9 months (Carver et al., 2011b) and there are also density dependent transmission patterns in the *Calomys laucha* – Laguna negra virus (LNV) system (Yahnke et al., 2001). In contrast, there is no association of the abundance of *Akodon montensis* and Jaborá and Ape Aime hantaviruses in this species (Owen et al., 2010).

Generally, density dependent hantavirus prevalence seems to be present for several but not all hantavirus species. There can be direct density dependence and delayed density dependence, which can be governed by a number of processes (Olsson et al., 2010). These include seasonal effects such as the influx of young (uninfected) or immune individuals that have better chances to mature and reproduce (Kallio et al., 2006b; Kallio et al., 2010) as well as other demographic effects (Mills et al., 1999a; Mills et al., 1999b; Adler et al., 2008).

Symptoms of hantavirus infection in rodent reservoir species and effects on population dynamics of the reservoir were thought to be minimal. However, several studies have demonstrated that there are consequences for rodent hosts. Weight gain (Douglass et al., 2007), immune response (Lehmer et al., 2007) and survival are reduced in SNV-positive male deer mice and in breeding deer mice of either sex compared to uninfected individuals (Luis et al., 2012). Winter survival of PUUV-positive bank voles is reduced (Kallio et al., 2007). In contrast, survival of nonbreeding but not breeding voles is lower in PUUV infected versus uninfected bank voles (Tersago et al., 2012) indicating differing demographic consequences of SNV *versus* PUUV infection in rodents. In addition, host behavior can be affected by hantavirus infection. Male Norway rats engage in more aggressive behavior the more SEOV is present in tissue (Klein et al., 2004).

Impact of rodent population dynamics on hantavirus seroprevalence in humans

Occasional hantavirus outbreaks in the human population (Heyman and Vaheri, 2008; Mills et al., 2010; Faber et al., 2010; Heyman et al., 2011) imply an association with population eruptions that occur in several small reservoir rodents. A thorough assessment of links between reservoir rodent population dynamics and human hantavirus infections requires information at sufficient spatial and temporal scale to accommodate the multiannual nature of rodent population fluctuation. However, human epidemiological data beyond annual case numbers (Heyman et al., 2011) are scarce in many European countries (Vaheri et al., 2013) regarding PUUV and DOBV infections. The situation for New World hantaviruses is similar (Mills et al., 2010).

Detailed information about human hantavirus cases and bank vole population dynamics is available only from a few European countries. In Belgium, the number of PUUV IgG seropositive bank voles correlates positively with the number of reported NE cases at the regional level (Tersago et al., 2011). In Finland, the number of human hantavirus infections depends on bank vole abundance with a time lag of several months (Kallio et al., 2009) and in Sweden as well as in Germany, bank vole abundance is also related to NE cases (Palo, 2009; Reil et al., 2011; and unpublished data).

Similarly, there seems to be a positive association of *Peromyscus maniculatus* abundance SNV seroprevalence and SNV related HCPS in humans (Calisher et al., 2011) at least at regional scale (Carver et al., 2011a).

Effect of diversity on transmission within rodent hosts

Hantavirus rodent host species are widely distributed habitat generalists, locally abundant, resilient to anthropogenic disturbance and usually occur in rodent communities of low diversity (Ostfeld and Keesing, 2012). At low rodent diversity hantavirus transmission is more successful than at high rodent diversity for PUUV (Tersago et al., 2008), LANV (Yahnke et al., 2001), ANDV (Piudo et al., 2011), Choclo and Calabazo hantaviruses (Ruedas et al., 2004; Suzan et al., 2009) and SNV (Carver et al., 2011a; Orrock et al., 2011), which seems to be a common pattern in epidemiology (Keesing et al., 2010).

This is probably due to a dilution effect causing an increase of interspecific encounters and a decrease in intraspecific encounters at growing small mammal biodiversity (Clay et al., 2009). Usually, only intraspecific host interaction will lead to hantavirus transmission and, therefore, a reduction of such encounters will reduce transmission rates (Dearing and Dizney, 2010), resulting in a lower proportion of host individuals being infected.

This pattern is not only valid for the small mammal community but also for predators. Their diversity is negatively correlated to SNV prevalence (Orrock et al., 2011) and abundant predators might limit reservoir host numbers sufficiently to reduce transmission risk (Ostfeld and Holt, 2004).

PREDICTIVE MODELING OF HUMAN HANTAVIRUS INFECTION RISK

Human PUUV infection risk in Europe is largely determined by landscape and habitat composition, climatic parameters, biodiversity and human exposure intensity (Reusken and Heyman, 2013). Human SNV infection risk can also depend on climate (Glass et al., 2002; Orrock et al., 2011) although this is equivocal. In either system a climatically operated increase in food availability for rodent hosts, build-up of rodent abundance and increasing hantavirus prevalence in rodents seems to be the mechanism behind increased human infection risk (Glass et al., 2002; Tersago et al., 2009).

Bottom-up regulation of food quantity and quality is probably the major driver of population increase in bank voles (Tersago et al., 2009) and deer mice (Calderon et al., 1999). Accordingly, results from field studies and modeling efforts mentioned above suggest that it might be sufficient to forecast the phase of the multiannual population cycle of rodent reservoir populations instead of hantavirus prevalence within reservoirs in order to predict human hantavirus infection risk - at least for the systems bank vole – PUUV and *Peromyscus maniculatus* – SNV. The sooner in trophic cascades the forecast can be made, the more useful the prediction will be because more lead-up time is achieved (Mills et al., 2010).

While basic forecast models yield useful results (Glass et al., 2002; Tersago et al., 2009) that can be used for alerting high-risk groups and authorities, models that can predict local human infection risk are not available. This may be due to the large number of factors beyond climate and rodent food that matter for human infection risk be they related to climate (e.g., survival of hantavirus outside host, habitat features, human outdoor activity) or not (e.g., biodiversity, landscape). Some of these factors interact and the patchiness of virus distribution and potential effects of hantavirus infection on host demography and dynamics add further complexity.

As a result, complex models for the prediction of human hantavirus infection risk may be required for small scaled forecasts that consider the most relevant factors and models might have to be scale dependent (regional/local) for optimal results. The development of such models will highly depend on the availability of long-term monitoring data for host abundance, climate, food, human infection that are only rarely available at sufficient temporal scale and spatial resolution. This is especially true for the less widely and/or patchy distributed hantaviruses and recently discovered hantaviruses that occur in insectivores or other reservoir small mammals whose abundance, let alone human infections, are not monitored.

CONCLUSIONS

As seen in the previous sections, there is an enormous and rapidly increasing knowledge about hantaviruses. This is most obviously driven by "**virus discovery**" efforts resulting in the identification of numerous hantaviruses in shrews, moles and recently in bats. Currently, about 50 hantaviruses with more than 400 strains or lineages were described and are currently listed in GenBank (http://www.ncbi.nlm.nih.gov; Plyusnin et al., 2011). However, the number of virus isolates allowing a functional characterization is still low, underlining the difficulties of virus isolation in cell culture. The recent isolation of DOBV strain "Greifswald" from a spillover-infected *A. flavicollis* may suggest that it might be promising to achieve **virus isolation** from spillover-infected nonreservoirs during the acute-phase of infection (Popugaeva et al., 2012). In addition, the majority of sequences in GenBank are still incomplete and only for a very few viruses a complete genome, including the noncoding regions, is available. **Complete genomes** are mainly derived from cell-culture isolates. The application of Next Generation Sequencing approaches will most likely help to significantly increase the number of complete hantavirus genomes (Plyusnin and Elliott, 2011). For future comprehensive phylogenetic investigations and functional genomics studies it is important to determine complete genome sequences and isolate virus strains directly from reservoirs. For this purpose, it might in the future be helpful to exploit permanent cell lines derived from natural reservoirs, as recently developed for bank voles (Essbauer et al., 2011; Stoltz et al., 2011). Finally, the determination of complete genome sequences may help also to improve the taxonomic classification of hantaviruses and the definition of hantavirus species and genotypes.

The determination of complete genomes will allow further **characterization of the genome organization**. Although through the pioneering work on HTNV and its confirmation for the later discovered rodent-borne pathogens, the genome organization of hantaviruses might be more complex than initially expected. Thus, in the S-segment of hantaviruses associated with Arvicolinae, Sigmodontinae and Neotominae rodents, a putative ORF was predicted on the S-segment overlapping the N-encoding ORF. The recent prediction of another ORF on the S-segment of shrew-borne SWSV strains from different trapping sites indicates the necessity of an open-minded proof of the putative ORFs in the genome of hantaviruses. The determination of complete genome sequences, including the NCR sequences in particular, might also be important for the development of a reverse-genetics system.

Studies on **molecular evolution** will also profit from the determination of the complete genome sequences. The accumulation of complete genome sequences may allow in the future to rationally select genome regions that will enable the resolution of the relationship of a certain hantavirus at different levels, i.e., between different reservoir populations, within a population and even within the same individual (quasispecies). Further, only the most comprehensive dataset of hantavirus sequences would allow a reliable analysis of reassortment and intragenomic recombination. Moreover, this will help to learn more about the phylogenetic origin, evolutionary history and host-association of the highly divergent hantaviruses found in rodents, insectivores and bats.

In contrast to the rapidly ongoing discovery of novel hantaviruses and their molecular genome characterization, the definition of their virulence and identification of virulence-mediating gene products are proceeding slowly. These **functional genomics** studies are currently mainly hampered by the lack of a suitable reverse-genetics system. The lack of a reverse-genetics system for hantaviruses is somewhat surprising as representatives of other bunyavirus genera reverse genetics have been established (Billecocq et al., 2008; Elliott et al., 2013). With the increasing knowledge of complete genome sequences and the development of novel *in vitro* cell-culture models (see above), the establishment of the reverse-genetics for hantaviruses might be achievable in the future. Until that, alternative "forward" genetics approaches might be applied to identify virulence and immune evasion factors or factors involved in the various stages of virus replication. Thus, by *in vitro*-reassortment progeny viruses might be obtained with properties differing from the parental viruses (Handke et al., 2010; Kirsanovs et al., 2010). In addition, recently the generation of different phenotypes of a PUUV strain was observed after cell-culture passaging (Sundström et al., 2011). Moreover, PUUV strain Sotkamo variants containing an intact NSs-ORF or a deficient NSs-ORF have been isolated by focus-assay-mediated virus purification (Rang et al., 2006). Besides the reverse-genetics system, it is important to select an animal model for the question to be answered. Infection models have been established for several hantaviruses and their reservoirs, i.e., HTNV and striped field mouse (Lee et al., 1978), PUUV and bank voles (Lundkvist et al., 1996), SEOV and Norway rats (Klein et al., 2000). Infection models are also vital to prevent cell-culture mediated attenuation of a hantavirus isolate resulting in the loss of the ability of the virus to re-infect the reservoir host (Lundkvist et al., 1997b). The number of suitable animal (disease) models is quite limited. The use of (suckling) laboratory mice might be limited to certain scientific questions as it does not reflect the natural situation. More suitable might be the PUUV/monkey model (Klingstrom et al., 2005; Sironen et al., 2008) and the ANDV and Maporal virus/Syrian hamster model (Safronetz et al., 2012). Thus, future investigations of spillover-infected nonreservoirs have to prove whether they are suited as an alternative novel animal disease model. The NGS-mediated metagenomics approach may help in the future to understand the complexity of the virus–reservoir interaction.

The **definition of a host species** is highly important to understand the evolutionary history of a hantavirus and its host adaptation. The definition of a reservoir is difficult and practically usually based on a multiple molecular detection in one species and the absence in sympatrically occurring other small mammal species (Hjelle and Yates, 2001). For several hantaviruses, such systematic studies with sympatrically occurring potential reservoirs are still needed. In addition, these comprehensive studies will help to document the frequency of spillover infections of nonreservoirs, which might be higher than previously expected, at least for certain hantaviruses, e.g., TULV (Schlegel et al., 2012a). Such systematic studies will help also to more precisely describe the geographical distribution of hantaviruses, and thereby allow the generation of risk maps. Although the knowledge of the **geographical distribution of hantaviruses** in Europe is quite good (Heyman et al., 2011), even for Europe the nonhomogeneous distribution of PUUV needs to be studied more precisely. Importantly, the recent detection of autochthonous hantavirus infections in wild and pet rats in the UK and their potential health impact on humans (Jameson et al., 2013a, b) raises the question about the presence of this hantavirus in Europe (and the whole world) and its potential as a human pathogen. The necessity for additional studies on the presence of hantaviruses is also obvious for Africa, where only few studies documented the presence of hantaviruses and human hantavirus infections (Klempa et al., 2010), but mainly for Australia with only a single study in potential reservoirs in the 1970s (LeDuc et al., 1986; Bi et al., 2005). To understand the current distribution of hantaviruses it would also be important to combine virus screening with phylogeography investigations of the reservoirs to understand their (re-)colonization processes and the "incursion" of the associated hantavirus lineages. This is an important issue also in the present when changes in the distribution range of potential reservoirs are ongoing.

Important prerequisites for these studies are well-established and standardized **diagnostics methods**. The validity and comparability of diagnostic assays should be demonstrated by quality assurance (Escadafal et al., 2012). In addition, the interpretation of results from seroepidemiological studies or the validity of the causative hantavirus of recorded human cases should be done critically, clearly keeping in mind the limitations of the diagnostic assay(s) used. Epidemiological studies in general would profit in the future from the use of an international joint study using harmonized protocols and assays. In addition, the selection of a molecular diagnostics assay should consider the high frequency of mutations and selection that might result in false-negative results and the necessity of having diagnostic assays for imported hantavirus infections, as recently evidenced by the exposure of tourists at the Yosemite National Park in the United States.

Bats have been in the focus of virus discovery approaches for several years. Bats have been found to harbor several viruses, with some being of (putative) zoonotic potential (Drexler et al., 2010; Canuti et al., 2011; Drexler et al., 2011; Drexler et al., 2012a, Drexler et al., 2012b). After the initial finding of short hantavirus-like sequences in two bat species (Sumibcay et al., 2012; Weiss et al., 2012), one may still be skeptical about whether bats really represent reservoirs for a novel group of hantaviruses. The recent finding of multiple novel and closely related sequences in bat species within the two families Vespertilinoidae and Rhinolophidae confirms that bats seem to be another reservoir taxon for hantaviruses (see Table 10.5). This is very important as the hantavirus evolution theory must now include three mammal orders. It can be expected that several additional hantaviruses will be identified in other bat species and other mammalian orders. The epidemiology of bat-associated hantaviruses might differ from that of rodent- and insectivore-associated viruses as some bat species are medium and long distance migrators (Hutterer et al., 2005).

The identification of the reasons for the **emergence of hantaviruses**, the driving forces for **hantavirus outbreaks** and the development of **early warning** tools need to be in the focus of future hantavirus research. These investigations should also incorporate proving the potential impact of global climate change (Klempa, 2009). To achieve this goal a **long-term rodent (small mammal) monitoring** under standardized conditions at representative monitoring sites is essential on a regional, national and perhaps global scale. To understand the multiple factors influencing the probability and frequency of human hantavirus infections, a more holistic and interdisciplinary approach targeting population dynamics, population (and landscape) genetics, phylogeography, infections (and co-infections) with different pathogens, immunology, ecology, climatology, epidemiology, etc., is needed.

After the enormous findings of novel hantaviruses in shrews, moles and most recently in bats, one may ask "What comes next?" Are there any unknown hantaviruses (or related pathogens) around in representatives of other rodent families (see Table 10.5), other mammal orders, or vertebrates other than mammals?

ACKNOWLEDGMENTS

The work in the laboratories of DHK, JJ and RGU was supported by the Federal Environment Agency within the Environment Research Plan of the German Federal Ministry for the Environment, Nature Conservation and Nuclear Safety (grants no. 370941401 and 371348401 to JJ and RGU), Deutsche Forschungsgemeinschaft (KR1293/9-1 and /12-1 to DHK), Robert Koch-Institut (grant no. 1369-382 to DHK) and the EU grants APHAEA (grant no. 2811ERA117) and FP7-261504 EDENext and is catalogued by the EDENext Steering Committee as EDENext 128 (http://www.edenext.eu). The contents of this publication are the sole responsibility of the authors and don't necessarily reflect the views of the European Commission.

REFERENCES

Aars, J., Dallas, J. F., Piertney, S. B., Marshall, F., Gow, J. L., Telfer, S., & Lambin, X. (2006). Widespread gene flow and high genetic variability in populations of water voles *Arvicola terrestris* in patchy habitats. *Mol. Ecol.*, *15*, 1455–1466.

Adler, F. R., Pearce-Duvet, J. M., & Dearing, M. D. (2008). How host population dynamics translate into time-lagged prevalence: An investigation of Sin Nombre virus in deer mice. *Bull Math. Biol.*, *70*, 236–252.

Alff, P. J., Gavrilovskaya, I. N., Gorbunova, E., Endriss, K., Chong, Y., Geimonen, E., Sen, N., Reich, N. C., & Mackow, E. R. (2006). The pathogenic NY-1 hantavirus G1 cytoplasmic tail inhibits RIG-I- and TBK-1-directed interferon responses. *J. Virol.*, *80*, 9676–9686.

Alff, P. J., Sen, N., Gorbunova, E., Gavrilovskaya, I. N., & Mackow, E. R. (2008). The NY-1 hantavirus Gn cytoplasmic tail coprecipitates TRAF3 and inhibits cellular interferon responses by disrupting TBK1-TRAF3 complex formation. *J. Virol.*, *82*, 9115–9122.

Andreassen, H. P., Glorvigen, P., Remy, A., & Ims, R. A. (2013). New views on how population-extrinsic and community-extrinsic processes interact during the vole population cycles. *Oikos*. In press.

Andreo, V., Glass, G., Shields, T., Provensal, C., & Polop, J. (2011). Modeling potential distribution of *Oligoryzomys longicaudatus*, the Andes virus (Genus: *Hantavirus*) reservoir, in Argentina. *Ecohealth*, *8*, 332–348.

Antic, D., Kang, C. Y., Spik, K., Schmaljohn, C., Vapalahti, O., & Vaheri, A. (1992). Comparison of the deduced gene products of the L, M and S genome segments of hantaviruses. *Virus Res.*, *24*, 35–46.

Arai, S., Bennett, S. N., Sumibcay, L., Cook, J. A., Song, J. W., Hope, A., Parmenter, C., Nerurkar, V. R., Yates, T. L., & Yanagihara, R. (2008). Phylogenetically distinct hantaviruses in the masked shrew (*Sorex cinereus*) and dusky shrew (*Sorex monticolus*) in the United States. *Am. J. Trop. Med. Hyg.*, *78*, 348–351.

Arai, S., Song, J. W., Sumibcay, L., Bennett, S. N., Nerurkar, V. R., Parmenter, C., Cook, J. A., Yates, T. L., & Yanagihara, R. (2007). Hantavirus in northern short-tailed shrew, United States. *Emerg. Infect. Dis.*, *13*, 1420–1423.

Araki, K., Yoshimatsu, K., Ogino, M., Ebihara, H., Lundkvist, A., Kariwa, H., Takashima, I., & Arikawa, J. (2001). Truncated hantavirus nucleocapsid proteins for serotyping Hantaan, Seoul, and Dobrava hantavirus infections. *J. Clin. Microbiol.*, *39*, 2397–2404.

Bagamian, K. H., Douglass, R., Alvarado, A., Kuenzi, A., Amman, B., Waller, L., & Mills, J. (2012). Population density and seasonality effects on Sin Nombre virus transmission in North American deermice (*Peromyscus maniculatus*) in outdoor enclosures. *PLOS One*, *7*, e37254.

Bennett, G. G. (1997). Norway rats' communication about foods and feeding sites. *Repellents in Wildlife Management*, 185–201.

Bi, P., Cameron, S., Higgins, G., & Burrell, C. (2005). Are humans infected by hantaviruses in Australia? *Int. Med. J.*, *35*, 672–674.

Billecocq, A., Gauliard, N., Le May, N., Elliott, R. M., Flick, R., & Bouloy, M. (2008). RNA polymerase I-mediated expression of viral RNA for the rescue of infectious virulent and avirulent Rift Valley fever viruses. *Virology*, *378*, 377–384.

Botten, J., Mirowsky, K., Kusewitt, D., Bharadwaj, M., Yee, J., Ricci, R., Feddersen, R. M., & Hjelle, B. (2000). Experimental infection model for Sin Nombre hantavirus in the deer mouse (Peromyscus maniculatus). *Proc. Natl. Acad. Sci. U S A*, *97*, 10578–10583.

Brown, J. H., & Ernest, S. K. M. (2002). Rain and rodents: complex dynamics of desert consumers. *Bioscience*, *52*, 979–987.

Brummer-Korvenkontio, M., Vaheri, A., Hovi, T., von Bonsdorff, C. H., Vuorimies, J., Manni, T., Penttinen, K., Oker-Blom, N., & Lahdevirta, J. (1980). Nephropathia epidemica: detection of antigen in bank voles and serologic diagnosis of human infection. *J. Infect. Dis.*, *141*, 131–134.

Calderon, G., Pini, N., Bolpe, J., Levis, S., Mills, J., Segura, E., Guthmann, N., Cantoni, G., Becker, J., Fonollat, A., Ripoll, C., Bortman, M., Benedetti, R., & Enria, D. (1999). Hantavirus reservoir hosts associated with peridomestic habitats in Argentina. *Emerg. Infect. Dis.*, *5*, 792–797.

Calisher, C. H., Mills, J. N., Root, J. J., Doty, J. B., & Beaty, B. J. (2011). The relative abundance of deer mice with antibody to Sin Nombre virus corresponds to the occurrence of hantavirus pulmonary syndrome in nearby humans. *Vector Borne Zoonotic Dis.*, *11*, 577–582.

Canuti, M., Eis-Huebinger, A. M., Deijs, M., de Vries, M., Drexler, J. F., Oppong, S. K., Müller, M. A., Klose, S. M., Wellinghausen, N., Cottontail, V. M., Kalko, E. K., Drosten, C., & van der Hoek, L. (2011). Two novel parvoviruses in frugivorous New and Old World bats. *PLoS One. 2011*, *6*(12), e29140. doi:10.1371/journal.pone.0029140. Epub 2011 Dec 27.

Carey, D. E., Reuben, R., Panicker, K. N., Shope, R. E., & Myers, R. M. (1971). Thottapalayam virus: a presumptive arbovirus isolated from a shrew in India. *The Indian Journal of Medical Research*, *59*, 1758–1760.

Carver, S., Kuenzi, A., Bagamian, K. H., Mills, J. N., Rollin, P. E., Zanto, S. N., & Douglass, R. (2011a). A temporal dilution effect: hantavirus infection in deer mice and the intermittent presence of voles in Montana. *Oecologia*, *166*, 713–721.

Carver, S., Trueax, J. T., Douglass, R., & Kuenzi, A. (2011b). Delayed density-dependent prevalence of Sin Nombre virus infection in deer mice (*Peromyscus maniculatus*) in central and western Montana. *J. Wildl. Dis.*, *47*, 56–63.

CDC. (2008). Ref Type: Internet Communication, http://www.cdc.gov/ncidod/diseases/hanta/hps/noframes/caseinfo.htm.

Chaparro, J., Vega, J., Terry, W., Vera, J. L., Barra, B., Meyer, R., Peters, C. J., Khan, A. S., & Ksiazek, T. G. (1998). Assessment of person-to-person transmission of hantavirus pulmonary syndrome in a Chilean hospital setting. *J. Hosp. Infect., 40*, 281–285.

Childs, J. E., Ksiazek, T. G., Spiropoulou, C. F., Krebs, J. W., Morzunov, S., Maupin, G. O., Gage, K. L., Rollin, P. E., Sarisky, J., Enscore, R. E., Russell, E., Frey, J. K., Peters, C. J., & Nichol, S. T. (1994). Serologic and genetic identification of *Peromyscus maniculatus* as the primary rodent reservoir for a new hantavirus in the southwestern United States. *J. Infect. Dis, 169*(6), 1271–1280.

Cueto, G. R., Cavia, R., Bellomo, C., Padula, P. J., & Suarez, O. V. (2008). Prevalence of hantavirus infection in wild Rattus norvegicus and R. rattus populations of Buenos Aires City, Argentina. *Trop. Med. Int. Health, 13*(1), 46–51.

Chumakov, M. P., Gavrilovskaya, I. N., Boiko, V. A., Zakharova, M. A., Myasnikov Yu, A., Bashkirev, T. A., Apekina, N. S., Safiullin, R. S., & Potapov, V. S. (1981). Detection of hemorrhagic fever with renal syndrome (HFRS) virus in the lungs of bank voles *Clethrionomys glareolus* and redbacked voles *Clethrionomys rutilus* trapped in HFRS foci in the European part of U.S.S.R., and serodiagnosis of this infection in man. *Arch. Virol, 69*, 295–300.

Clay, C. A., Lehmer, E. M., St Jeor, S., & Dearing, M. D. (2009). Testing mechanisms of the dilution effect: deer mice encounter rates, Sin Nombre virus prevalence and species diversity. *Ecohealth, 6*, 250–259.

Clement, J., Frans, J., & Van Ranst, M. (2003). Human Tula virus infection or rat-bite fever? *Eur. J. Clin. Microbiol. Infect. Dis., 22*, 332–333.

Clement, J., Maes, P., Lagrou, K., Van Ranst, M., & Lameire, N. (2012). A unifying hypothesis and a single name for a complex globally emerging infection: hantavirus disease. *Eur. J. Clin. Microbiol. Infect. Dis., 31*, 1–5.

Cornulier, T., Yoccoz, N. G., Bretagnolle, V., Brommer, J. E., Butet, A., Ecke, F., Elston, D. A., Framstad, E., Henttonen, H., Hornfeldt, B., Huitu, O., Imholt, C., Ims, R. A., Jacob, J., Jedrzejewska, B., Millon, A., Petty, S. J., Pietiainen, H., Tkadlec, E., Zub, K., & Lambin, X. (2013). Europe-wide dampening of population cycles in keystone herbivores. *Science, 340*, 63–66.

Coulaud, X., Chouaib, E., Georges, A. J., Rollin, P., & Gonzalez, J. P. (1987). First human case of haemorrhagic fever with renal syndrome in the Central African Republic. *Trans. Royal Soc. Trop. Med. Hyg., 81*, 686.

Daud, N. H., Kariwa, H., Tanikawa, Y., Nakamura, I., Seto, T., Miyashita, D., Yoshii, K., Nakauchi, M., Yoshimatsu, K., Arikawa, J., & Takashima, I. (2007). Mode of infection of Hokkaido virus (Genus *Hantavirus*) among grey red-backed voles, *Myodes rufocanus*, in Hokkaido, Japan. *Microbiol. Immunol., 51*, 1081–1090.

Davis, S., Calvet, E., & Leirs, H. (2005). Fluctuating rodent populations and risk to humans from rodent-borne zoonoses. *Vector Borne Zoonotic Dis., 5*, 305–314.

Dearing, M. D., & Dizney, L. (2010). Ecology of hantavirus in a changing world. *Ann. N. Y. Acad. Sci., 1195*, 99–112.

Deter, J., Bryja, J., Chaval, Y., Galan, M., Henttonen, H., Laakkonen, J., Voutilainen, L., Vapalahti, O., Vaheri, A., Salvador, A. R., Morand, S., Cosson, J. F., & Charbonnel, N. (2008). Association between the DQA MHC class II gene and Puumala virus infection in *Myodes glareolus*, the bank vole. *Infect. Genet. Evol., 8*, 450–458.

Dizney, L. J., & Ruedas, L. A. (2009). Increased host species diversity and decreased prevalence of Sin Nombre virus. *Emerg. Infect. Dis., 15*, 1012–1018.

Dobly, A., Yzoard, C., Cochez, C., Ducoffre, G., Aerts, M., Roels, S., & Heyman, P. (2012). Spatio-temporal dynamics of Puumala hantavirus in suburban reservoir rodent populations. *J. Vector. Ecol.*, *37*, 276–283.

Douglass, R. J., Calisher, C. H., Wagoner, K. D., & Mills, J. N. (2007). Sin Nombre virus infection of deer mice in Montana: characteristics of newly infected mice, incidence, and temporal pattern of infection. *J. Wildl. Dis.*, *43*, 12–22.

Douron, E., Moriniere, B., Matheron, S., Girard, P. M., Gonzalez, J. P., Hirsch, F., & McCormick, J. B. (1984). HFRS after a wild rodent bite in the Haute-Savoie—and risk of exposure to Hantaan-like virus in a Paris laboratory. *Lancet*, *1*, 676–677.

Drexler, J. F., Gloza-Rausch, F., Glende, J., Corman, V. M., Muth, D., Goettsche, M., Seebens, A., Niedrig, M., Pfefferle, S., Yordanov, S., Zhelyazkov, L., Hermanns, U., Vallo, P., Lukashev, A., Müller, M. A., Deng, H., Herrler, G., & Drosten, C. (2010). Genomic characterization of severe acute respiratory syndrome-related coronavirus in European bats and classification of coronaviruses based on partial RNA-dependent RNA polymerase gene sequences. *J. Virol.*, *2010 Nov*, *84*(21), 11336–11349. doi:10.1128/JVI.00650-10. Epub 2010 Aug 4.

Drexler, J. F., Corman, V. M., Muller, M. A., Maganga, G. D., Vallo, P., Binger, T., Gloza-Rausch, F., Rasche, A., Yordanov, S., Seebens, A., Oppong, S., Adu Sarkodie, Y., Pongombo, C., Lukashev, A. N., Schmidt-Chanasit, J., Stocker, A., Carneiro, A. J., Erbar, S., Maisner, A., Fronhoffs, F., Buettner, R., Kalko, E. K., Kruppa, T., Franke, C. R., Kallies, R., Yandoko, E. R., Herrler, G., Reusken, C., Hassanin, A., Krüger, D. H., Matthee, S., Ulrich, R. G., Leroy, E. M., & Drosten, C. (2012a). Bats host major mammalian paramyxoviruses. *Nat. Commun.*, *3*, 796.

Drexler, J. F., Seelen, A., Corman, V. M., Fumie Tateno, A., Cottontail, V., Melim Zerbinati, R., Gloza-Rausch, F., Klose, S. M., Adu-Sarkodie, Y., Oppong, S. K., Kalko, E. K., Osterman, A., Rasche, A., Adam, A., Müller, M. A., Ulrich, R. G., Leroy, E. M., Lukashev, A. N., & Drosten, C. (2012b). Bats worldwide carry hepatitis E virus-related viruses that form a putative novel genus within the family Hepeviridae. *J. Virol.*, *2012 Sep*, *86*(17), 9134–9147. doi:10.1128/JVI.00800-12. Epub 2012 Jun 13.

Dupont, A., Gonzalez, J. P., Georges, A., & Ivanoff, B. (1987). Seroepidemiology of Hantaan-related virus in Gabon. *Trans. R. Soc. Trop. Med. Hyg.*, *81*, 519.

Easterbrook, J. D., Kaplan, J. B., Vanasco, N. B., Reeves, W. K., Purcell, R. H., Kosoy, M. Y., Glass, G. E., Watson, J., & Klein, S. L. (2007). A survey of zoonotic pathogens carried by Norway rats in Baltimore, Maryland, USA. *Epidemiol. Infect.*, *135*, 1192–1199.

Ebihara, H., Yoshimatsu, K., Ogino, M., Araki, K., Ami, Y., Kariwa, H., Takashima, I., Li, D., & Arikawa, J. (2000). Pathogenicity of Hantaan virus in newborn mice: genetic reassortant study demonstrating that a single amino acid change in glycoprotein G1 is related to virulence. *J. Virol*, *74*(19), 9245–9255.

Ehlers, B., Kuchler, J., Yasmum, N., Dural, G., Schmidt-Chanasit, J., Jäkel, T., Matuschka, F. -R., Richter, D., Essbauer, S., Hughes, D. J., Summers, C., Bennett, M., Stewart, J. P., & Ulrich, R. G. (2007). Identification of novel rodent herpesviruses, including the first gammaherpesvirus of Mus musculus. *J. Virol.*, *81*, 8091–8100.

Elliott, R. M., Blakqori, G., van Knippenberg, I. C., Koudriakova, E., Li, P., McLees, A., Shi, X., & Szemiel, A. M. (2013). Establishment of a reverse genetics system for Schmallenberg virus, a newly emerged orthobunyavirus in Europe. *J. Gen. Virol.*, *94*, 851–859.

Elliott, R., & Weber, F. (2009). Bunyaviruses and the Type I Interferon System. *Viruses*, *1*, 1003–1021.

Elliott, R. M., Schmaljohn, C. S., & Collett, M. S. (1991). Bunyaviridae genome structure and gene expression. *Curr. Top. Microbiol. Immunol.*, *169*, 91–141.

Elton, C. S. (1924). Periodic fluctuations in the numbers of animals: their causes and effects. *British Journal of Experimental Biology*, *2*, 119–163.

Elton, C. S. (1942). *Voles, mice and lemmings: problems in population dynamics*. Oxford: Clarendon Press.

Escadafal, C., Avsic-Zupanc, T., Vapalahti, O., Niklasson, B., Teichmann, A., Niedrig, M., & Donoso-Mantke, O. (2012). Second external quality assurance study for the serological diagnosis of hantaviruses in Europe. *PLoS Negl. Trop. Dis.*, *6*, e1607.

Essbauer, S. S., Krautkrämer, E., Herzog, S., & Pfeffer, M. (2011). A new permanent cell line derived from the bank vole (Myodes glareolus) as cell culture model for zoonotic viruses. *Virol. J.*, *2011 Jul 5*, *8*, 339. doi:10.1186/1743-422X-8-339.

Ettinger, J., Hofmann, J., Enders, M., Tewald, F., Oehme, R. M., Rosenfeld, U. M., Ali, H. S., Schlegel, M., Essbauer, S., Osterberg, A., Jacob, J., Reil, D., Klempa, B., Ulrich, R. G., & Krüger, D. H. (2012). Multiple synchronous outbreaks of Puumala virus, Germany, 2010. *Emerg. Infect. Dis.*, *18*, 1461–1464.

Faber, M., Wollny, T., Schlegel, M., Wanka, K. M., Thiel, J., Frank, C., Rimek, D., Ulrich, R. G., & Stark, K. (2013). Puumala Virus Outbreak in Western Thuringia, Germany, 2010: Epidemiology and Strain Identification. Zoonoses Public Health. in press.

Faber, M. S., Ulrich, R. G., Frank, C., Brockmann, S. O., Pfaff, G. M., Jacob, J., Krüger, D. H., & Stark, K. (2010). Steep rise in notified hantavirus infections in Germany, April 2010. *Euro surveillance*. *15*(20), pii=19574.

Feuer, R., Boone, J. D., Netski, D., Morzunov, S. P., & St Jeor, S. C. (1999). Temporal and spatial analysis of Sin Nombre virus quasispecies in naturally infected rodents. *J. Virol.*, *73*, 9544–9554.

Gajdusek, D. C. (1962). Virus hemorrhagic fevers. Special reference to hemorrhagic fever with renal syndrome (epidemic hemorrhagic fever). *J. Pediatr.*, *60*, 841–857.

Garanina, S. B., Platonov, A. E., Zhuravlev, V. I., Murashkina, A. N., Yakimenko, V. V., Korneev, A. G., & Shipulin, G. A. (2009). Genetic diversity and geographic distribution of hantaviruses in Russia. *Zoonoses and public health*, *56*, 297–309.

Garcin, D., Lezzi, M., Dobbs, M., Elliott, R. M., Schmaljohn, C., Kang, C. Y., & Kolakofsky, D. (1995). The 5′ ends of Hantaan virus (Bunyaviridae) RNAs suggest a prime-and-realign mechanism for the initiation of RNA synthesis. *J. Virol.*, *69*, 5754–5762.

Glass, G. E., Childs, J. E., Korch, G. W., & LeDuc, J. W. (1988). Association of intraspecific wounding with hantaviral infection in wild rats (*Rattus norvegicus*). *Epidemiol. Infect.*, *101*, 459–472.

Glass, G. E., Yates, T. L., Fine, J. B., Shields, T. M., Kendall, J. B., Hope, A. G., Parmenter, C. A., Peters, C. J., Ksiazek, T. G., Li, C. S., Patz, J. A., & Mills, J. N. (2002). Satellite imagery characterizes local animal reservoir populations of Sin Nombre virus in the southwestern United States. *Proc. Natl. Acad. Sci. U S A*, *99*, 16817–16822.

Goldsmith, C. S., Elliott, L. H., Peters, C. J., & Zaki, S. R. (1995). Ultrastructural characteristics of Sin Nombre virus, causative agent of hantavirus pulmonary syndrome. *Arch. Virol.*, *140*, 2107–2122.

Gonzalez Della Valle, M., Edelstein, A., Miguel, S., Martinez, V., Cortez, J., Cacace, M. L., Jurgelenas, G., Sosa Estani, S., & Padula, P. (2002). Andes virus associated with hantavirus pulmonary syndrome in northern Argentina and determination of the precise site of infection. *Am. J. Trop. Med. Hyg.*, *66*, 713–720.

Gonzalez, J. P., Josse, R., Johnson, E. D., Merlin, M., Georges, A. J., Abandja, J., Danyod, M., Delaporte, E., Dupont, A., Ghogomu, A., Kouka-Bemba, D., Madelon, M. C., Sima, A., & Meunier, D. M. Y. (1989). Antibody prevalence against haemorrhagic fever viruses in randomized representative Central African populations. *Res. Virol.*, *140*, 319–331.

Gonzalez, J. P., McCormick, J. B., Baudon, D., Gautun, J. P., Meunier, D. Y., Dournon, E., & Georges, A. J. (1984). Serological evidence for Hantaan-related virus in Africa. *Lancet*, *2*, 1036–1037.

Graham, T. B., & Chomel, B. B. (1997). Population dynamics of the deer mouse (*Peromyscus maniculatus*) and Sin Nombre virus, California Channel Islands. *Emerg. Infect. Dis.*, *3*, 367–370.

Groen, J., Gerding, M., Koeman, J. P., Roholl, P. J., van Amerongen, G., Jordans, H. G., Niesters, H. G., & Osterhaus, A. D. (1995). A macaque model for hantavirus infection. *J. Infect. Dis.*, *172*, 38–44.

Guivier, E., Galan, M., Salvador, A. R., Xuéreb, A., Chaval, Y., Olsson, G. E., Essbauer, S., Henttonen, H., Voutilainen, L., Cosson, J. F., & Charbonnel, N. (2010). Tnf-α expression and promoter sequences reflect the balance of tolerance/resistance to Puumala hantavirus infection in European bank vole populations. *Infect. Genet. Evol.*, *10*, 1208–1217.

Guo, W. P., Lin, X. D., Wang, W., Tian, J. H., Cong, M. L., Zhang, H. L., Wang, M. R., Zhou, R. H., Wang, J. B., Li, M. H., Xu, J., Holmes, E. C., & Zhang, Y. Z. (2013). Phylogeny and origins of hantaviruses harbored by bats, insectivores, and rodents. *PLoS Pathog.*, *9*, e1003159.

Handke, W., Oelschlegel, R., Franke, R., Wiedemann, L., Krüger, D. H., & Rang, A. (2010). Generation and characterisation of genetic reassortants between Puumala and Prospect Hill hantavirus in vitro. *J. Gen.Virol.*, *91*, 2351–2359.

Hardestam, J., Petterson, L., Ahlm, C., Evander, M., Lundkvist, A., & Klingstrom, J. (2008). Antiviral effect of human saliva against hantavirus. *J. Med. Virol.*, *80*, 2122–2126.

Hardestam, J., Simon, M., Hedlund, K. O., Vaheri, A., Klingstrom, J., & Lundkvist, A. (2007). Ex vivo stability of the rodent-borne Hantaan virus in comparison to that of arthropod-borne members of the Bunyaviridae family. *Applied Environmental Microbiology*, *73*, 2547–2551.

Hedman, K., Vaheri, A., & Brummer-Korvenkontio, M. (1991). Rapid diagnosis of hantavirus disease with an IgG-avidity assay. *Lancet*, *338*, 1353–1356.

Henttonen, H., Buchy, P., Suputtamongkol, Y., Jittapalapong, S., Herbreteau, V., Laakkonen, J., Chaval, Y., Galan, M., Dobigny, G., Charbonnel, N., Michaux, J., Cosson, J. F., Morand, S., & Hugot, J. P. (2008). Recent discoveries of new hantaviruses widen their range and question their origins. *Annals of the New York Academy of Sciences*, *1149*, 84–89.

Hepojoki, J., Strandin, T., Lankinen, H., & Vaheri, A. (2012). Hantavirus structure—molecular interactions behind the scene. *J. Gen. Virol.*, *93*, 1631–1644.

Heyman, P., Ceianu, C. S., Christova, I., Tordo, N., Beersma, M., Joao Alves, M., Lundkvist, A., Hukic, M., Papa, A., Tenorio, A., Zelena, H., Essbauer, S., Visontai, I., Golovljova, I., Connell, J., Nicoletti, L., Van Esbroeck, M., Gjeruldsen Dudman, S., Aberle, S. W., Avsic-Zupanc, T., Korukluoglu, G., Nowakowska, A., Klempa, B., Ulrich, R. G., Bino, S., Engler, O., Opp, M., & Vaheri, A. (2011). A five-year perspective on the situation of haemorrhagic fever with renal syndrome and status of the hantavirus reservoirs in Europe, 2005–2010. *Euro Surveill*, *16*(36), pii=19961.

Heyman, P., Plyusnina, A., Berny, P., Cochez, C., Artois, M., Zizi, M., Pirnay, J. P., & Plyusnin, A. (2004). Seoul hantavirus in Europe: first demonstration of the virus genome in wild Rattus norvegicus captured in France. *Eur. J. Clin. Microbiol. Inf. Dis.*, *23*, 711–717.

Heyman, P., & Vaheri, A. (2008). Situation of hantavirus infections and haemorrhagic fever with renal syndrome in European countries as of December 2006. *Euro Surveill*, *13*(28).

Heyman, P., Baert, K., Plyusnina, A., Cochez, C., Lundkvist, A., Esbroeck, M. V., Goossens, E., Vandenvelde, C., Plyusnin, A., & Stuyck, J. (2009). Serological and genetic evidence for the presence of Seoul hantavirus in Rattus norvegicus in Flanders. *Belgium. Scand. J. Infect. Dis.*, *41*, 51–56.

Hujakka, H., Koistinen, V., Kuronen, I., Eerikäinen, P., Parviainen, M., Lundkvist, A., Vaheri, A., Vapalahti, O., & Närvänen, A. (2003). Diagnostic rapid tests for acute hantavirus infections: specific tests for Hantaan, Dobrava and Puumala viruses versus a hantavirus combination test. *J. Virol. Methods*, *108*, 117–122.

Hjelle, B., Anderson, B., Torrez-Martinez, N., Song, W., Gannon, W. L., & Yates, T. L. (1995). Prevalence and geographic genetic variation of hantaviruses of New World harvest mice (*Reithrodontomys*): identification of a divergent genotype from a Costa Rican Reithrodontomys mexicanus. *Virology*, *207*, 452–459.

Hjelle, B., & Glass, G. E. (2000). Outbreak of hantavirus infection in the Four Corners region of the United States in the wake of the 1997–1998 El Nino-southern oscillation. *J. Infect. Dis.*, *181*, 1569–1573.

Hjelle, B., & Yates, T. (2001). Modeling hantavirus maintenance and transmission in rodent communities. *Curr. Top. Microbiol. Immunology*, *256*, 77–90.

Hofmann, J., Fuhrer, A., Bolz, M., Waldschlager-Terpe, J., Meier, M., Ludders, D., Enders, M., Oltmann, A., Meisel, H., & Krüger, D. H. (2012). Hantavirus infections by Puumala or Dobrava-Belgrade virus in pregnant women. *J. Clin. Virol.*, *55*, 266–269.

Horling, J., Chizhikov, V., Lundkvist, A., Jonsson, M., Ivanov, L., Dekonenko, A., Niklasson, B., Dzagurova, T., Peters, C. J., Tkachenko, E., & Nichol, S. (1996). Khabarovsk virus: a phylogenetically and serologically distinct hantavirus isolated from *Microtus fortis* trapped in Far-East Russia. *J. Gen. Virol.*, *77*, 687–694.

Horling, J., Lundkvist, A., Persson, K., Mullaart, M., Dzagurova, T., Dekonenko, A., Tkachenko, E., & Niklasson, B. (1995). Detection and subsequent sequencing of Puumala virus from human specimens by PCR. *J. Clin. Microbiol.*, *33*, 277–282.

Hörnfeldt, B. (2004). Long-term decline in numbers of cyclic voles in boreal Sweden: analysis and presentation of hypotheses. *Oikos*, *107*(2), 376–392.

Houpikian, P., & Raoult, D. (2001). Molecular phylogeny of the genus Bartonella: what is the current knowledge? *FEMS Microbiol. Lett.*, *200*, 1–7.

Hussein, I. T., Haseeb, A., Haque, A., & Mir, M. A. (2011). Recent advances in hantavirus molecular biology and disease. *Adv. Applied Microbiol.*, *74*, 35–75.

Hutterer, R., Ivanova, T., Meyer-Cord, C., & Rodrigues, L. M. (2005). *Bat Migrations in Europe: A Review of Banding Data and Literature. – Naturschutz und Biologische Vielfalt 28, Bonn (Bundesamt für Naturschutz)*.

Jaaskelainen, K. M., Kaukinen, P., Minskaya, E. S., Plyusnina, A., Vapalahti, O., Elliott, R. M., Weber, F., Vaheri, A., & Plyusnin, A. (2007). Tula and Puumala hantavirus NSs ORFs are functional and the products inhibit activation of the interferon-beta promoter. *J. Med. Virol.*, *79*, 1527–1536.

Jaaskelainen, K. M., Plyusnina, A., Lundkvist, A., Vaheri, A., & Plyusnin, A. (2008). Tula hantavirus isolate with the full-length ORF for nonstructural protein NSs survives for more consequent passages in interferon-competent cells than the isolate having truncated NSs ORF. *Virology Journal*, *5*, 3.

Jacob, J., & Tkadlec, E. (2010). Rodent outbreaks in Europe: dynamics and damage. In G. R. Singleton, S. Belmain, P. R. Brown & B. Hardy (Eds.), *Rodent outbreaks – Ecology and impacts* (pp. 207–223). Los Baños, Philippines: International Rice Research Institute.

Jameson, L., Taori, S., Atkinson, B., Levick, P., Featherstone, C., van der Burgt, G., McCarthy, N., Hart, J., Osborne, J., Walsh, A., Brooks, T., & Hewson, R. (2013a). Pet rats as a source of hantavirus in England and Wales, 2013. *Euro Surveill*, *18*(9).

Jameson, L. J., Logue, C. H., Atkinson, B., Baker, N., Galbraith, S. E., Carroll, M. W., Brooks, T., & Hewson, R. (2013b). The continued emergence of hantaviruses: isolation of a Seoul virus implicated in human disease, United Kingdom, October 2012. *Euro Surveill*, *18*(1), 4–7.

Jin, M., Park, J., Lee, S., Park, B., Shin, J., Song, K. J., Ahn, T. I., Hwang, S. Y., Ahn, B. Y., & Ahn, K. (2002). Hantaan virus enters cells by clathrin-dependent receptor-mediated endocytosis. *Virology*, *294*, 60–69.

Johansson, P., Yap, G., Low, H. T., Siew, C. C., Kek, R., Ng, L. C., & Bucht, G. (2010). Molecular characterization of two hantavirus strains from different rattus species in Singapore. *Virology Journal, 7*, 15.

Johnson, K. M. (2001). Hantaviruses: history and overview. *Curr. Topics Microbiol. Immunol., 256*, 1–14.

Jonsson, C. B., & Schmaljohn, C. S. (2001). Replication of hantaviruses. *Curr. Top. Microbiol. Immunol., 256*, 15–32.

Kallio-Kokko, H., Lundkvist, A., Plyusnin, A., Avsic-Zupanc, T., Vaheri, A., & Vapalahti, O. (2000). Antigenic properties and diagnostic potential of recombinant Dobrava virus nucleocapsid protein. *J. Med. Virol., 61*, 266–274.

Kallio, E. R., Begon, M., Henttonen, H., Koskela, E., Mappes, T., Vaheri, A., & Vapalahti, O. (2009). Cyclic hantavirus epidemics in humans—predicted by rodent host dynamics. *Epidemics, 1*, 101–107.

Kallio, E. R., Begon, M., Henttonen, H., Koskela, E., Mappes, T., Vaheri, A., & Vapalahti, O. (2010). Hantavirus infections in fluctuating host populations: the role of maternal antibodies. *Proc. Biol. Sci., 277*, 3783–3791.

Kallio, E. R., Klingstrom, J., Gustafsson, E., Manni, T., Vaheri, A., Henttonen, H., Vapalahti, O., & Lundkvist, A. (2006a). Prolonged survival of Puumala hantavirus outside the host: evidence for indirect transmission via the environment. *J. Gen. Virol., 87*, 2127–2134.

Kallio, E. R., Poikonen, A., Vaheri, A., Vapalahti, O., Henttonen, H., Koskela, E., & Mappes, T. (2006b). Maternal antibodies postpone hantavirus infection and enhance individual breeding success. *Proceedings. Biological Sciences / The Royal Society, 273*, 2771–2776.

Kallio, E. R., Voutilainen, L., Vapalahti, O., Vaheri, A., Henttonen, H., Koskela, E., & Mappes, T. (2007). Endemic hantavirus infection impairs the winter survival of its rodent host. *Ecology, 88*, 1911–1916.

Kang, H. J., Bennett, S. N., Dizney, L., Sumibcay, L., Arai, S., Ruedas, L. A., Song, J. W., & Yanagihara, R. (2009a). Host switch during evolution of a genetically distinct hantavirus in the American shrew mole (*Neurotrichus gibbsii*). *Virology, 388*, 8–14.

Kang, H. J., Bennett, S. N., Sumibcay, L., Arai, S., Hope, A. G., Mocz, G., Song, J. W., Cook, J. A., & Yanagihara, R. (2009b). Evolutionary insights from a genetically divergent hantavirus harbored by the European common mole (Talpa europaea). *PloS one, 4*, e6149.

Kang, H. J., Bennett, S. N., Hope, A. G., Cook, J. A., & Yanagihara, R. (2011). Shared ancestry between a newfound mole-borne hantavirus and hantaviruses harbored by cricetid rodents. *J. Virol., 85*(15), 7496–7503.

Kariwa, H., Yoshimatsu, K., Sawabe, J., Yokota, E., Arikawa, J., Takashima, I., Fukushima, H., Lundkvist, A., Shubin, F. N., Isachkova, L. M., Slonova, R. A., Leonova, G. N., & Hashimoto, N. (1999). Genetic diversities of hantaviruses among rodents in Hokkaido, Japan and Far East Russia. *Virus Res., 59*, 219–228.

Kariwa, H., Yoshizumi, S., Arikawa, J., Yoshimatsu, K., Takahashi, K., Takashima, I., & Hashimoto, N. (1995). Evidence for the existence of Puumula-related virus among *Clethrionomys rufocanus* in Hokkaido, Japan. *Am. J. Trop. Med. Hyg., 53*, 222–227.

Keesing, F., Belden, L. K., Daszak, P., Dobson, A., Harvell, C. D., Holt, R. D., Hudson, P., Jolles, A., Jones, K. E., Mitchell, C. E., Myers, S. S., Bogich, T., & Ostfeld, R. S. (2010). Impacts of biodiversity on the emergence and transmission of infectious diseases. *Nature, 468*, 647–652.

King, A. M. Q., Adams, M. J., Carstens, E. B., & Lefkowitz, E. J. (2011). Virus Taxonomy, Ninth Report of the International Committee on Taxonomy of Viruses. In A. M. Q. King, M. J. Adams, E. B. Carstens & E. J. Lefkowitz (Eds.), *Virus Taxonomy* (1 ed.). San Diego: Elsevier Academic Press.

Kirsanovs, S., Klempa, B., Franke, R., Lee, M., Schönrich, G., Rang, A., & Krüger, D. H. (2010). Genetic reassortment between high-virulent and low-virulent Dobrava-Belgrade virus strains. *Virus Genes, 41*(3), 319–328.

Klein, S. L., Zink, M. C., & Glass, G. E. (2004). Seoul virus infection increases aggressive behaviour in male Norway rats. *Animal Behaviour, 67*, 421–429.

Klein, S. L., Bird, B. H., & Glass, G. E. (2000). Sex differences in Seoul virus infection are not related to adult sex steroid concentrations in Norway rats. *J. Virol., 74*, 8213–8217.

Klempa, B., Schmidt, H. A., Ulrich, R., Kaluz, S., Labuda, M., Meisel, H., Hjelle, B., & Krüger, D. H. (2003). Genetic interaction between distinct Dobrava hantavirus subtypes in *Apodemus agrarius* and *A. flavicollis* in nature. *J. Virol., 77*, 804–809.

Klempa, B., Fichet-Calvet, E., Lecompte, E., Auste, B., Aniskin, V., Meisel, H., Denys, C., Koivogui, L., ter Meulen, J., & Krüger, D. H. (2006). Hantavirus in African wood mouse, Guinea. *Emerg. Infect. Dis., 12*, 838–840.

Klempa, B., Fichet-Calvet, E., Lecompte, E., Auste, B., Aniskin, V., Meisel, H., Barriere, P., Koivogui, L., ter Meulen, J., & Krüger, D. H. (2007). Novel hantavirus sequences in shrew, Guinea. *Emerg. Infect. Dis., 13*, 520–522.

Klempa, B. (2009). Hantaviruses and climate change. *Clin. Microbiol. Infect., 15*, 518–523.

Klempa, B., Krüger, D. H., Auste, B., Stanko, M., Krawczyk, A., Nickel, K. F., Uberla, K., & Stang, A. (2009). A novel cardiotropic murine adenovirus representing a distinct species of mastadenoviruses. *J Virol., 83*, 5749–5759.

Klempa, B., Koivogui, L., Sylla, O., Koulemou, K., Auste, B., Krüger, D. H., & ter Meulen, J. (2010). Serological evidence of human hantavirus infections in Guinea, West Africa. *J. Infect. Dis., 201*, 1031–1034.

Klempa, B., Witkowski, P. T., Popugaeva, E., Auste, B., Koivogui, L., Fichet-Calvet, E., Strecker, T., Ter Meulen, J., & Kruger, D. H. (2012). Sangassou virus, the first hantavirus isolate from Africa, displays distinct genetic and functional properties in the group of Murinae-associated hantaviruses. *J.Virol., 86*(7), 3819–3827.

Klempa, B., Avsic-Zupanc, T., Clement, J., Dzagurova, T. K., Henttonen, H., Heyman, P., Jakab, F., Krüger, D. H., Maes, P., Papa, A., Tkachenko, E. A., Ulrich, R. G., Vapalahti, O., & Vaheri, A. (2013). Complex evolution and epidemiology of Dobrava-Belgrade hantavirus: definition of genotypes and their characteristics. *Arch. Virol., 158*, 521–529.

Klingstrom, J., Falk, K. I., & Lundkvist, A. (2005). Delayed viremia and antibody responses in Puumala hantavirus challenged passively immunized cynomolgus macaques. *Arch. Virol, 150*, 79–92.

Korpimäki, E., Brown, P. R., Jacob, J., & Pech, R. P. (2004). The puzzles of population cycles and outbreaks of small mammals solved? *Bioscience, 54*, 1071–1079.

Kramski, M., Meisel, H., Klempa, B., Krüger, D. H., Pauli, G., & Nitsche, A. (2007). Detection and typing of human pathogenic hantaviruses by real-time reverse transcription-PCR and pyrosequencing. *Clin. Chem., 53*, 1899–1905.

Kramski, M., Achazi, K., Klempa, B., & Krüger, D. H. (2009). Nephropathia epidemica with a 6-week incubation period after occupational exposure to Puumala hantavirus. *J. Clin. Virol., 44*, 99–101.

Krebs, C. J., & Myers, J. H. (1974). Population cycles in small mammals. *Advances in Ecological Research, 8*, 267–399.

Krüger, D. H., Schonrich, G., & Klempa, B. (2011). Human pathogenic hantaviruses and prevention of infection. *Human Vaccines, 7*, 685–693.

Krüger, D. H., Ulrich, R., & Lundkvist, A. (2001). Hantavirus infections and their prevention. *Microbes Infect., 3*, 1129–1144.

Kukkonen, S. K., Vaheri, A., & Plyusnin, A. (2005). L protein, the RNA-dependent RNA polymerase of hantaviruses. *Arch. Virol.*, *150*, 533–556.

LeDuc, J. W., Smith, G. A., Childs, J. E., Pinheiro, F. P., Maiztegui, J. I., Niklasson, B., Antoniades, A., Robinson, D. M., Khin, M., Shortridge, K. F., et al. (1986). Global survey of antibody to *Hantaan*-related viruses among peridomestic rodents. *Bulletin of the World Health Organization*, *64*(1), 139–144.

Lee, H. W., Baek, L. J., & Johnson, K. M. (1982). Isolation of Hantaan virus, the etiologic agent of Korean hemorrhagic fever, from wild urban rats. *J. Infect. Dis.*, *146*, 638–644.

Lee, H. W., Lee, P. W., Baek, L. J., Song, C. K., & Seong, I. W. (1981). Intraspecific transmission of Hantaan virus, etiologic agent of Korean hemorrhagic fever, in the rodent *Apodemus agrarius*. *Am. J. Trop. Med. Hyg.*, *30*, 1106–1112.

Lee, H. W., Lee, P. W., & Johnson, K. M. (1978). Isolation of the etiologic agent of Korean Hemorrhagic fever. *J. Infect. Dis.*, *137*, 298–308.

Lee, H. W., & van der Groen, G. (1989). Hemorrhagic fever with renal syndrome. *Prog. Med. Viro.*, *36*, 62–102.

Lee, P. W., Amyx, H. L., Yanagihara, R., Gajdusek, D. C., Goldgaber, D., & Gibbs, C. J., Jr. (1985). Partial characterization of Prospect Hill virus isolated from meadow voles in the United States. *J. Infect. Dis.*, *152*, 826–829.

Lehmer, E. M., Clay, C. A., Wilson, E., St Jeor, S., & Dearing, M. D. (2007). Differential resource allocation in deer mice exposed to Sin Nombre virus. *Physiol. Biochem. Zool.*, *80*, 514–521.

Levine, J. R., Prescott, J., Brown, K. S., Best, S. M., Ebihara, H., & Feldmann, H. (2010). Antagonism of type I interferon responses by new world hantaviruses. *J. Virol.*, *84*, 11790–11801.

Levis, S., Garcia, J., Pini, N., Calderon, G., Ramirez, J., Bravo, D., St Jeor, S., Ripoll, C., Bego, M., Lozano, E., Barquez, R., Ksiazek, T. G., & Enria, D. (2004). Hantavirus pulmonary syndrome in northwestern Argentina: circulation of Laguna Negra virus associated with *Calomys callosus*. *Am. J. Trop. Med. Hyg.*, *71*, 658–663.

Li, D., Schmaljohn, A. L., Anderson, K., & Schmaljohn, C. S. (1995). Complete nucleotide sequences of the M and S segments of two hantavirus isolates from California: evidence for reassortment in nature among viruses related to hantavirus pulmonary syndrome. *Virology*, *206*, 973–983.

Lin, X. D., Guo, W. P., Wang, W., Zou, Y., Hao, Z. Y., Zhou, D. J., Dong, X., Qu, Y. G., Li, M. H., Tian, H. F., Wen, J. F., Plyusnin, A., Xu, J., & Zhang, Y. Z. (2012). Migration of Norway rats resulted in the worldwide distribution of Seoul hantavirus today. *J. Virol.*, *86*, 972–981.

Lindström, J., Ranta, E., Kokko, H., Lundberg, P., & Kaitala, V. (2001). From arctic lemmings to adaptive dynamics: Charles Elton's legacy in population ecology. *Biol. Rev. Camb. Philos. Soc.*, *76*(1), 129–158.

Lober, C., Anheier, B., Lindow, S., Klenk, H. D., & Feldmann, H. (2001). The Hantaan virus glycoprotein precursor is cleaved at the conserved pentapeptide WAASA. *Virology*, *289*, 224–229.

Loehman, R. A., Elias, J., Douglass, R. J., Kuenzi, A. J., Mills, J. N., & Wagoner, K. (2012). Prediction of *Peromyscus maniculatus* (deer mouse) population dynamics in Montana, USA, using satellite-driven vegetation productivity and weather data. *J. Wildl. Dis.*, *48*, 348–360.

Luis, A. D., Douglass, R. J., Hudson, P. J., Mills, J. N., & Bjornstad, O. N. (2012). Sin Nombre hantavirus decreases survival of male deer mice. *Oecologia*, *169*, 431–439.

Lundkvist, A., Kallio-Kokko, H., Sjölander, K. B., Lankinen, H., Niklasson, B., Vaheri, A., & Vapalahti, O. (1996). Characterization of Puumala virus nucleocapsid protein: identification of B-cell epitopes and domains involved in protective immunity. *Virology. 1996 Feb 15*, *216*(2), 397–406.

Lundkvist, A., Hukic, M., Hörling, J., Gilljam, M., Nichol, S., & Niklasson, B. (1997a). Puumala and Dobrava viruses cause hemorrhagic fever with renal syndrome in Bosnia-Herzegovina: evidence of highly cross-neutralizing antibody responses in early patient sera. *J. Med. Virol., 53,* 51–59.

Lundkvist, A., Cheng, Y., Sjolander, K. B., Niklasson, B., Vaheri, A., & Plyusnin, A. (1997b). Cell culture adaptation of Puumala hantavirus changes the infectivity for its natural reservoir, *Clethrionomys glareolus,* and leads to accumulation of mutants with altered genomic RNA S segment. *J. Virol., 71,* 9515–9523.

Mackow, E. R., & Gavrilovskaya, I. N. (2001). Cellular receptors and hantavirus pathogenesis. *Curr. Top. Microbiol. Immunol., 256,* 91–115.

Mackow, E. R., & Gavrilovskaya, I. N. (2009). Hantavirus regulation of endothelial cell functions. *Thromb. Haemost., 102,* 1030–1041.

Maes, P., Klempa, B., Clement, J., Matthijnssens, J., Gajdusek, D. C., Krüger, D. H., & Van Ranst, M. (2009). A proposal for new criteria for the classification of hantaviruses, based on S and M segment protein sequences. *Infection Genetic Evolution, 9,* 813–820.

Mayer-Scholl, A., Draeger, A., Luge, E., Ulrich, R., & Nöckler, K. (2011). Comparison of two PCR systems for the rapid detection of *Leptospira* spp. from kidney tissue. *Curr. Microbiol., 62,* 1104–1106.

Matthys, V., Gorbunova, E. E., Gavrilovskaya, I. N., Pepini, T., & Mackow, E. R. (2011). The C-terminal 42 residues of the Tula virus Gn protein regulate interferon induction. *J. Virol., 85,* 4752–4760.

McCormick, J. B., Sasso, D. R., Palmer, E. L., & Kiley, M. P. (1982). Morphological identification of the agent of Korean haemorrhagic fever (Hantaan virus) as a member of the Bunyaviridae. *Lancet, 1,* 765–768.

McElroy, A. K., Smith, J. M., Hooper, J. W., & Schmaljohn, C. S. (2004). Andes virus M genome segment is not sufficient to confer the virulence associated with Andes virus in Syrian hamsters. *Virology, 326,* 130–139.

Medina, R. A., Torres-Perez, F., Galeno, H., Navarrete, M., Vial, P. A., Palma, R. E., Ferres, M., Cook, J. A., & Hjelle, B. (2009). Ecology, genetic diversity, and phylogeographic structure of Andes virus in humans and rodents in Chile. *J. Virol., 83,* 2446–2459.

Meerburg, B. G., Singleton, G. R., & Kijlstra, A. (2009). Rodent-borne diseases and their risks for public health. *Crit. Rev. Microbiol., 35,* 221–270.

Meisel, H., Wolbert, A., Razanskiene, A., Marg, A., Kazaks, A., Sasnauskas, K., Pauli, G., Ulrich, R., & Krüger, D. H. (2006). Development of novel immunoglobulin G (IgG), IgA, and IgM enzyme immunoassays based on recombinant Puumala and Dobrava hantavirus nucleocapsid proteins. *Clin. Vaccine Immunol., 13,* 1349–1357.

Mertz, G. J., Hjelle, B., Crowley, M., Iwamoto, G., Tomicic, V., & Vial, P. A. (2006). Diagnosis and treatment of new world hantavirus infections. *Curr. Opin. Infect. Dis., 19,* 437–442.

Mertz, G. J., Hjelle, B. L., & Bryan, R. T. (1997). Hantavirus infection. *Adv. Intern. Med., 42,* 369–421.

Mills, J. N., Amman, B. R., & Glass, G. E. (2010). Ecology of hantaviruses and their hosts in North America. *Vector Borne Zoonotic Dis., 10,* 563–574.

Mills, J. N., Ksiazek, T. G., Peters, C. J., & Childs, J. E. (1999a). Long-term studies of hantavirus reservoir populations in the southwestern United States: a synthesis. *Emerg. Infect. Dis., 5,* 135–142.

Mills, J. N., Yates, T. L., Ksiazek, T. G., Peters, C. J., & Childs, J. E. (1999b). Long-term studies of hantavirus reservoir populations in the southwestern United States: rationale, potential, and methods. *Emerg. Infect. Dis., 5,* 95–101.

Mohamed, N., Nilsson, E., Johansson, P., Klingström, J., Evander, M., Ahlm, C., & Bucht, G. (2013). Development and evaluation of a broad reacting SYBR-green based quantitative real-time PCR for the detection of different hantaviruses. *J. Clin. Virol.*, *56*, 280–285.

Murphy, F. A. (2008). Emerging zoonoses: the challenge for public health and biodefense. *Prev. Vet. Med.*, *86*, 216–223.

Nemirov, K., Lundkvist, A., Vaheri, A., & Plyusnin, A. (2003). Adaptation of Puumala hantavirus to cell culture is associated with point mutations in the coding region of the L segment and in the noncoding regions of the S segment. *J. Virol.*, *77*, 8793–8800.

Nichol, S. T., Spiropoulou, C. F., Morzunov, S., Rollin, P. E., Ksiazek, T. G., Feldmann, H., Sanchez, A., Childs, J., Zaki, S., & Peters, C. J. (1993). Genetic identification of a hantavirus associated with an outbreak of acute respiratory illness. *Science*, *262*, 914–917.

Nordström, H., Johansson, P., Li, Q. G., Lundkvist, A., Nilsson, P., & Elgh, F. (2004). Microarray technology for identification and distinction of hantaviruses. *J. Med. Virol.*, *72*, 646–655.

Olsson, G. E., Hjertqvist, M., Ahlm, C., Evander, M., & Hornfeldt, B. (2010). Nephropathia epidemica: Data on voles indicate new, extensive outbreak. *Lakartidningen*, *107*, 1769–1770.

Olsson, G. E., White, N., Ahlm, C., Elgh, F., Verlemyr, A. C., Juto, P., & Palo, R. T. (2002). Demographic factors associated with hantavirus infection in bank voles (*Clethrionomys glareolus*). *Emerg. Infect. Dis.*, *8*, 924–929.

Orrock, J. L., Allan, B. F., & Drost, C. A. (2011). Biogeographic and ecological regulation of disease: prevalence of Sin Nombre virus in island mice is related to island area, precipitation, and predator richness. *Am. Nat.*, *177*, 691–697.

Ostfeld, R. S., & Holt, R. D. (2004). Are predators good for your health? Evaluating evidence for top-down regulation of zoonotic disease reservoirs. *Frontiers in Ecology and the Environment*, *2*, 13–20.

Ostfeld, R. S., & Keesing, F. (2012). Effects of Host Diversity on Infectious Disease. *Annual Review of Ecology, Evolution, and Systematics*, *43*, 157–182.

Owen, R. D., Goodin, D. G., Koch, D. E., Chu, Y. -K., & Jonsson, C. B. (2010). Spatiotemporal variation in *Akodon montensis* (Cricetidae: Sigmodontinae) and hantaviral seroprevalence in a subtropical forest ecosystem. *Journal of Mammalogy*, *91*(2), 467–481.

Padula, P. J., Edelstein, A., Miguel, S. D., Lopez, N. M., Rossi, C. M., & Rabinovich, R. D. (1998). Hantavirus pulmonary syndrome outbreak in Argentina: molecular evidence for person-to-person transmission of Andes virus. *Virology*, *241*, 323–330.

Palo, R. T. (2009). Time series analysis performed on Nephropathia epidemica in humans of northern Sweden in relation to bank vole population dynamic and the NAO index. *Zoonoses and Public Health*, *56*, 150–156.

Peters, C. J. (2006). Emerging infections: lessons from the viral hemorrhagic fevers. *Trans. Am. Clin. Climatol. Assoc.*, *117*, 189–196.

Peters, C. J., Simpson, G. L., & Levy, H. (1999). Spectrum of hantavirus infection: hemorrhagic fever with renal syndrome and hantavirus pulmonary syndrome. *Annu. Rev. Med.*, *50*, 531–545.

Phan, T. G., Kapusinszky, B., Wang, C., Rose, R. K., Lipton, H. L., & Delwart, E. L. (2011). The fecal viral flora of wild rodents. *PLoS Pathog.*, *7*, e1002218.

Pinna, D. M., Martinez, V. P., Bellomo, C. M., Lopez, C., & Padula, P. (2004). New epidemiologic and molecular evidence of person to person transmission of hantavirus Andes Sout. *Medicina (B Aires)*, *64*, 43–46.

Piudo, L., Monteverde, M. J., Walker, R. S., & Douglass, R. J. (2011). Rodent community structure and Andes virus infection in sylvan and peridomestic habitats in northwestern Patagonia, Argentina. *Vector Borne Zoonotic Dis.*, *11*, 315–324.

Plyusnin, A., Vapalahti, O., Lankinen, H., Lehvaslaiho, H., Apekina, N., Myasnikov, Y., Kallio-Kokko, H., Henttonen, H., Lundkvist, A., Brummer-Korvenkontio, M., Gavrilovskaya, I., & Vaheri, A. (1994). Tula virus: a newly detected hantavirus carried by European common voles. *J. Virol.*, *68*, 7833–7839.

Plyusnin, A., Cheng, Y., Lehvaslaiho, H., & Vaheri, A. (1996). Quasispecies in wild-type Tula hantavirus populations. *J. Virol.*, *70*, 9060–9063.

Plyusnin, A., Vapalahti, O., & Vaheri, A. (1996). Hantaviruses: genome structure, expression and evolution. *J. Gen. Virol.*, *77*, 2677–2687.

Plyusnin, A., Vapalahti, O., Vasilenko, V., Henttonen, H., & Vaheri, A. (1997). Dobrava hantavirus in Estonia: does the virus exist throughout Europe? *Lancet.*, *349*(9062), 1369–1370.

Plyusnin, A., Mustonen, J., Asikainen, K., Plyusnina, A., Niemimaa, J., Henttonen, H., & Vaheri, A. (1999). Analysis of puumala hantavirus genome in patients with nephropathia epidemica and rodent carriers from the sites of infection. *J. Med. Virol.*, *59*, 397–405.

Plyusnin, A., & Morzunov, S. P. (2001). Virus evolution and genetic diversity of hantaviruses and their rodent hosts. *Curr. Topics Microbiol. Immunol.*, *256*, 47–75.

Plyusnin, A., Vaheri, A., & Lundkvist, A. (2003). Genetic interaction between Dobrava and Saaremaa hantaviruses: now or millions of years ago? *J. Virol.*, *77*, 7156–7157.

Plyusnina, A., Laakkonen, J., Niemimaa, J., Nemirov, K., Muruyeva, G., Pohodiev, B., Lundkvist, A., Vaheri, A., Henttonen, H., Vapalahti, O., & Plyusnin, A. (2008). Genetic analysis of hantaviruses carried by *Myodes* and *Microtus* rodents in Buryatia. *Virology Journal*, *5*, 4.

Plyusnin, A., Beaty, B. J., Elliott, R. M., Goldbach, R., Kormelink, R., Lundkvist, A., Schmaljohn, C. S., & Tesh, R. B. (2011). Bunyaviridae. In A. M. Q. King, M. J. Adams, E. B. Carstensen & E. J. Lefkowitz (Eds.), *Virus Taxonomy: Ninth report of the International Committee on Taxonomy of Viruses*. San Diego, USA: Elsevier Inc.

Plyusnin, A., & Elliott, R. (2011). Concluding remarks. In *Bunyaviridae. Molecular and Cellular Biology*. Caister Academic Press.

Popugaeva, E., Witkowski, P. T., Schlegel, M., Ulrich, R. G., Auste, B., Rang, A., Krüger, D. H., & Klempa, B. (2012). Dobrava-Belgrade hantavirus from Germany shows receptor usage and innate immunity induction consistent with the pathogenicity of the virus in humans. *PLoS One*, *7*, e35587.

Radosa, L., Schlegel, M., Gebauer, P., Ansorge, H., Heroldová, M., Jánová, E., Stankof, M., Mošanský, L., Fričová, J., Pejčoch, M., Suchomel, J., Purchart, L., Groschup, M. H., Krüger, D. H., Ulrich, R. G., & Klempa, B. (2013). Detection of shrew-borne hantavirus in Eurasian pygmy shrew (*Sorex minutus*) in Central Europe. *Infection, Genetics and Evolution.* (in press).

Rang, A., Heider, H., Ulrich, R., & Krüger, D. H. (2006). A novel method for cloning of noncytolytic viruses. *J. Virol. Methods*, *135*, 26–31.

Ravkov, E. V., Nichol, S. T., & Compans, R. W. (1997). Polarized entry and release in epithelial cells of Black Creek Canal virus, a New World hantavirus. *J. Virol.*, *71*, 1147–1154.

Rawlings, J. A., Torrez-Martinez, N., Neill, S. U., Moore, G. M., Hicks, B. N., Pichuantes, S., Nguyen, A., Bharadwaj, M., & Hjelle, B. (1996). Cocirculation of multiple hantaviruses in Texas, with characterization of the small (S) genome of a previously undescribed virus of cotton rats (*Sigmodon hispidus*). *Am. J. Trop. Med. Hyg.*, *55*, 672–679.

Razanskiene, A., Schmidt, J., Geldmacher, A., Ritzi, A., Niedrig, M., Lundkvist, A., Krüger, D. H., Meisel, H., Sasnauskas, K., & Ulrich, R. (2004). High yields of stable and highly pure nucleocapsid proteins of different hantaviruses can be generated in the yeast *Saccharomyces cerevisiae*. *J. Biotechnol.*, *111*, 319–333.

Razzauti, M., Plyusnina, A., Henttonen, H., & Plyusnin, A. (2008). Accumulation of point mutations and reassortment of genomic RNA segments are involved in the microevolution of Puumala hantavirus in a bank vole (*Myodes glareolus*) population. *J. Gen. Virol.*, *89*, 1649–1660.

Razzauti, M., Plyusnina, A., Sironen, T., Henttonen, H., & Plyusnin, A. (2009). Analysis of Puumala hantavirus in a bank vole population in northern Finland: evidence for co-circulation of two genetic lineages and frequent reassortment between strains. *J. Gen. Virol.*, *90*, 1923–1931.

Reil, D., Imholt, C., Schmidt, S., Rosenfeld, U. M., Ulrich, R. G., Eccard, J. A., & Jacob, J. (2011). Relationship between bank vole abundance, seroprevalence and human hantavirus infections. *Julius-Kühn-Archiv* (432), 197.

Reusken, C., & Heyman, P. (2013). Factors driving hantavirus emergence in Europe. *Curr. Opin. Virol.*, *3*, 92–99.

Rodier, G., Soliman, A. K., Bouloumie, J., & Kremer, D. (1993). Presence of antibodies to Hantavirus in rat and human populations of Djibouti *Trans. Royal Soc. Trop. Med. Hyg.*, *87*, 160–161.

Rodriguez, L. L., Owens, J. H., Peters, C. J., & Nichol, S. T. (1998). Genetic reassortment among viruses causing hantavirus pulmonary syndrome. *Virology*, *242*, 99–106.

Roehr, B. (2012). US officials warn 39 countries about risk of hantavirus among travellers to Yosemite. *BMJ.*, *345*, e6054.

Rosa, E. S., Mills, J. N., Padula, P. J., Elkhoury, M. R., Ksiazek, T. G., Mendes, W. S., Santos, E. D., Araujo, G. C., Martinez, V. P., Rosa, J. F., Edelstein, A., & Vasconcelos, P. F. (2005). Newly recognized hantaviruses associated with hantavirus pulmonary syndrome in northern Brazil: partial genetic characterization of viruses and serologic implication of likely reservoirs. *Vector Borne Zoonotic Dis.*, *5*, 11–19.

Ruedas, L. A., Salazar-Bravo, J., Tinnin, D. S., Armien, B., Caceres, L., Garcia, A., Diaz, M. A., Gracia, F., Suzan, G., Peters, C. J., Yates, T. L., & Mills, J. N. (2004). Community ecology of small mammal populations in Panama following an outbreak of Hantavirus pulmonary syndrome. *J. Vector Ecol.*, *29*, 177–191.

Rüdisser, E. (1942). Beobachtungen zur Pathogenese und Aetiologie von Nierenentzündungen im Feldzug im Osten Winter 1941/42, Münch. *Med. Wochenschr.*, *89*, 863–866.

Sackl, P., Tiefenbach, M., Tajmel, J., & Spitzenberger, F. (2007). Weitere Ausbreitung der Brandmaus *Apodemus agrarius* (PALLAS, 1771) in Österreich (Mammalia). *Joannea Zool.*, *9*, 5–13.

Safronetz, D., Ebihara, H., Feldmann, H., & Hooper, J. W. (2012). The Syrian hamster model of hantavirus pulmonary syndrome. *Antiviral Res.*, *95*, 282–292.

Scharninghausen, J. J., Pfeffer, M., Meyer, H., Davis, D. S., Honeycutt, R. L., & Faulde, M. (2002). Genetic evidence for Tula virus in *Microtus arvalis* and *Microtus agrestis* populations in Croatia. *Vector Borne Zoonotic Dis.*, *2*, 19–27.

Schex, S., Dobler, G., Riehm, J., Müller, J., & Essbauer, S. (2011). *Rickettsia* spp. in wild small mammals in Lower Bavaria, South-Eastern Germany. *Vector Borne Zoonotic Dis.*, *11*, 493–502.

Schlegel, M., Klempa, B., Auste, B., Bemmann, M., Schmidt-Chanasit, J., Buchner, T., Groschup, M. H., Meier, M., Balkema-Buschmann, A., Zoller, H., Krüger, D. H., & Ulrich, R. G. (2009). Dobrava-Belgrade virus spillover infections, Germany, *Emerg. Infect. Dis.*, *15*, 2017–2020.

Schlegel, M., Kindler, E., Essbauer, S. S., Wolf, R., Thiel, J., Groschup, M. H., Heckel, G., Oehme, R. M., & Ulrich, R. G. (2012a). Tula virus infections in the Eurasian water vole in Central Europe. *Vector Borne Zoonotic Dis.*, *12*, 503–513.

Schlegel, M., Radosa, L., Rosenfeld, U. M., Schmidt, S., Triebenbacher, C., Lohr, P. W., Fuchs, D., Heroldova, M., Janova, E., Stanko, M., Mosansky, L., Fricova, J., Pejcoch, M., Suchomel, J., Purchart, L., Groschup, M. H., Krüger, D. H., Klempa, B., & Ulrich, R. G. (2012b). Broad geographical distribution and high genetic diversity of shrew-borne Seewis hantavirus in Central Europe. *Virus Genes*, *45*, 48–55.

Schlegel, M. (2013). Host range and spillover infections of rodent- and insectivore-borne hantaviruses. *Dissertation, Ernst-Moritz-Arndt-Universität Greifswald.* URL http://ub-ed.ub.uni-greifswald.de/opus/volltexte/2013/1453/.

Schmaljohn, C., & Hjelle, B. (1997). Hantaviruses: a global disease problem. *Emerg. Infect. Dis.*, *3*, 95–104.

Schmaljohn, C. S., & Dalrymple, J. M. (1983). Analysis of Hantaan virus RNA: evidence for a new genus of Bunyaviridae. *Virology*, *131*, 482–491.

Schmaljohn, C. S., Hasty, S. E., Dalrymple, J. M., LeDuc, J. W., Lee, H. W., von Bonsdorff, C. H., Brummer-Korvenkontio, M., Vaheri, A., Tsai, T. F., Regnery, H. L., et al. (1985). Antigenic and genetic properties of viruses linked to hemorrhagic fever with renal syndrome. *Science*, *227*, 1041–1044.

Schmidt-Chanasit, J., Essbauer, S., Petraityte, R., Yoshimatsu, K., Tackmann, K., Conraths, F. J., Sasnauskas, K., Arikawa, J., Thomas, A., Pfeffer, M., Scharninghausen, J. J., Splettstoesser, W., Wenk, M., Heckel, G., & Ulrich, R. G. (2010). Extensive host sharing of central European Tula virus. *J. Virol.*, *84*, 459–474.

Schonrich, G., Rang, A., Lutteke, N., Raftery, M. J., Charbonnel, N., & Ulrich, R. G. (2008). Hantavirus-induced immunity in rodent reservoirs and humans. *Immunological Reviews*, *225*, 163–189.

Schulz, E., Gottschling, M., Ulrich, R. G., Richter, D., Stockfleth, E., & Nindl, I. (2012). Isolation of three novel rat and mouse papillomaviruses and their genomic characterization. *PLoS One*, *7*, e47164.

Schultze, D., Lundkvist, A., Blauenstein, U., & Heyman, P. (2002). Tula virus infection associated with fever and exanthema after a wild rodent bite. *Eur. J. Clin. Microbiol. Infect. Dis.*, *21*, 304–306.

Shi, X., & Elliott, R. M. (2004). Analysis of N-linked glycosylation of hantaan virus glycoproteins and the role of oligosaccharide side chains in protein folding and intracellular trafficking. *J. Virol.*, *78*, 5414–5422.

Shi, X., McCaughey, C., & Elliott, R. M. (2003). Genetic characterisation of a Hantavirus isolated from a laboratory-acquired infection. *J. Med. Virol.*, *71*, 105–109.

Sibold, C., Sparr, S., Schulz, A., Labuda, M., Kozuch, O., Lysy, J., Krüger, D. H., & Meisel, H. (1995). Genetic characterization of a new hantavirus detected in *Microtus arvalis* from Slovakia. *Virus Genes*, *10*, 277–281.

Sibold, C., Meisel, H., Krüger, D. H., Labuda, M., Lysy, J., Kozuch, O., Pejcoch, M., Vaheri, A., & Plyusnin, A. (1999). Recombination in Tula hantavirus evolution: analysis of genetic lineages from Slovakia. *J Virol.*, *73*, 667–675.

Siński, E., Bajer, A., Welc, R., Pawełczyk, A., Ogrzewalska, M., & Behnke, J. M. (2006). Babesia microti: prevalence in wild rodents and *Ixodes ricinus* ticks from the Mazury Lakes District of North-Eastern Poland. *Int. J. Med. Microbiol.*, *296*, 137–143.

Sirola, H., Kallio, E. R., Koistinen, V., Kuronen, I., Lundkvist, A., Vaheri, A., Vapalahti, O., Henttonen, H., & Närvänen, A. (2004). Rapid field test for detection of hantavirus antibodies in rodents. *Epidemiol. Infect.*, *132*, 549–553.

Sironen, T., Klingstrom, J., Vaheri, A., Andersson, L. C., Lundkvist, A., & Plyusnin, A. (2008). Pathology of Puumala hantavirus infection in macaques. *PLoS One*, *3*, e3035.

Sjolander, K. B., Elgh, F., Kallio-Kokko, H., Vapalahti, O., Hagglund, M., Palmcrantz, V., Juto, P., Vaheri, A., Niklasson, B., & Lundkvist, A. (1997). Evaluation of serological methods for diagnosis of Puumala hantavirus infection (nephropathia epidemica). *J. Clin. Microbiol.*, *35*, 3264–3268.

Smith, M. G., White, A., Lambin, X., Sherratt, J. A., & Begon, M. (2006). Delayed density-dependent season length alone can lead to rodent population cycles. *American Naturalist*, *167*, 695–704.

Song, J. W., Gligic, A., & Yanagihara, R. (2002). Identification of Tula hantavirus in Pitymys subterraneus captured in the Cacak region of Serbia-Yugoslavia. *Int. J. Infect. Dis.*, *6*, 31–36.

Song, J. W., Baek, L. J., Schmaljohn, C. S., & Yanagihara, R. (2007a). Thottapalayam virus, a prototype shrewborne hantavirus. *Emerg. Infect. Dis.*, *13*, 980–985.

Song, J. W., Gu, S. H., Bennett, S. N., Arai, S., Puorger, M., Hilbe, M., & Yanagihara, R. (2007b). Seewis virus, a genetically distinct hantavirus in the Eurasian common shrew (*Sorex araneus*). *Virology Journal*, *4*, 114.

Song, J. W., Kang, H. J., Song, K. J., Truong, T. T., Bennett, S. N., Arai, S., Truong, N. U., & Yanagihara, R. (2007c). Newfound hantavirus in Chinese mole shrew. *Vietnam. Emerg. Infect. Dis.*, *13*, 1784–1787.

Song, K. J., Baek, L. J., Moon, S., Ha, S. J., Kim, S. H., Park, K. S., Klein, T. A., Sames, W., Kim, H. C., Lee, J. S., Yanagihara, R., & Song, J. W. (2007d). Muju virus, a novel hantavirus harboured by the arvicolid rodent *Myodes regulus* in Korea. *J. Gen. Virol.*, *88*, 3121–3129.

Song, J. W., Kang, H. J., Gu, S. H., Moon, S. S., Bennett, S. N., Song, K. J., Baek, L. J., Kim, H. C., O'Guinn, M. L., Chong, S. T., Klein, T. A., & Yanagihara, R. (2009). Characterization of Imjin virus, a newly isolated hantavirus from the Ussuri white-toothed shrew (*Crocidura lasiura*). *J. Virol.*, *83*, 6184–6191.

Song, W., Torrez-Martinez, N., Irwin, W., Harrison, F. J., Davis, R., Ascher, M., Jay, M., & Hjelle, B. (1995). Isla Vista virus: a genetically novel hantavirus of the California vole *Microtus californicus*. *J. Gen. Virol.*, *76*, 3195–3199.

Spiropoulou, C. F., Albarino, C. G., Ksiazek, T. G., & Rollin, P. E. (2007). Andes and Prospect Hill hantaviruses differ in early induction of interferon although both can downregulate interferon signaling. *J. Virol.*, *81*, 2769–2776.

Spiropoulou, C. F. (2001). Hantavirus maturation. *Curr. Top. Microbiol. Immunol.*, *256*, 33–46.

Spiropoulou, C. F. (2011). Molecular Biology of Hantavirus Infection. In A. Plyusnin & R. M. Elliot (Eds.), *Bunyaviridae, Molecular and Cellular Biology* (Vol. 1, pp. 41–60). Norfolk, UK: Caister Academic Press.

Stoltz, M., Sundström, K. B., Hidmark, Å., Tolf, C., Vene, S., Ahlm, C., Lindberg, A. M., Lundkvist, Å., & Klingström, J. (2011). A model system for in vitro studies of bank vole borne viruses. *PLoS One*, *6*(12), e28992. doi:10.1371/journal.pone.0028992. Epub 2011 Dec 16.

Stuhlfault, K. (1943). Bericht über ein neues schlammfieberähnliches Krankheitsbild bei Deutschen Truppen in Lappland. *Dtsch. Med. Wochenschr.*, *69*, 439–443.

Sumibcay, L., Kadjo, B., Gu, S. H., Kang, H. J., Lim, B. K., Cook, J. A., Song, J. W., & Yanagihara, R. (2012). Divergent lineage of a novel hantavirus in the banana pipistrelle (*Neoromicia nanus*) in Cote d'Ivoire. *Virology Journal*, *9*, 34.

Sundström, K. B., Stoltz, M., Lagerqvist, N., Lundkvist, Å., Nemirov, K., & Klingström, J. (2011). Characterization of two substrains of Puumala virus that show phenotypes that are different from each other and from the original strain. *J. Virol.*, *85*, 1747–1756.

Suzan, G., Marce, E., Giermakowski, J. T., Mills, J. N., Ceballos, G., Ostfeld, R. S., Armien, B., Pascale, J. M., & Yates, T. L. (2009). Experimental evidence for reduced rodent diversity causing increased hantavirus prevalence. *PloS One*, *4*, e5461.

Tao, H., Xia, S. M., Chan, Z. Y., Song, G., & Yanagihara, R. (1987). Morphology and morphogenesis of viruses of hemorrhagic fever with renal syndrome. II. Inclusion bodies—ultrastructural markers of hantavirus-infected cells. *Intervirology*, *27*, 45–52.

Taylor, S. L., Frias-Staheli, N., Garcia-Sastre, A., & Schmaljohn, C. S. (2009). Hantaan virus nucleocapsid protein binds to importin alpha proteins and inhibits tumor necrosis factor alpha-induced activation of nuclear factor kappa B. *J. Virol.*, *83*, 1271–1279.

Telfer, S., Lambin, X., Birtles, R., Beldomenico, P., Burthe, S., Paterson, S., & Begon, M. (2010). Species interactions in a parasite community drive infection risk in a wildlife population. *Science*, *330*, 243–246.

Tersago, K., Crespin, L., Verhagen, R., & Leirs, H. (2012). Impact of Puumala virus infection on maturation and survival in bank voles: a capture-mark-recapture analysis. *J. Wildl. Dis., 48,* 148–156.

Tersago, K., Schreurs, A., Linard, C., Verhagen, R., Van Dongen, S., & Leirs, H. (2008). Population, environmental, and community effects on local bank vole (*Myodes glareolus*) Puumala virus infection in an area with low human incidence. *Vector Borne Zoonotic Dis., 8,* 235–244.

Tersago, K., Verhagen, R., Servais, A., Heyman, P., Ducoffre, G., & Leirs, H. (2009). Hantavirus disease (nephropathia epidemica) in Belgium: effects of tree seed production and climate. *Epidemiol. Infect., 137,* 250–256.

Tersago, K., Verhagen, R., & Leirs, H. (2011). Temporal variation in individual factors associated with hantavirus infection in bank voles during an epizootic: implications for Puumala virus transmission dynamics. *Vector Borne Zoonotic Dis., 11,* 715–721.

Thrall, P. H., Antonovics, J., & Hall, D. W. (1993). Host and pathogen coexistence in sexually transmitted and vector-borne diseases characterized by frequency-dependent disease transmission. *The American Naturalist, 142,* 543–552.

Torrez-Martinez, N., Bharadwaj, M., Goade, D., Delury, J., Moran, P., Hicks, B., Nix, B., Davis, J. L., & Hjelle, B. (1998). Bayou virus-associated hantavirus pulmonary syndrome in Eastern Texas: identification of the rice rat, *Oryzomys palustris*, as reservoir host. *Emerg. Infect. Dis., 4,* 105–111.

Traavik, T., Sommer, A. I., Mehl, R., Berdal, B. P., Stavem, K., Hunderi, O. H., & Dalrymple, J. M. (1984). Nephropathia epidemica in Norway: antigen and antibodies in rodent reservoirs and antibodies in selected human populations. *J. Hyg., 93,* 139–146.

Ulrich, R., Hjelle, B., Pitra, C., & Krüger, D. H. (2002). Emerging viruses: the case 'hantavirus'. *Intervirology, 45,* 318–327.

Ulrich, R., Meisel, H., Schütt, M., Schmidt, J., Kunz, A., Klempa, B., Niedrig, M., Kimmig, P., Pauli, G., Krüger, D. H., & Koch, J. (2004). Verbreitung von Hantavirusinfektionen in Deutschland. *Bundesgesundheitsblatt Gesundheitsforschung Gesundheitsschutz, 47,* 661–670.

Ulrich, R. G., Heckel, G., Pelz, H. J., Wieler, L. H., Nordhoff, M., Dobler, G., Freise, J., Matuschka, F. R., Jacob, J., Schmidt-Chanasit, J., Gerstengarbe, F. W., Jakel, T., Suss, J., Ehlers, B., Nitsche, A., Kallies, R., Johne, R., Gunther, S., Henning, K., Grunow, R., Wenk, M., Maul, L. C., Hunfeld, K. P., Wolfel, R., Schares, G., Scholz, H. C., Brockmann, S. O., Pfeffer, M., & Essbauer, S. S. (2009). Rodents and rodent associated disease vectors: the network of "rodent carrying pathogens" introduces itself. *Bundesgesundheitsblatt Gesundheitsforschung Gesundheitsschutz, 52,* 352–369.

Vaheri, A., Vapalahti, O., & Plyusnin, A. (2008). How to diagnose hantavirus infections and detect them in rodents and insectivores. *Rev. Med. Virol., 18,* 277–288.

Vaheri, A., Henttonen, H., Voutilainen, L., Mustonen, J., Sironen, T., & Vapalahti, O. (2013). Hantavirus infections in Europe and their impact on public health. *Rev. Med. Virol., 23,* 35–49.

Vahlenkamp, M., Muller, T., Tackmann, K., Loschner, U., Schmitz, H., & Schreiber, M. (1998). The muskrat (*Ondatra zibethicus*) as a new reservoir for Puumala-like hantavirus strains in Europe. *Virus Res., 57,* 139–150.

Vapalahti, O., Lundkvist, A., Fedorov, V., Conroy, C. J., Hirvonen, S., Plyusnina, A., Nemirov, K., Fredga, K., Cook, J. A., Niemimaa, J., Kaikusalo, A., Henttonen, H., Vaheri, A., & Plyusnin, A. (1999). Isolation and characterization of a hantavirus from *Lemmus sibiricus*: evidence for host switch during hantavirus evolution. *J. Virol., 73,* 5586–5592.

Vapalahti, O., Lundkvist, A., Kallio-Kokko, H., Paukku, K., Julkunen, I., Lankinen, H., & Vaheri, A. (1996a). Antigenic properties and diagnostic potential of Puumala virus nucleocapsid protein expressed in insect cells. *J. Clin. Microbiol., 34,* 119–125.

Vapalahti, O., Lundkvist, A., Kukkonen, S. K., Cheng, Y., Gilljam, M., Kanerva, M., Manni, T., Pejcoch, M., Niemimaa, J., Kaikusalo, A., Henttonen, H., Vaheri, A., & Plyusnin, A. (1996b). Isolation and characterization of Tula virus, a distinct serotype in the genus Hantavirus, family Bunyaviridae. *J. Gen. Virol.*, *77*, 3063–3067.

Vapalahti, O., Mustonen, J., Lundkvist, A., Henttonen, H., Plyusnin, A., & Vaheri, A. (2003). Hantavirus infections in Europe. *Lancet Infectious Diseases*, *3*, 653–661.

van Regenmortel, M. H., Mayo, M. A., Fauquet, C. M., & Maniloff, J. (2000). Virus nomenclature: consensus versus chaos. *Arch. Virol.*, *145*, 2227–2232.

Vorou, R. M., Papavassiliou, V. G., & Tsiodras, S. (2007). Emerging zoonoses and vector-borne infections affecting humans in Europe. *Epidemiol. Infect.*, *135*, 1231–1247.

Weidmann, M., Schmidt, P., Vackova, M., Krivanec, K., Munclinger, P., & Hufert, F. T. (2005). Identification of genetic evidence for Dobrava virus spillover in rodents by nested reverse transcription (RT)-PCR and TaqMan RT-PCR. *J. Clin. Microbiol.*, *43*, 808–812.

Weiss, S., Witkowski, P. T., Auste, B., Nowak, K., Weber, N., Fahr, J., Mombouli, J. V., Wolfe, N. D., Drexler, J. F., Drosten, C., Klempa, B., Leendertz, F. H., & Krüger, D. H. (2012). Hantavirus in bat, Sierra Leone. *Emerg. Infect. Dis.*, *18*, 159–161.

Wilson, D. E., & Reeder, D. M. (2005a). Mammal species of the world: a taxonomic and geographic reference. Third Edition. Volume 1. 3 ed., 1. 2 vols. Baltimore, Maryland, USA: The Johns Hopkins University Press.

Wilson, D. E., & Reeder, D. M. (2005b). Mammal species of the world: a taxonomic and geographic reference. Third Edition. Volume 2. 3 ed., 2. 2 vols. Baltimore, Maryland, USA: The Johns Hopkins University Press.

Woods, C., Palekar, R., Kim, P., Blythe, D., de Senarclens, O., Feldman, K., Farnon, E. C., Rollin, P. E., Albarino, C. G., Nichol, S. T., & Smith, M. (2009). Domestically acquired Seoul virus causing hemorrhagic fever with renal syndrome–Maryland, 2008. *Clin. Infect. Dis.*, *49*, e109–e112.

World Health Organization. (1983). Haemorrhagic fever with renal syndrome: memorandum from a WHO meeting. *Bull. WHO*, *61*, 269–275.

Yadav, P. D., Vincent, M. J., & Nichol, S. T. (2007). Thottapalayam virus is genetically distant to the rodent-borne hantaviruses, consistent with its isolation from the Asian house shrew (*Suncus murinus*). *Virology Journal*, *4*, 80.

Yahnke, C. J., Meserve, P. L., Ksiazek, T. G., & Mills, J. N. (2001). Patterns of infection with Laguna Negra virus in wild populations of *Calomys laucha* in the central Paraguayan chaco. *Am. J. Trop. Med. Hyg.*, *65*, 768–776.

Yanagihara, R., & Gajdusek, D. C. (1987). Hemorrhagic fever with renal syndrome: global epidemiology and ecology of hantavirus infections. In L. M. de la Maza & E. M. Peterson (Eds.), *Medical Virology VI* (pp. 171–214). Amsterdam: Elsevier.

Yanagihara, R., Daum, C. A., Lee, P. W., Baek, L. J., Amyx, H. L., Gajdusek, D. C., & Gibbs, C. J., Jr (1987). Serological survey of Prospect Hill virus infection in indigenous wild rodents in the USA. *Trans. R. Soc. Trop. Med. Hyg.*, *81*, 42–45.

Yashina, L. N., Abramov, S. A., Gutorov, V. V., Dupal, T. A., Krivopalov, A. V., Panov, V. V., Danchinova, G. A., Vinogradov, V. V., Luchnikova, E. M., Hay, J., Kang, H. J., & Yanagihara, R. (2010). Seewis virus: phylogeography of a shrew-borne hantavirus in Siberia, Russia. *Vector Borne Zoonotic Dis.*, *10*, 585–591.

Yates, T. L., Mills, J. N., Parmenter, C. A., Ksiazek, T. G., Parmenter, R. R., Vande Castle, J. R., Calisher, C. H., Nichol, S. T., Abbott, K. D., & Young, J. C. (2002). The ecology and evolutionary history of an emergent disease: hantavirus pulmonary syndrome. *Bioscience*, *52*, 989–998.

Yoshimatsu, K., Arikawa, J., & Kariwa, H. (1993). Application of a recombinant baculovirus expressing hantavirus nucleocapsid protein as a diagnostic antigen in IFA test: cross reactivities among 3 serotypes of hantavirus which causes hemorrhagic fever with renal syndrome (HFRS). *The Journal of Veterinary Medical Science, 55,* 1047–1050.

Zavasky, D. M., Hjelle, B., Peterson, M. C., Denton, R. W., & Reimer, L. (1999). Acute infection with Sin Nombre hantavirus without pulmonary edema. *Clin. Infect. Dis., 29,* 664–666.

Zhang, F. X., Zou, Y., Chen, H. X., Yan, S. H., Hasen, G., Wang, J. B., Li, C. F., Zhang, S. Y., Zhao, Z. W., & Zhang, Y. Z. (2007a). Study on the epidemiological characteristics of hemorrhagic fever with renal syndrome in Inner Mongolia. *Zhonghua Liu Xing Bing Xue Za Zhi, 28,* 1101–1104.

Zhang, Y. Z., Zou, Y., Yao, L. S., Hu, G. W., Du, Z. S., Jin, L. Z., Liu, Y. Y., Wang, H. X., Chen, X., Chen, H. X., & Fu, Z. F. (2007b). Isolation and characterization of hantavirus carried by Apodemus peninsulae in Jilin, China. *J. Gen. Virol., 88,* 1295–1301.

Zhuravlev, V. I., Garanina, S. B., Kabin, V. V., Shipulin, G. A., & Platonov, A. E. (2008). Detection of a new natural virus focus Dobrava in the Astrakhan Region. *Voprosy virusologii, 53,* 37–40.

Zoller, L., Faulde, M., Meisel, H., Ruh, B., Kimmig, P., Schelling, U., Zeier, M., Kulzer, P., Becker, C., Roggendorf, M., et al. (1995). Seroprevalence of hantavirus antibodies in Germany as determined by a new recombinant enzyme immunoassay. *Eur. J. Clin. Microbiol. Infect. Dis., 14,* 305–313.

Zou, Y., Hu, J., Wang, Z. X., Wang, D. M., Yu, C., Zhou, J. Z., Fu, Z. F., & Zhang, Y. Z. (2008). Genetic characterization of hantaviruses isolated from Guizhou, China: evidence for spillover and reassortment in nature. *J. Med. Virol., 80,* 1033–1041.

Chapter 11

Nipah Virus

A Virus with Multiple Pathways of Emergence

David T.S. Hayman[1,2] and Nicholas Johnson[3]

[1]*Department of Biology, Colorado State University, Fort Collins, Colorado, USA*, [2]*Department of Biology, University of Florida, Gainesville, FL, USA*, [3]*Animal Health and Veterinary Laboratories Agency, Surrey, United Kingdom*

INTRODUCTION: NIPAH VIRUS OUTBREAKS

Nipah virus, NiV, has caused several intense outbreaks of disease in Asia. However, to provide perspective for the NiV outbreaks, a brief introduction to Hendra virus (HeV) emergence that preceded it is useful. Hendra virus was first reported to have caused disease in horses and humans in 1994 in Australia, after a sudden outbreak of acute respiratory disease in thoroughbred horses (Murray et al., 1995). The etiological agent was initially undetermined after a range of noninfectious and exotic infectious diseases to Australia were excluded (Douglas et al., 1997; Douglas, 1998). The previously unknown virus, initially named equine morbillivirus and subsequently Hendra virus, of the family *Paramyxoviridae,* was subsequently identified as the causal agent. Two HeV outbreaks were reported in 1994, the first of which was diagnosed retrospectively in Mackay after the second outbreak in Brisbane. The characterization of HeV by researchers was to be especially pertinent when NiV emerged five years later in another part of the world.

The epidemic caused by the virus subsequently named Nipah virus is believed to have originated in the northwestern state of Perak in northern Malaysia. The first indication of the appearance of a new disease was an epidemic of encephalitis among farm workers, particularly those with contact with pigs. Infection typically caused encephalitis, with many patients going into coma prior to death (Goh et al., 2000; Wong et al., 2002a). The human epidemic started in October 1998 and lasted until May 1999 (Mohd Nor et al., 2000). Initially the disease was reported as an outbreak of Japanese encephalitis virus (ProMed, 1998). Despite exposure to pigs as the primary risk factor for infection (Parashar et al., 2000), the outbreak had features unlike the mosquito-borne JEV. In 1999 NiV was identified as the etiological agent of the outbreak (Amal et al., 2000; Chua et al., 2000a) and retrospective investigations suggest that the initial transmission from the wildlife reservoir into the pig population occurred as early as January 1997 (Pulliam et al., 2012). The first virus isolation was made from an infected human from the village of Sungai Nipah, hence the naming of the virus Nipah (Chua et al., 1999). Electron microscopy confirmed that the virus was a paramyxovirus, and partial genomic sequencing demonstrated that it was closely related to HeV, recently isolated from Australia (Chua et al., 2000a). The emergence of the virus in pigs was not immediately noticed as the disease caused was not sufficiently distinct from existing diseases of pigs to warrant attention and pig mortality was not high. However, mortality in man was particularly high with 105 human deaths resulting from 265 reported cases. Due to the delay in linking human encephalitis cases to what would be described as porcine respiratory and neurological syndrome, the virus was spread south in Malaysia to other farms and abattoirs through livestock movements. Human NiV in Malaysia was invariably due to close contact with NiV infected pigs, with human-to-human transmission rare (Amal et al., 2000; Anonymous, 1999a,b; Parashar et al., 2000; Mounts et al., 2001), and so the outbreak was ultimately stopped by a massive cull of affected pig farms (Kamil et al., 2001; Lam and Chua, 2002). The cull of pigs from the affected farms led to the slaughter of over a million animals, almost one third of the total standing stock within the country. The cull and introduction of biosecurity measures to prevent further NiV emergence appears to have succeeded with no further outbreaks being reported to the present day (Mohd Nor et al., 2000; Hayman et al., 2012a).

One of the most striking examples of NiV spread through livestock movements from the Malaysian outbreak was the spread of disease to abattoirs in Singapore on the southern tip of Peninsular Malaysia. In March 1999, a number of cases of encephalitis and pneumonia were reported in staff working in a single abattoir (Paton et al., 1999). The response in Singapore was a complete closure of abattoirs. Serological screening of at-risk persons was implemented, and included abattoir workers, meat inspectors, public butchers, zoo workers and customs inspectors who discovered illegal pork shipments from Malaysia after an import ban on Malaysian pigs (Chan et al., 2002). A final total of 22 workers from two abattoirs were considered to have been infected, all having had close contact with pigs or pig carcasses, although only 12 of the 22 were symptomatic (Chan et al., 2002). All who died or were seropositive against NiV in Singapore had direct contact with pigs imported from Malaysia (Paton et al., 1999, Chan et al., 2002).

Following the explosive appearance of NiV, the virus re-emerged but with a different pathway to infection of humans (Figure 11.1). In 2001, several NiV encephalitis cases were discovered in Bangladesh (Anonymous, 2002, 2004a, 2004b; Hsu et al., 2004; Luby et al., 2007). The next example of re-emergence was reported in Bangladesh in 2003 (Hsu et al., 2004). Retrospective analysis identified 13 potential cases from one location and 12 from a second. There

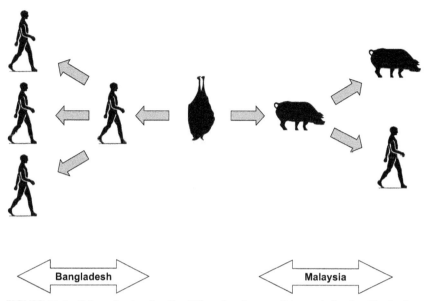

FIGURE 11.1 Schematic showing the different pathways of transmission for Nipah virus from Pteropid bats to humans in Bangladesh (left) and pigs to humans in Malaysia (right).

appeared to be no involvement of an intermediate host such as a domestic animal, and strong evidence that human-to-human contact had resulted in transmission of disease. Further outbreaks of NiV infection in humans has been demonstrated in Bangladesh and India (Homaira et al., 2010; Chadha et al., 2006). Although these south Asian outbreaks involve fewer individuals than the Malaysian outbreak, the mortality rate is consistently higher at approximately 75%. One of the key routes of infection has been the consumption of raw date palm sap (Luby et al., 2006). Date palm sap is collected from across Bangladesh and is a common target of nocturnal fruit bats, which steal the sap from collecting vessels and contaminate the remaining sap with body fluids such as urine and saliva. Subsequent consumption of contaminated sap is thought to lead to human infection (Rahman et al., 2012). The outbreaks in Bangladesh occur on the border with India, and it was a retrospective study that showed that in 2001 an outbreak of NiV occurred in the city of Siliguri, northeast India, which borders Bangladesh. In Siliguri, human-to-human transmission appears to have occurred and both IgM antibodies to NiV and NiV viral RNA have been detected in samples from patients (Chadha et al., 2006).

Apart from these locations, there have been no reported cases of NiV infection in humans in other areas of the world. It is striking that the routes of transmission from the Pteropid reservoir (discussed below) and humans are different (Figure 11.1). There was little evidence during the Malaysian outbreak to suggest human-to-human transmission, whereas in the repeated outbreaks in Bangladesh there is clear evidence of human-to-human transmission. Direct

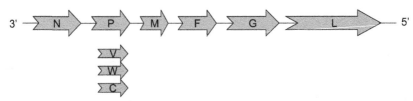

FIGURE 11.2   Schematic of the Nipah virus genome showing the sequence of open-reading frames within the genome. The alternative gene products of the phosphoprotein gene are also shown.

comparison between NiV strains isolated from the Malaysian and Bangladeshi outbreaks in experimental infections in ferrets suggests that the Bangladesh isolate is shed at higher levels (Clayton et al., 2012). This factor alone may be sufficient to enable the virus to transmit more efficiently between humans, although the mechanism through which this isolate drives higher virus shedding has not been identified.

NIPAH VIRUS VIROLOGY AND DISEASE

Nipah virus is classified within the genus *Henipavirus* within the family Paramyxoviridae and subfamily Paramyxovirinae (Chua et al., 2000a). Paramyxoviruses have nonsegmented, negative-stranded (-ve) ribonucleic acid (RNA) genomes (Bellini et al., 1998). The NiV genome has been fully sequenced and consists of negative-stranded RNA, 18,246 base pairs in length (Chua et al., 2000a; Harcourt et al., 2001; Chan et al., 2001). This RNA genome codes for six open-reading frames flanked by 3′ and 5′ noncoding regions (Figure 11.2). The gene order from 3′ to 5′ is: nucleoprotein (N), phosphoprotein (P), matrix (M), fusion (F), glycoprotein (G) and polymerase (L) (Vigant and Lee, 2011). The phosphoprotein gene of NiV, in common with many paramyxoviruses, can express multiple proteins, and NiV appears to replicate like other paramyxoviruses (Jack et al., 2005; Li et al., 2006; Halpin et al., 2004; Harcourt et al., 2000, 2001; Wang et al., 2000). The 3′ leader and 5′ trailer regions of paramyxovirus genomes act as transcription and replication promoters (Lamb and Parks, 2007). Each gene coding region is between untranslated regions (UTRs) containing start and stop sequences, with short intergenomic regions. Expression of the complete gene produces P, a component of the nucleocapsid. Shorter gene products produced by RNA editing or alternative translational initiation produce the V, W and C proteins that antagonize interferon signaling by binding and sequestering STAT proteins. This antagonism is not species restricted and appears to inhibit human, bat and pig interferon production when tested in cellular models (Hagmaier et al., 2006). The nucleoprotein binds tightly to the RNA genome. The F and G proteins insert within the host-derived envelope and enable binding and entry into the target cell. The host receptor is the ephrin-B2 protein, a transmembrane protein kinase found on many cell types (Bonaparte et al.,

2005; Negrete et al., 2005). This protein is evolutionarily conserved within mammals, which may be a reason for the wide host-range of NiV. The distribution of NiV also correlates well with Ephrin-B2 receptors in human disease. The polymerase copies the RNA genome into a positive sense intermediate and then progeny genomes.

NiV virions are larger (500 nm diameter) than other paramyxoviruses (150–400 nm diameter), with large variations in size (180–1900 nm) (Goldsmith et al., 2003), but share most morphological features. Negative-contrast electron microscopy may distinguish NiV from HeV by the presence of a double-fringed surface projection on HeV, compared to the single fringe observed on NiV (Hyatt et al., 2001).

Clinical signs of NiV encephalitis in humans are not pathognomonic for NiV. The incubation period for clinical signs varied from 2 to 30 days in the Malaysian outbreak, and was typically 1–2 weeks. Prodromal signs of NiV infection are nonspecific and include fever, headache, myalgia, and respiratory signs (Chong et al., 2008; Goh et al., 2000). As disease progresses, further signs of neurological disease develop, including hyporeflexia or areflexia, reduced consciousness, segmental myoclonus, gaze palsy and limb weakness, before coma and death. Diagnosis of NiV is by viral RNA detection by reverse transcription (RT)-polymerase chain reaction (PCR) and virus isolation from urine, respiratory secretions, and cerebrospinal fluid (CSF) of patients. Detection of virus in the CSF is associated with high mortality (Chua et al., 2000b; Chua et al., 2001; Tan et al., 1999). Anti-NiV IgM antibody can be detected in blood from the majority of patients after a week and CSF 10–15 days after clinical signs appear (Chua et al., 2001). Anti-NiV IgG can be detected in blood by day 17–18 in all patients (Ramasundrum et al., 2000). Brain lesions, such as focal lesions in subcortical and deep white matter, caused by NiV encephalitis correlate with disease severity (Goh et al., 2000) and neurological deterioration and coma precedes death in 32–73% of NiV patients (Goh et al., 2000; Sim et al., 2002, Hossain et al., 2008). Mortality is most likely in cases with rapidly developing and severe neurological signs, and many of these patients do not survive long enough for a premortem diagnosis to be made. In these cases, autopsies show systemic vasculitis with extensive thrombosis, parenchymal necrosis in gray and white matter of the central nervous system (CNS) and widespread viral antigen in endothelial and smooth muscle cells of blood vessels, CNS parenchymal cells and neurons. In visceral organs intranuclear inclusions are found in multinucleated giant cells (Wong et al., 2002b). Encephalitic patients that do survive often have mild to severe residual neurologic deficits of the patients (Goh et al., 2000; Sejvar et al., 2007). As in the second reported fatal HeV case (O'Sullivan et al. 1997) relapsing encephalitis after apparent survival has been reported months after apparent initial recovery. Twelve of 160 encephalitic cases and five non-encephalitic Malaysian patients had relapsing encephalitis in Malaysia (Chong and Tan, 2003; Tan et al., 2002). To date, none have been reported in Bangladesh (Sejvar et al., 2007). Overall, however, case fatality rates reported from Bangladesh are higher than they were for Malaysia, which is perhaps due to viral traits (Clayton et al., 2012), but which may reflect differences in care (Hossain et al., 2008). Treatment of NiV encephalitis, like most viral encephalitis, is supportive therapy, including mechanical ventilation in comatose patients. Anti-viral ribavirin has been used in patients, but without clear success (Chong et al., 2000; Goh et al., 2000; Chong et al., 2001). A similar lack of efficacy of ribavirin is shown by experimental studies in animal models (Georges-Courbot et al., 2006). More recently experimental studies have shown monoclonal antibodies have potential to protect from disease by post-exposure passive antibody therapy in animal models (Bossart et al., 2009; Zhu et al., 2006a, b).

Disease in pigs differs from that in man. Natural and experimental infection of pigs with NiV causes different symptoms depending on the age the pigs are infected (Hooper, 2000; Middleton et al., 2002; Mohd Nor et al., 2000). Mortality is typically low, with the predominant clinical signs being respiratory, although neurological symptoms may also be seen. Those animals that become symptomatic show clinical signs after 7–14 days. Young animals (0.5–6 months old) typically develop febrile respiratory illness. Clinically, their respiratory signs may vary, but may include a harsh nonproductive coughing through to hemoptysis in severe cases. More prominent in adult pigs are neurological signs, including myoclonus, paresis, frothy salivation, difficulty swallowing, seizures, paralysis, and lateral recumbency with paddling movements. Pathological signs in pigs include nonspecific edema, congestion, and petechial to ecchymotic hemorrhages in visceral organs, such as lungs and kidneys, and in the brain. Histological changes include vasculitis and syncytial cell formations in endothelial cells of affected tissues and NiV antigen can be detected by immunohistochemistry (IHC).

Studies in Malaysia suggest multiple factors could have been necessary for NiV to be transmitted to commercial piggeries (Chua et al., 2002a), and modeling studies suggest multiple introductions were needed before infection could persist within the piggeries that were the major sources of human infection (Pulliam et al., 2012). The ongoing transmission of NiV was only possible in piggeries because of direct pig contact with infected excretions in high intensity pig farms with insufficient biosecurity. Experimental studies confirmed urine, saliva, and pharyngeal secretions could act as sources of infection (Middleton et al., 2002). Both oral and parental experimental inoculation of NiV via oronasal routes led to infection of pigs, and transmission to in-contact pigs was rapid (Middleton et al., 2002). Other mechanisms for transmission could include a range of biological or mechanical modes, but the spread of NiV within Malaysia was due to movement of live pigs with farms in close proximity to infected farms remaining uninfected if no pigs moved between them (Mohd Nor, 2000). Most recently, Hayman et al. (2011) detected non-neutralizing antibodies in domestic pigs in West Africa. The significance of this is uncertain, given the non-neutralizing nature of the antibody and the diverse array of paramyxoviruses detected in African bats (Drexler et al., 2012; Baker et al., 2012; Baker et al., 2013).

BAT RESERVOIR HOSTS

One of the most urgent concerns with the NiV outbreak in Malaysia was to identify the source of this new infection and again the emergence of HeV from Pteropid bats (Young et al., 1996; Halpin et al., 2000) pointed to the likely source of NiV. The presence of bats of the *Pteropus* genus at the infected pig farms meant *Pteropus* bats were quickly identified as the putative viral reservoir (Chua et al., 2002b; Johara et al., 2001). A preliminary study of 324 Malaysian bats from 14 species failed to detect virus but did identify neutralizing antibodies to NiV in 5 species (Johara et al., 2001). Malaysia has 13 species of fruit-eating bats and the highest seroprevalance was found in two pteropid species, the Island flying fox (*Pteropus hypomelanus*) and the Malaysian flying fox (*Pteropus vampyrus*). Further investigation of a colony of *P. hypomelanus* led to the isolation of virus from both urine samples and from a partially eaten fruit discarded by a fruit bat (Chua et al., 2002b). Subsequently NiV was also isolated from bats in Cambodia (Reynes et al., 2005), further strengthening the hypothesis that pteropid bats are reservoirs for NiV. Experimental infection of *Pteropus* bats with NiV does not cause illness in the bats, but bats do shed NiV (Middleton et al., 2007), suggesting once again that these bats are suitable hosts for NiV. In Bangladesh there was strong evidence for an initial transmission event from bats to humans and then onward transmission by direct contact for each case (Luby et al., 2009), and serological surveys have provided no evidence for other wildlife reservoirs, so investigations remain focused on *Pteropus* bats. A complete list of bats linked to NiV is shown in Table 11.1.

Bats belong to the mammalian order Chiroptera, the second largest mammalian order. The Chiroptera are divided into two suborders; Vespertilioniformes and Pteropodiformes. The pteropid bats, the putative hosts for henipaviruses, are within the suborder Pteropodiformes, family Pteropodidae, and genus *Pteropus* (Hutcheon and Kirsch, 2006; Teeling et al., 2005). The Pteropodidae are divided into seven subfamilies, 46 genera and 186 species. Pteropid bats are frugivorous, deriving nutrition from fruit-bearing trees. Pteropid bats are found in Australia, the tropical regions of Asia and in locations around the Indian Ocean, including Madagascar, although not in mainland Africa. The large or Malaysian flying fox (*P. vampyrus*) is one of the largest species within the genus *Pteropus* with an adult wing span of 1.5 meters and weighing over 1 kilogram. Many species form large colonies and will travel long distances in search of fruit (Breed et al., 2010; Epstein et al., 2009). In Africa, other frugivorous bat species occupy the ecological niche that *Pteropus* bats fill in Asia. In particular, species such as the straw-colored fruit bat (*Eidolon helvum*) and the Egyptian fruit bat (*Roussettus aegyptiacus*) are of similar size and form large colonies. *Eidolon helvum* shares a close genetic relationship to pteropid bats and roosts

TABLE 11.1 Bat Species Linked to Nipah Virus (NiV) Through Virus Isolation, Detection of Viral RNA by Polymerase Chain Reaction, or the Detection of Circulating Antibodies Against NiV by the Use of Serological Assays

Species	Distribution	Country of Nipah virus detection	Type of detection method	References
Frugivorous bats				
Pteropus vampyrus	Continental and insular southeast Asia, from southern Myanmar and Vietnam to throughout much of Indonesia and the Philippines	Malaysia; Thailand; Indonesia	Virus isolation; serological antibody detection; serological antibody detection	Sohayati et al. 2011; Wacharapluesadee et al. 2005; Sendow et al. 2006
Pteropus lylei	Yunnan, China, Cambodia, Thailand and Vietnam	Thailand; Cambodia	Polymerase chain reaction detection of nucleic acids; virus isolation	Wacharapluesadee et al. 2005, Wacharapluesadee et al. 2009; Reynes et al. 2005
Pteropus hypomelanus	Patchy distribution including the Maldives, Andaman and Nicobar Islands (India) east to Melanesia, the Philippines and Indonesia (Sulawesi)	Malaysia; Thailand	Virus isolation; serological antibody detection	Chua et al. 2002b; Wacharapluesadee et al. 2005
Pteropus giganteus	South Asia, into China and southeast Asia as far as Myanmar	Bangladesh; India	Serological antibody detection; serological antibody detection	Hsu et al. 2004; Epstein et al. 2008
Pteropus rufus	Madagascar	Madagascar	Serological antibody detection	Lehle et al. 2007
Rousettus leschenaultia	South Asia, into China and southeast Asia as far as Bali, Indonesia	China; Vietnam	Serological antibody detection	Li et al. 2008; Hasebe et al. 2012

Cynopterus brachyotis	South Asia, through to southern China and southeast Asia	Malaysia	Serological antibody detection	Yob et al. 2001
Cynopterus sphinx	South Asia, through to southern China and southeast Asia	Vietnam	Serological antibody detection	Hasebe et al. 2012
Eidolon dupreanum	Madagascar	Madagascar	Serological antibody detection	Lehle et al. 2007
Eidolon helvum	Sub-Saharan Africa	Ghana	Serological antibody detection	Hayman et al. 2008
Epomophorus gambianus	Central and west Africa through to Ethiopia in the east	Ghana	Serological antibody detection	Hayman et al. 2008
Hypsignathus monstrosus	Central and west Africa	Ghana	Serological antibody detection	Hayman et al. 2008
Insectivorous bats				
Hipposideros larvatus	Southeast Asia, and parts of northeastern south Asia and southern China	Thailand	Serological antibody detection	Wacharapluesadee et al. 2005
Hipposideros pomona	Southeast Asia, and parts of south Asia and southern China	China	Serological antibody detection	Li, et al. 2008
Rhinolophus affinis	Southeast Asia, and parts of south Asia and southern China	China	Serological antibody detection	Li et al. 2008
Rhinolophus sinicus	Southeast Asia, and parts of south Asia and southern China	China	Serological antibody detection	Li et al. 2008

Continued

TABLE 11.1 Bat Species Linked to Nipah Virus (NiV) Through Virus Isolation, Detection of Viral RNA by Polymerase Chain Reaction, or the Detection of Circulating Antibodies Against NiV by the Use of Serological Assays—cont'd

Species	Distribution	Country of Nipah virus detection	Type of detection method	References
Insectivorous bats				
Myotis daubentonii	Southeast Asia, and parts of south Asia and southern China	China	Serological antibody detection	Li et al. 2008
Myotis ricketti	Southeast Asia, and parts of south Asia and southern China	China	Serological antibody detection	Li et al. 2008
Miniopterus sp.	Southeast Asia, and parts of south Asia and southern China	China	Serological antibody detection	Li et al. 2008
Eonycteris spelaea	Southeast Asia, and parts of south Asia and southern China	Malaysia	Serological antibody detection	Yob et al. 2001
Scotophilus kuhlii	Southeast Asia, and parts of south Asia and southern China	Malaysia	Serological antibody detection	Yob et al. 2001

in trees much as fruit bats from the *Pteropus* genus do (Hayman et al., 2012b; Giannini and Simmons, 2003). Typically Pteropodidae bats only reproduce once a year and only produce one young (Hayman et al., 2012b). Reproduction, however, may play a key role in virus transmission in natural colonies for fruit bats, as serological surveys suggest that pregnancy and lactation in female bats may be associated with higher HeV seroprevalence (Plowright et al., 2008) and HeV spillover (Plowright et al., 2011). Longitudinal studies of NiV RNA in bat urine in Thailand suggest that two NiV strains (Bangladesh and Malaysian) are shed during the reproductive period, although peak NiV shedding of the Bangladesh strain was after the pups separated from the dams (Wacharapluesadee et al., 2010). Further evidence for the importance of reproduction in NiV transmission has been suggested, because HeV is detected in placental and fetal tissues (Williamson and Torres-Velez, 2010) and NiV in the uterus (Middleton et al. 2007) of bats after experimental infection.

Experimental infection of pteropid bats with NiV has demonstrated that infected bats excrete NiV and develop anti-NiV antibody without symptomatic illness (Middleton et al., 2007). The mechanism of NiV transmission between bats, however, is less clear. Serological studies suggest horizontal transmission of henipaviruses (Plowright et al., 2008) and NiV has been isolated from urine of wild roosting bats (Chua et al., 2002b, Reynes et al., 2005). Vertical transmission is suggested by the detection of HeV in placental and fetal tissues (Williamson and Torres-Velez, 2010) and NiV in the uterus (Middleton et al. 2007). Further studies are also needed to clarify the role of recrudescence after a captive study on naturally seropositive bats suggested NiV recrudesced (Sohayati et al., 2011). Nipah virus is capable of infecting a wide range of mammalian species in nature, including pigs, dogs, horses, and humans, in addition to bats, and under experimental conditions. However, questions remain on the relative transmissibility observed between species compared to its ability to infect conspecifics within bat species.

Isolation of NiV from bats has been successfully performed in Malaysia from pooled urine of *P. hypomelanus* (Chua et al., 2001), *P. vampyrus* (Rahman et al., 2010) and from a captive *P. vampyrus* (Sohayati et al., 2011), and NiV has been isolated from the urine *P. lylei* at roost in Cambodia (Reynes et al., 2005). Most recently, Cedar virus (CedPV), a new putative henipavirus has been isolated from fruit bats in Australia (Marsh et al., 2012). In all cases virus has been detected from urine samples from wild bats of unknown health status. Serological observational and surveillance studies in bats have been numerous and suggest NiV is widespread (Anonymous, 2003; Epstein et al., 2008; Lehle et al., 2007; Li et al., 2008; Olson et al., 2002; Reynes et al., 2005; Sendow et al., 2006). However, interpretation of anti-NiV antibodies has been somewhat difficult. Experimental challenge studies suggest seroconversion is consistent 10–14 days after challenge, but that antibody titers vary and individuals seropositive 14 days after challenge may lose detectable antibody by day 20 (Middleton et al., 2007). Furthermore, the captive study of *P. vampyrus* that demonstrated possible recrudescence in a bat showed that the bat lost detectable antibody to become serologically negative and showed equivocal evidence for seroconversion after virus isolation occurred (Sohayati et al., 2011), although other individuals in the colony did seroconvert.

EVIDENCE FOR A WIDER DISTRIBUTION OF NIPAH VIRUS

In Australia all four species of pteropid fruit bats in Australia (*P. alecto, P. poliocephalus, P. scapulatus, P. conspicillatus*) appear to act as reservoir hosts of HeV (Halpin et al., 2000; Young et al., 1996). Evidence for NiV infection in Australasia is complicated by HeV cross-reactivity although antibodies to NiV have been detected in bats from Papua New Guinea (Breed et al., 2010). In Malaysia *P. vampyrus* and *P. hypomelanus* act as reservoirs (Chua et al., 2002b; Yob et al., 2001). Anti-NiV antibodies were detected in three further bat species (Johara et al., 2001), including two other genera of Pteropodidae (*Cynopterus brachyotis, and Eonycteris spelaea)* and one Vespertilioniformes, *Scotophilus kuhlii,* but at very low seroprevalence (Table 11.1). In the rest of southeast Asia evidence of NiV exists across the region from Indonesia (Sendow et al., 2006; Sendow et al., 2010), Thailand (Wacharapluesadee et al., 2005; 2010), Cambodia (Olson et al., 2002; Reynes et al., 2005) and Vietnam (Hasebe et al., 2012), through east (Li et al., 2008) and south Asia as far as India (Epstein et al., 2008; Yadav et al., 2012; Epstein et al., 2008; Hsu et al., 2004, Arankalle et al., 2011). In Madagascar serological evidence of NiV in *Pteropus rufus* has been discovered; however, non-pteropid but sympatric fruit bat species *Eidolon dupreanum* and *Rousettus madagascariensis* were also seropositive (Lehle et al., 2007). Antibody to NiV and a very large number of diverse henipavirus-like RNA virus sequences have been detected in Africa recently (Drexler et al., 2009; Hayman et al., 2008; Drexler et al., 2012) suggesting that henipa-like viruses are present throughout Africa, in particular in *E. helvum* (Figure 11.3).

CONTROL OF OUTBREAKS

A review of the different pathways through which NiV has emerged in Malaysia, Singapore and Bangladesh demonstrates that wildlife, mainly but perhaps not exclusively pteropid bats, is the source of NiV, but that under certain circumstances domestic animals act as a link between the reservoir and humans (Figure 11.1). However, human-to-human transmission is also a pathway to spread of virus. This has led us and other authors to advocate a One Health approach to research on NiV (Hayman et al., 2012a; Wood et al., 2012; Daszak et al., 2012). This extends to control of outbreaks. The unpredictable nature of outbreaks suggests that sporadic incidence of disease will continue to occur. While much effort has been expended in identifying factors that could have triggered the emergence of NiV, ranging from deforestation, environmental factors (Chua et al., 2002b) to agricultural intensification (Pulliam et al., 2012), it is unlikely

FIGURE 11.3   Roosting *Eidolon helvum*, an African bat from which Nipah-like viral RNA sequences and anti-NiV antibodies have been detected. *Photo courtesy of David Hayman.*

that human activities will be curtailed purely to prevent NiV emergence. However, the outbreaks that have occurred in Bangladesh have been self-limiting or relatively easy to contain.

Both the Malaysian and Bangladeshi scenarios share a number of common features from the perspective of control. The source is clearly bats, so any activity that removes contact between pteropid bats and humans or livestock should be implemented. These can be simple, locally applied methods that can be implemented at low cost to the farmers involved (Stone, 2011). This also extends to potentially contaminated fruit entering the food chain and may involve removal of fruit trees that overhang animal holding facilities or redesign of buildings that hold livestock to prevent fruit that could have been partially eaten by bats or bat excreta falling into the holding pen (Chua et al. 2002b). Partially eaten fruit that had fallen from overhanging trees was observed in the index farm in Malaysia (Chua et al., 2002b). Critically, rapid diagnostic tests for NiV infection capable of detecting virus in a range of samples (blood, sputum, cerebrospinal fluid) should be available. Furthermore, these should be effective and available for both human and veterinary testing. Serological tests should also be applied to enable retrospective testing and screening of pig herds and other domestic animals in the event of a suspected outbreak. Education is a key method of preparing human populations to act as early monitors for disease outbreak. This could be targeted at the livestock industry and those that harvest palm sap (Stone, 2011; Rahman et al., 2012, Nahar et al., 2010). It could also be extended to the families of those that work within these industries. The general population should also be informed of high risk activities, such as consumption of raw palm sap, and early signs of infection.

Health monitoring of livestock for respiratory and/or neurological disease and sudden increases in juvenile mortality may help in detecting introduction of NiV into herds. Suspected outbreaks should be investigated rapidly and contact tracing of individuals who are suspected of disease performed. There is no licensed vaccine for NiV that could be deployed and no effective treatment at the present time (Vigant and Lee, 2011), so only palliative care can be offered to infected individuals. However, rapid detection leading to containment and barrier nursing will reduce human-to-human transmission probability. When an outbreak is confirmed within livestock, movement between different locations should be restricted and culling of infected herds should be introduced. Biosecurity should be heightened to reduce direct contact with animals and animal bodily fluids. This extends from sites where livestock are resident, livestock transportation through to abattoir facilities. A range of additional methods may be considered to prevent NiV infection. Proximate measures include vaccine development and antiviral therapy, but more long-lasting ultimate measures must include spillover prevention, and these are discussed next.

FUTURE DIRECTIONS

Vaccines and antiviral therapies are necessary, not only for general public health concerns, but also to reduce the risk to researchers working with NiV. For vaccine and antiviral therapy development, a number of experimental models exist. Golden hamsters (Wong et al., 2003) and guinea pigs (Torres-Velez et al., 2008) are antiviral drug testing models, as they show similar histological lesions (Williamson and Torres-Velez, 2010), whereas cats have been used as a model for a successful G protein subunit vaccine experiment (Middleton et al., 2002; Mungall et al., 2006). Indeed, vaccine development does not appear to be a technical challenge, as a number of experimental vaccines have been demonstrated to be effective in model animals (Guillaume et al., 2004a; Mungall et al., 2006; Weingartl et al., 2006). Furthermore, ferrets have been used for a successful post-exposure passive antibody therapy, in which NiV G protein was targeted by a human monoclonal antibody (Bossart et al., 2009) and non-human primate models have been developed (Geisbert et al., 2010; Marianneau et al., 2010; Rockx et al., 2010). The key issues with vaccines are delivery and costs. Logistically, livestock are likely to be most reachable, but biosecurity measures, surveillance and culling may be more cost-effective.

To address the ultimate causes of NiV spillover, a better understanding of bat reservoirs and the ecological and sociological drivers for NiV emergence is necessary (Chua et al., 2002a; Hayman et al., 2012a; Plowright et al., 2011; Pulliam et al., 2012). Improved medical facilities and education are important throughout the regions where NiV exists, and in India and Bangladesh where human-to-human transmission has occurred this is essential (Gurley et al., 2007; Homira et al., 2007). In Malaysia, it is believed that the ban on commercial piggeries having mango and other fruit trees that attract *Pteropus* bats may have prevented further epidemics. Such simple control measures may be useful for HeV control in Australia, though horse vaccination is now also considered an alternative. Throughout the tropics, including in Australia where HeV and CedPV occur in *Pteropus* bats, ecological changes are affecting bat ecology, and of particular importance Pteropodidae bats are increasingly roosting in urban areas (Hayman et al., 2012b; Plowright et al., 2011). In Bangladesh, reducing contamination of palm sap by using bat exclusion methods has been proposed as a means of reducing NiV outbreaks from contaminated sap (Luby et al., 2006; Nahar et al., 2010; Khan et al., 2010; Khan et al., 2012). Further epidemiological, ecological and sociological studies, however, are necessary throughout the areas where NiV may be endemic, in order for socially acceptable and sustainable prevention to be achieved (Hayman et al., 2012a; Wood et al., 2012; Blum et al. 2009).

CONCLUSIONS

The high fatality rates in the NiV outbreaks and repeated epidemics with human-to-human transmission in Bangladesh mean that NiV infection of humans has become a serious threat to human health in Asia. Measures necessary to prevent further NiV cases in human and domestic animal populations will likely need holistic One Health approaches, and include education and healthcare improvements in Bangladesh, in addition to more fundamental research on the ecological, environmental, and sociological factors leading to spillover of NiV from bats. Increasing NiV, and more generally paramyxovirus, surveillance and laboratory capacity for rapid diagnosis of NiV in high-risk areas is essential, but these must be done in a way that ensures they are sustainable and be integrated with other general health sector improvements.

ACKNOWLEDGMENTS

DTSH acknowledges funding from the David H. Smith Fellowship and the Research and Policy for Infectious Disease Dynamics (RAPIDD) program of the Science and Technology Directorate, Department of Homeland Security and Fogarty International Center, National Institutes of Health.

REFERENCES

Amal, N. M., Lye, M. S., Ksiazek, T. G., Kitsutani, P. D., Hanjeet, K. S., Kamaluddin, M. A., Ong, F., Devi, S., Stockton, P. C., Ghazali, O., Zainab, R., & Taha, M. A. (2000). Risk factors for Nipah virus transmission, Port Dickson, Negeri Sembilan, Malaysia: results from a hospital-based case—control study. *Southeast Asian Journal of Tropical Medicine and Public Health*, *31*, 301–306.

Anonymous. (1999a). Outbreak of Hendra-like virus—Malaysia and Singapore, 1998–1999. *MMWR*, *48*, 265–269.

Anonymous. (1999b). Update: outbreak of Nipah virus—Malaysia and Singapore, 1999. *MMWR*, *48*, 335–337.

Anonymous, W.H.O. (2002). Acute neurological syndrome. Bangladesh. *Weekly Epidemiological Record, 77*, 297.

Anonymous. (2003). Outbreaks of encephalitis due to Nipah/Hendra-like viruses, Western Bangladesh. *Health and Science Bulletin (English), 1*, 1–6.

Anonymous. (2004a). Person-to-person transmission of Nipah virus during outbreak in Faridpur District, 2004. *Health and Science Bulletin, 2*, 5–9.

Anonymous, W.H.O. (2004b). Nipah virus outbreak(s) in Bangladesh, January–April 2004. *Weekly Epidemiological Record, 79*, 168–171.

Arankalle, V. A., Bandyopadhyay, B. T., Ramdasi, A. Y., Jadi, R., Patil, D. R., Rahman, M., Majumdar, M., Banerjee, P. S., Hati, A. K., Goswami, R. P., Neogi, D. K., & Mishra, A. C. (2011). Genomic characterization of Nipah virus, West Bengal, India. *Emerg. Infect. Dis., 17*, 907–909.

Baker, K. S., Todd, S., Marsh, G., Fernandez-Loras, A., Suu-Ire, R., Wood, J. L. N., Wang, L. F., Murcia, P. R., & Cunningham, A. A. (2012). Co-circulation of diverse paramyxoviruses in an urban African fruit bat population. *J. Gen. Virol., 93*, 850–856.

Baker, K. S., Marsh, G. A., Todd, S., Crameri, G., Barr, J., Kamins, A. O., Peel, A. C., Yu, M., Hayman, D. T. S., Nagjm, B., Mtove, G., Amos, B., Reyburn, H., Nyarko, E. O., Suu-Ire, R., Murcia, P. R., Cunningham, A. A., Wood, J. L. N., & Wang, L. F. (2013). Novel potentially-zoonotic paramyxoviruses from the African straw-coloured fruit bat, *Eidolon helvum*. *J. Virol., 87*, 1348–1358.

Bellini, W. J., Rota, P. A., & Anderson, L. J. (1998). Paramyxoviruses. In B. W. J. Mahy & L. Collier (Eds.), *Topley & Wilson's Microbiology and Microbial Infections, Virology* (Vol. 1, 9th ed) (pp. 435–462). London: Edward Arnold.

Blum, L. S., Khan, R., Nahar, N., & Breiman, R. F. (2009). In-depth assessment of an outbreak of Nipah encephalitis with person-to-person transmission in Bangladesh: implications for prevention and control strategies. *Am. J. Trop. Med. Hyg., 80*, 96–102.

Bonaparte, M. I., Dimitrov, A. S., Bossart, K. N., Crameri, G., Mungall, B. A., Bishop, K. A., Choudry, V., Dimitrov, D. S., Wang, L. F., Eaton, B. T., & Broder, C. C. (2005). Ephrin-B2 ligand is the functional receptor for Hendra and Nipah virus. *PNAS. USA, 30*, 10652–10657.

Bossart, K. N., Zhu, Z., Middleton, D., Klippel, J., Crameri, G., Bingham, J., McEachern, J. A., Green, D., Hancock, T. J., Chan, Y. P., Hickey, A. C., Dimitrov, D. S., Wang, L. F., & Broder, C. C. (2009). A neutralizing human monoclonal antibody protects against lethal disease in a new ferret model of acute Nipah virus infection. *PLoS Pathogens, 5*, e1000642.

Breed, A. C., Field, H. E., Smith, C. S., Edmonston, J., & Meers, J. (2010). Bats Without Borders: Long-Distance Movements and Implications for Disease Risk Management. *EcoHealth, 7*, 204–212.

Chadha, M. S., Comer, J. A., Lowe, L., Rota, P. A., Rollin, P. E., Bellini, W. J., Ksiezek, T. G., & Mishra, A. (2006). Nipah virus-associated encephalitis outbreak, Siliguri, India. (2006). *Emerg, Infect. Dis., 12*, 235–240.

Chan, Y. P., Chua, K. B., Koh, C. L., Lim, M. E., & Lam, S. K. (2001). Complete nucleotide sequences of Nipah virus isolates from Malaysia. *J. Gen. Virol., 829*, 2151–2155.

Chan, K. P., Rollin, P. E., Ksiazek, T. G., Leo, Y. S., Goh, K. T., Paton, N. I., Sng, E. H., & Ling, A. E. (2002). A survey of Nipah virus infection among various risk groups in Singapore. *Epidemiol. Infect., 128*, 93–98.

Chong, H. T., Kunjapan, S. R., Thayaparan, T., Tong, J. M. G., Patharunam, V., Jusoh, M. R., & Tan, C. T. (2000). Nipah encephalitis outbreak in Malaysia, clinical features in patients from Seremban. *Neurological Journal of Southeast Asia, 5*, 61–67.

Chong, H. -T., Kamarulzaman, A., Tan, C. -T., Goh, K. -J., Thayaparan, T., Kunjapan, S. R., Chew, N. -K., Chua, K. -B., & Lam, S. -K. (2001). Treatment of acute Nipah encephalitis with ribavirin. *Annals of Neurology, 49*, 810–813.

Chong, H. T., & Tan, C. T. (2003). Relapsed and late-onset Nipah encephalitis, a report of three cases. *Neurological Journal of Southeast Asia, 8*, 109–112.

Chong, H. T., Jahangir Hossain, M., & Tan, C. T. (2008). Differences in epidemiologic and clinical features of Nipah virus encephalitis between the Malaysian and Bangladesh outbreaks. *Neurology Asia, 13*, 23–26.

Chua, K. B., Goh, K. J, Wong, K. T,, Kamarulzaman, A., Tan, P. S., Ksiazek, T. G., Zaki, S. R., Paul, G., Lam, S. K., & Tan, C. T. (1999). Fatal encephalitis due to Nipah virus among pig-farmers in Malaysia. *The Lancet, 354*, 1257–1259.

Chua, K. B., Bellini, W. J., Rota, P. A., Harcourt, B. H., Tamin, A., Lam, S. K., Ksiazek, T. G., Rollin, P. E., Zaki, S. R., Sheih, W. J., Goldsmith, C. S., Gubler, D. J., Roehrig, J. T., Eaton, B., Gould, A. R., Olson, J., Field, H., Daniels, P., Ling, A. E., Peters, C. J., Anderson, L. J., & Mahy, B. W. J. (2000a). Nipah virus: A recently emergent deadly paramyxovirus. *Science, 288*, 1432–1435.

Chua, K. B., Lam, S. K., Goh, K. J., Hooi, P. S., Ksiazek, T. G., Kamarulzaman, A., Olson, J., & Tan, C. T. (2001). The presence of Nipah virus in respiratory secretions and urine of patients during an outbreak of Nipah virus encephalitis in Malaysia. *J. Infect., 42*, 40–43.

Chua, K. B., Chua, B. H., & Wang, C. W. (2002a). Anthropogenic deforestation, El Niño and the emergence of Nipah virus in Malaysia. *Malaysian J. Pathol., 24*, 15–21.

Chua, K. B., Koh, C. L., Hooi, P. S., Wee, K. F., Khong, J. H., Chua, B. H., Chan, Y. P., Lim, M. E., & Lam, S. K. (2002b). Isolation of Nipah virus from Malaysian Island flying-foxes. *Microbes and Infection, 4*, 145–151.

Chua, K. B., Lam, S. K., Tan, C. T., Hooi, P. S., Goh, K. J., Chew, N. K., Tan, K. S., Kamarulzaman, A., & Wong, K. T. (2000b). High mortality in Nipah encephalitis is associated with presence of virus in cerebrospinal fluid. *Annals of Neurology, 48*, 802–805.

Clayton, B. A., Middleton, D., Bergfeld, J., Haining, J., Arkinstall, R., Wang, L., & Marsh, G. A. (2012). Transmission routes for Nipah virus from Malaysia and Bangladesh. *Emerg. Infect. Dis., 18*, 1983–1993.

Daszak, P., Zambrana-Torrelio, C., Bogich, T. L., Fernandez, M., Epstein, J. H., Murray, K. A., & Hamilton, H. (2012). Interdisciplinary approaches to understanding disease emergence: The past, present, and future drivers of Nipah virus emergence. *PNAS, USA*. In Press.

Douglas, I. C., Baldock, F., & Black, P. (1997). Outbreak investigation of an emerging disease (equine morbillivirus). In *In Epidemiol Sante Anim - Proc of 8th ISVEE conference*, pp. 04.081-004.008.083. Paris.

Douglas, I. C. (1998). Equine morbillivirus: the search for the agent. In *Australian Assoc Cattle Veterinarians*. Sydney, Australia, .

Drexler, J. F., Corman, V. M., Gloza-Rausch, F., Seebens, A., Annan, A., Ipsen, A., Kruppa, T., Müller, M. A., Kalko, E. K., Adu-Sarkodie, Y., Oppong, S., & Drosten, C. (2009). Henipavirus RNA in African bats. *PLoS One, 4*, e6367.

Drexler, J. F., Corman, V. M., Müller, M. A., Maganga, G. D., Vallo, P., Binger, T., Gloza-Rausch, F., Rasche, A., Yordanov, S., Seebens, A., Oppong, S., Sarkodie, Y. A., Pongombo, C., Lukashev, A. N., Schmidt-Chanasit, J., Stöcker, A., Carneiro, A. J. B., Erbar, S., Maisner, A., Fronhoffs, F., Buettner, R., Kalko, E. K. V., Kruppa, T., Franke, C. R., Kallies, R., Yandoko, E. R. N., Herrler, G., Reusken, C., Hassanin, A., Krüger, D. H., Matthee, S., Ulrich, R. G., Leroy, E. M., & Drosten, C. (2012). Bats host major mammalian paramyxoviruses. *Nature Communications, 3*, 796.

Epstein, J. H., Prakash, V., Smith, C. S., Daszak, P., McLaughlin, A. B., Meehan, G., Field, H. E., & Cunningham, A. A. (2008). Henipavirus infection in fruit bats (*Pteropus giganteus*), India. *Emerg. Infect. Dis.*, *14*, 1309–1311.

Epstein, J. H., Olival, K. J., Pulliam, J., Smith, C., Westrum, J., Hughes, T., Dobson, A. P., Zubaid, A., Rahman, S. A., Basir, M. M., Field, H. E., & Daszak, P. (2009). *Pteropus vampyrus*, a hunted migratory species with a multinational home-range and a need for regional management. *J. Applied Ecol.*, *46*, 991–1002.

Field, H., Schaaf, K., Kung, N., Simon, C., Waltisbuhl, D., Hobert, H., Moore, F., Middleton, D., Crook, A., Smith, G., Daniels, P., Glanville, R., & Lovell, D. (2010). Hendra virus outbreak with novel clinical features, Australia. *Emerg. Infect. Dis.*, *16*, 338–340.

Geisbert, T. W., Daddario-DiCaprio, K. M., Hickey, A. C., Smith, M. A., Chan, Y. -P., Wang, L. -F., Mattapallil, J. J., Geisbert, J. B., Bossart, K. N., & Broder, C. C. (2010). Development of an acute and highly pathogenic nonhuman primate model of Nipah virus infection. *PLoS One*, *5*, e10690.

Georges-Courbot, M. C., Contamin, H., Faure, C., Loth, P., Baize, S., Leyssen, P., Neyts, J., & Deubel, V. (2006). Poly(I)–Poly(C12u) but not ribavirin prevents death in a hamster model of Nipah virus infection. *Antimicrobial Agents and Chemotherapy*, *50*, 1768–1772.

Giannini, N. P., & Simmons, N. B. (2003). A phylogeny of megachiropteran bats (Mammalia: Chiroptera: Pteropodidae) based on direct optimization analysis of one nuclear and four mitochondrial genes. *Cladistics*, *19*, 496–511.

Goh, K. J., Tan, C. T., Chew, N. K., Tan, P. S., Kamarulzaman, A., Sarji, S. A., Wong, K. T., Abdullah, B. J. J., Bing, C., & Lam, S. K. (2000). Clinical features of Nipah virus encephalitis among pig farmers in Malaysia. *New England Journal of Medicine*, *342*, 1229–1235.

Goldsmith, C. S., Whistler, T., Rollin, P. E., Ksiazek, T. G., Rota, P. A., Bellini, W. J., Daszak, P., Wong, K. T., Shieh, W. J., & Zaki, S. R. (2003). Elucidation of Nipah virus morphogenesis and replication using ultrastructural and molecular approaches. *Virus Res.*, *92*, 89–98.

Guillaume, V., Contamin, H., Loth, P., Georges-Courbot, M. -C., Lefeuvre, A., Marianneau, P., Chua, K. B., Lam, S. K., Buckland, R., Deubel, V., & Wild, T. F. (2004). Nipah virus: vaccination and passive protection studies in a hamster model. *J. Virol.*, *78*, 834–840.

Gurley, E. S., Montgomery, J. M., Hossain, M. J., Bell, M., Azad, A. K., Islam, M. R., Molla, M. A., Carroll, D. S., Ksiazek, T. G., Rota, P. A., Lowe, L., Comer, J. A., Rollin, P., Czub, M., Grolla, A., Feldmann, H., Luby, S. P., Woodward, J. L., & Breiman, R. F. (2007). Person-to-person transmission of Nipah virus in a Bangladeshi community. *Emerg. Infect. Dis.*, *13*, 1031–1037.

Hagmaier, K., Stock, N., Goodbourn, S., Wang, L. -F., & Randall, R. (2006). A single amino acid substitution in the V protein of Nipah virus alters its ability to block interferon signaling in cells from different species. *J. Gen. Virol.*, *87*, 3649–3653.

Halpin, K., Young, P. L., Field, H. E., & Mackenzie, J. S. (2000). Isolation of Hendra virus from pteropid bats: a natural reservoir of Hendra virus. *J. Gen. Virol.*, *81*, 1927–1932.

Halpin, K., Bankamp, B., Harcourt, B. H., Bellini, W. J., & Rota, P. A. (2004). Nipah virus conforms to the rule of six in a minigenome replication assay. *J. Gen.Virol.*, *85*, 701–707.

Harcourt, B. H., Tamin, A., Ksiazek, T. G., Rollin, P. E., Anderson, L. J., Bellini, W. J., & Rota, P. A. (2000). Molecular characterization of Nipah virus, a newly emergent paramyxovirus. *Virology*, *271*, 334–349.

Harcourt, B. H., Tamin, A., Halpin, K., Ksiazek, T. G., Rollin, P. E., Bellini, W. J., & Rota, P. A. (2001). Molecular characterization of the polymerase gene and genomic termini of Nipah virus. *Virology*, *287*, 192–201.

Hasebe, F., Thuy, N. T., Inoue, S., Yu, F., Kaku, Y., Watanabe, S., Akashi, H., Dat, D. T., Mai le, T. Q., & Morita, K. (2012). Serologic evidence of Nipah virus infection in bats. *Vietnam. Emerg. Inf. Dis.*, *18*, 536–537.

Hayman, D. T., Suu-Ire, R., Breed, A. C., McEachern, J. A., Wang, L., Wood, J. L., & Cunningham, A. A. (2008). Evidence of henipavirus infection in West African fruit bats. *PLoS One*, *3*, e32739.

Hayman, D. T. S., Wang, L. -F., Barr, J., Baker, K. S., Suu-Ire, R., Broder, C. C., Cunningham, A. A., & Wood, J. L. N. (2011). Antibodies to henipavirus or henipa-like viruses in domestic pigs in Ghana, West Africa. *PLoS ONE*, *6*, e25256.

Hayman, D. T. S., Gurley, E. S., Pulliam, J. R. C., & Field, H. E. (2012a). The application of One Health approaches to Henipavirus research. *Current Topics in Microbiology and Immunology*, DOI:10.1007/82_2012_276.

Hayman, D. T. S., McCrea, R., Suu-Ire, R., Wood, J. L. N., Cunningham, A. A., & Rowcliffe, J. R. (2012b). Straw-coloured fruit bat demography in Ghana. *J. Mammol.*, *93*, 1393–1404.

Homira, N., Rahman, M., Hossain, M. J., Chowdhury, I., Sultana, R., Khan, R., Ahmed, B. -N., Banu, S., Nahar, K., Poddar, G., Gurley, E., Comer, J. A., Rollin, P. E., Rota, P., Ksiazek, T. G., & Luby, S. (2007). Nipah outbreak with person-to-person transmission in Bangladesh, 2007. *Am. J. Trop. Med. Hyg.*, *77*(5, Suppl. S), 80.

Homaira, N., Rahman, M., Hossain, M. J., Nahar, N., Khan, R., Rahman, M., Podder, G., Nahar, K., Khan, D., Gurley, E. S., Rollin, P. E., Comer, J. A., Ksiazek, T. G., & Luby, S. P. (2010). Cluster of Nipah virus infection, Kushtia District, Bangladesh, 2007. *PLoS One*, *5*, e 13570.

Hooper, P. T. (2000). New fruit bat viruses affecting horses, pigs, and humans. In C. Brown & C. Bolin (Eds.), *Emerging Diseases of Animals* (pp. 85–99). Washington, DC: ASM Press.

Hossain, M. J., Gurley, E. S., Montgomery, J. M., Bell, M., Carroll, D. S., Hsu, V. P., Formenty, P., Croisier, A., Bertherat, E., Faiz, M. A., Azad, A. K., Islam, R., Molla, M. A., Ksiazek, T. G., Rota, P. A., Comer, J. A., Rollin, P. E., Luby, S. P., & Breiman, R. F. (2008). Clinical presentation of nipah virus infection in Bangladesh. *Clin. Infect. Dis.*, *46*, 977–984.

Hsu, V. P., Hossain, M. J., Partashar, U. D., Ali, M. M., Ksiazek, T. G., Kuzmin, I., Niezgoda, M., Rupprecht, C., Bresee, J., & Breiman, R. F. (2004). Nipah virus encephalitis reemergence. *Bangladesh. Emerg. Infect. Dis.*, *10*, 2082–2087.

Hutcheon, J. A., & Kirsch, J. A. W. (2006). A moveable face: deconstructing the Microchiroptera and a new classification of extant bats. *Acta Chiropterologica*, *8*, 1–10.

Hyatt, A. D., Zaki, S. R., Goldsmith, C. S., Wise, T. G., & Hengstberger, S. G. (2001). Ultrastructure of Hendra virus and Nipah virus within cultured cells and host animals. *Microbes and Infection*, *3*, 297–306.

Jack, P. J., Boyle, D. B., Eaton, B. T., & Wang, L. F. (2005). The complete genome sequence of J virus reveals a unique genome structure in the family Paramyxoviridae. *J. Virol.*, *79*, 10690–10700.

Johara, M. Y., Field, H., Rashdi, A. M., Morrissy, C., van der Heide, B., Rota, P., Adzhar, A. B., White, J., Daniels, P., Jamaluddin, A., & Ksiazek, T. (2001). Nipah virus infection in Bats (Order Chiroptera) in Peninsular Malaysia. *Emerg. Infect. Dis.*, *7*, 439–441.

Kamil, W. M., Rahmat, S. M., & Kadir, O. A. (2001). *JE/Nipah virus outbreak in Perak: the economics of the control program*. Retrieved 2002 (May 23), from http://www.jphpk.gov.my/English/JE2.html.

Khan, M. S., Hossain, J., Gurley, E. S., Nahar, N., Sultana, R., & Luby, S. P. (2010). Use of infrared camera to understand bats' access to date palm sap: implications for preventing Nipah virus transmission. *Ecohealth*, *7*, 517–525.

Khan, S. U., Gurley, E. S., Hossain, M. J., Nahar, N., Sharker, M. A., & Luby, S. P. (2012). A randomized controlled trial of interventions to impede date palm sap contamination by bats to prevent nipah virus transmission in Bangladesh. *PLoS One*, *7*, e42689.

Lam, S. K., & Chua, K. B. (2002). Nipah virus encephalitis outbreak in Malaysia. *Clin. Infect. Dis.*, *34*(Suppl. 2), S48–S51.

Lamb, R. A., & Parks, G. D. (2007). Paramyxoviridae: the viruses and their replication. In D. M. H.P. Knipe (Ed.), *Fields Virology,* (vol. 1, 2 vols, 5th ed) (pp. 1449–1496). Philadelphia: Lippincott Williams & Wilkins.

Lehle, C., Razafitrimo, G., Razainirina, J., Andriaholinirina, N., Goodman, S. M., Faure, C., Georges-Courbot, M. -C., Rousset, D., & Reynes, J. -M. (2007). Henipavirus and Tioman virus antibodies in pteropodid bats. *Madagascar. Emerg. Infect. Dis.*, *13*, 159–161.

Li, Z., Yu, M., Zhang, H., Magoffin, D. E., Jack, P. J., Hyatt, A., Wang, H. Y., & Wang, L. F. (2006). Beilong virus, a novel paramyxovirus with the largest genome of non-segmented negative-stranded RNA viruses. *Virology*, *346*, 219–228.

Li, Y., Wang, J., Hickey, A. C., Zhang, Y., Li, Y., Wu, Y., Zhang, H., Yuan, J., Han, Z., McEachern, J., Broder, C. C., Wang, L. -F., & Shi, Z. (2008). Antibodies to Nipah or Nipah-like viruses in bats, China. *Emerg. Infect. Dis.*, *14*, 1974–1976.

Luby, S. P., Rahman, M., Hossain, M. J., Blum, L. S., Husain, M. M., Gurley, E., Khan, R., Ahmed, B. -A., Rahman, S., Nahar, N., Kenah, E., Comer, J. A., & Ksiazek, T. G. (2006). Foodborne transmission of Nipah virus. *Bangladesh. Emerg. Infect. Dis.*, *12*, 1888–1894.

Luby, S. P., Gurley, E. S., & Hossain, M. J. (2009). Transmission of human infection with Nipah virus. *Clin. Infect. Dis.*, *49*, 1743–1748.

Luby, S., Rahman, M., Hossain, M. J., Ahmed, B. -N., Gurley, E., Banu, S., Homira, N., Rollin, P. E., Comer, J. A., Rota, P., Montgomery, J., & Ksiazek, T. G. (2007). Recurrent Nipah virus outbreaks in Bangladesh, 2001–2007. *Am. J. Trop. Med. Hyg.*, *77*(5, Suppl. S), 273.

Marianneau, P., Guillaume, V., Wong, T., Badmanathan, M., Looi, R. Y., Murri, S., Loth, P., Tordo, N., Wild, F., Horvat, B., & Contamin, H. (2010). Experimental infection of squirrel monkeys with Nipah virus. *Emerg. Infect. Dis.*, *16*, 507–510.

Marsh, G. A., de Jong, C., Barr, J. A., Tachedjian, M., Smith, C., Middleton, D., Yu, M., Foord, A. J., Haring, V., Payne, J., Robinson, R., Broz, I., Crameri, G., Field, H. E., & Wang, L. F. (2012). Cedar Virus: A Novel Henipavirus Isolated from Australian Bats. *PLoS Pathog.*, *8*, e1002836.

Middleton, D. J., Westbury, H. A., Morrissy, C. J., van der Heide, B. M., Russell, G. M., Braun, M. A., & Hyatt, A. D. (2002). Experimental Nipah virus infection in pigs and cats. *J.Comp. Pathol.*, *126*, 124–136.

Middleton, D. J., Morrissy, C. J., van der Heide, B. M., Russell, G. M., Braun, M. A., Westbury, H. A., Halpin, K., & Daniels, P. W. (2007). Experimental Nipah virus infection in pteropid bats (*Pteropus poliocephalus*). *J. Comp. Pathol.*, *136*, 266–272.

Mohd Nor, M. N., Gan, C. H., & Ong, B. L. (2000). Nipah virus infection of pigs in peninsular Malaysia. *Rev.. sci. tech. Off. Int. Epiz.*, *19*, 160–165.

Mounts, A. W., Kaur, H., Parashar, U. D., Ksiazek, T. G., Cannon, D., Arokiasamy, J. T., Anderson, L. J., & Lye, M. S. (2001). A cohort study of health care workers to assess nosocomial transmissibility of Nipah virus, Malaysia, 1999. *J. Infect. Dis.*, *183*, 810–813.

Mungall, B. A., Middleton, D., Crameri, G., Bingham, J., Halpin, K., Russell, G., Green, D., McEachern, J., Pritchard, L. I., Eaton, B. T., Wang, L. -F., Bossart, K. N., & Broder, C. C. (2006). Feline model of acute Nipah virus infection and protection with a soluble glycoprotein-based subunit vaccine. *J. Virol.*, *80*, 12293–12302.

Murray, K., Selleck, P., Hooper, P., Hyatt, A., Gould, A., Gleeson, L., Westbury, H., Hiley, L., Selvey, L., Rodwell, B., & Ketterer, P. (1995). A morbillivirus that caused fatal disease in horses and humans. *Science*, *268*, 94–97.

Nahar, N., Sultana, R., Gurley, E. S., Hossain, M. J., & Luby, S. P. (2010). Date palm sap collection: exploring opportunities to prevent Nipah transmission. *Ecohealth*, *7*, 196–203.

Negrete, O. A., Levroney, E. L., Aguilar, H. C., Bertolotti-Ciarlet, A., Nazarin, R., Tajyar, S., & Lee, B. (2005). EphrinB2 is the entry receptor for Nipah virus, an emergent deadly paramyxovirus. *Nature*, *7049*, 401–405.

O'Sullivan, J. D., Allworth, A. M., Paterson, D. L., Snow, T. M., Boots, R., Gleeson, L. J., Gould, A. R., Hyatt, A. D., & Bradfield, J. (1997). Fatal encephalitis due to novel paramyxovirus transmitted from horses. *The Lancet*, *349*, 93–95.

Olson, J. G., Rupprecht, C., Rollin, P. E., An, U. S., Niezgoda, M., Clemins, T., Walston, J., & Ksiazek, T. G. (2002). Antibodies to Nipah-like virus in bats (*Pteropus lylei*), Cambodia. *Emerg. Infect. Dis.*, *8*, 987 988.

Parashar, U. D., Sunn, L. M., Ong, F., Mounts, A. W., Arif, M. T., Ksiazek, T. G., Kamaluddin, M. A., Mustafa, A. N., Kaur, H., Ding, L. M., Othman, G., Radzi, H. M., Kitsutani, P. T., Stockton, P. C., Arokiasamy, J., Gary, H. E., Jr., & Anderson, L. J. (2000). Case–control study of risk factors for human infection with a new zoonotic paramyxovirus, Nipah virus, during a 1998–1999 outbreak of severe encephalitis in Malaysia. *J. Infect. Dis.*, *181*, 1755–1759.

Paton, N. I., Leo, Y. S., Zaki, S. R., Auchus, A. P., Lee, K. E., Ling, A. E., Chew, S. K., Ang, B., Rollin, P. E., Umapathi, T., Sng, I., Lee, C. C., Lim, E., & Ksiazek, T. G. (1999). Outbreak of Nipah-virus infection among abattoir workers in Singapore. *The Lancet*, *354*, 1253–1257.

Plowright, R. K., Field, H. E., Smith, C., Divljan, A., Palmer, C., Tabor, G. M., Daszak, P., & Foley, J. E. (2008). Reproduction and nutritional stress are risk factors for Hendra virus infection in little red flying foxes (*Pteropus scapulatus*). *Proceedings of the Royal Society B*, *275*, 861–869.

Plowright, R. K., Foley, P., Field, H. E., Dobson, A. P., Foley, J. E., Eby, P., & Daszak, P. (2011). Urban habituation, ecological connectivity and epidemic dampening: the emergence of Hendra virus from flying foxes (*Pteropus* spp.). *Proceedings of the Royal Society B*, *278*, 3703–3712.

ProMed. (1998). *Japanese Encephalitis, Suspected-Malaysia, Archive Number: 19981124.2269*.

Pulliam, J. R. C., Epstein, J. H., Dushoff, J., Rahman, S. A., Bunning, M., Jamaluddin, A. A., Hyatt, A. D., Field, H. E., Dobson, A. P., Daszak, P.and the Henipavirus Ecology Research Group (HERG) (2012). Agricultural intensification, priming for persistence and the emergence of Nipah virus: a lethal bat-borne zoonosis. *J. R. Soc. Interface*, *9*, 89–101.

Ramasundrum, V., Tan, C. T., Chua, K. B., Chong, H. T., Goh, K. J., Chew, N. K., Tan, K. S., Thayaparan, T., Kunjapan, S. R., Petharunam, V., Loh, Y. L., Ksiazek, T. G., & Lam, S. K. (2000). Kinetics of IgM and IgG seroconversion in Nipah virus infection. *Neurological Journal of Southeast Asia*, *5*, 23–28.

Rahman, S. A., Hassan, S. S., Olival, K. J., Mohamed, M., Chang, L. Y., Hassan, L., Saad, N. M., Shohaimi, S. A., Mamat, Z. C., Naim, M. S., Epstein, J. H., Suri, A. S., Field, H. E., Daszak, P.Henipavirus Ecology Research Group (2010). Characterization of Nipah virus from naturally infected *Pteropus vampyrus* bats, Malaysia. *Emerg. Infect. Dis.*, *16*, 1990–1993.

Rahman, M. A., Hossain, M. J., Sultana, S., Homaira, N., Khan, S. U., Rahman, M., Gurley, E. S., Rollin, P. E., Lo, M. K., Comer, J. A., Lowe, L., Rota, P. A., Ksiazek, T. G., & Kenah, E. (2012). Date palm sap linked to Nipah virus outbreak in Bangladesh, 2008. *Vector Borne Zoonotic Diseases*, *12*. Epub.

Reynes, J. -M., Counor, D., Ong, S., Faure, C., Seng, V., Molia, S., Walston, J., Georges- Courbot, M. C., Deubel, V., & Sarthou, J. -L. (2005). Nipah virus in Lyle's Flying Foxes, Cambodia. *Emerg. Infect. Dis.*, *11*, 1042–1047.

Rockx, B., Bossart, K. N., Feldmann, F., Geisbert, J. B., Hickey, A. C., Brining, D., Callison, J., Safronetz, D., Marzi, A., Kercher, L., Long, D., Broder, C. C., Feldmann, H., & Geisbert, T. W. (2010). A novel model of lethal Hendra virus infection in African green monkeys and the effectiveness of ribavirin treatment. *J. Virol.*, *84*, 9831–9839.

Sejvar, J. J., Hossain, J., Saha, S. K., Gurley, E. S., Banu, S., Hamadani, J. D., Faiz, M. A., Siddiqui, F. M., Mohammad, Q. D., Mollah, A. H., Uddin, R., Alam, R., Rahman, R., Tan, C. T., Bellini, W., Rota, P., Breiman, R. F., & Luby, S. P. (2007). Long-term neurological and functional outcome in Nipah virus infection. *Annals of Neurology*, *62*, 235–242.

Sendow, I., Field, H. E., Curran, J., Darminto, Morrissy, C., Meehan, G., Buick, T., & Daniels, P. (2006). Henipavirus in *Pteropus vampyrus* bats, Indonesia. *Emerg. Infect. Dis.*, *12*, 711–712.

Sendow, I., Field, H. E., Adjid, A., Ratnawati, A., Breed, A. C., Darminto, Morrissy, C., & Daniels, P. (2010). Screening for Nipah Virus Infection in West Kalimantan Province, Indonesia. *Zoonoses and Public Health*, *57*, 499–503.

Sim, B. -F., Jusoh, M. R., Chang, C. -C., & Khalid, R. (2002). Nipah encephalitis: a report of 18 patients from Kuala Lumpur Hospital. *Neurological Journal of Southeast Asia*, *7*, 13–18.

Sohayati, A. R., Hassan, L., Sharifah, S. H., Lazarus, K., Zaini, C. M., Epstein, J. H., Shamsyul Naim, N., Field, H. E., Arshad, S. S., Abdul Aziz, J., & Daszak, P. (2011). Evidence for Nipah virus recrudescence and serological patterns of captive *Pteropus vampyrus*. *Epidemiol. Infect.*, *139*, 1570–1579.

Stone, R. (2011). Breaking the chain in Bangladesh. *Science*, *331*, 1128–1131.

Tan, C. T., Goh, K. J., Wong, K. T., Sarji, S. A., Chua, K. B., Chew, N. K., Murugasu, P., Loh, Y. L., Chong, H. T., Sin, T. K., Thayaparan, T., Kumar, S., & Jusoh, M. R. (2002). Relapsed and late-onset Nipah encephalitis. *Annals of Neurology*, *51*, 703–708.

Tan, K. -S., Tan, C. -T., & Goh, K. -J. (1999). Epidemiological aspects of Nipah virus infection. *Neurological Journal of Southeast Asia*, *4*, 77–81.

Teeling, E. C., Springer, M. S., Madsen, O., Bates, P., O'Brien, S. J., & Murphy, W. J. (2005). A molecular phylogeny of bats illuminates biogeography and the fossil record. *Science*, *307*, 580–584.

Torres-Velez, F. J., Shieh, W. J., Rollin, P. E., Morken, T., Brown, C., Ksiazek, T. G., & Zaki, S. R. (2008). Histopathologic and immunohistochemical characterization of Nipah virus infection in the guinea pig. *Veterinary Pathology*, *45*, 576–585.

Vigant, F., & Lee, B. (2011). Hendra and Nipah infection: Pathology, models and potential therapies. *Infect. Disord. Drug Targets*, *11*, 315–336.

Wacharapluesadee, S., Lumlertdacha, B., Boongird, K., Wanghongsa, S., Chanhome, L., Rollin, P., Stockton, P., Rupprecht, C. E., Ksiazek, T. G., & Hemachudha, T. (2005). Bat Nipah virus. *Thailand. Emerg. Infect. Dis.*, *11*, 1949–1951.

Wacharapluesadee, S., Boongird, K., Wanghongsa, S., Ratanasetyuth, N., Supavonwong, P., Saengsen, D., Gongal, G. N., & Hemachudha, T. (2010). A longitudinal study of the prevalence of Nipah virus in *Pteropus lylei* bats in Thailand: evidence for seasonal preference in disease transmission. *Vector Borne and Zoonotic Diseases*, *10*, 183–190.

Wang, L. F., Yu, M., Hansson, E., Pritchard, L. I., Shiell, B., Michalski, W. P., & Eaton, B. T. (2000). The exceptionally large genome of Hendra virus: support for creation of a new genus within the family Paramyxoviridae. *J. Virol.*, *74*, 9972–9979.

Weingartl, H. M., Berhane, Y., Caswell, J. L., Loosmore, S., Audonnet, J. -C., Roth, J. A., & Czub, M. (2006). Recombinant nipah virus vaccines protect pigs against challenge. *J. Virol.*, *80*, 7929–7938.

Williamson, M. M., & Torres-Velez, F. J. (2010). Henipavirus pathology: a review of laboratory animal pathology. *Veterinary Pathology*, *47*, 871–880.

Wong, K. T., Shieh, W. -J., Kumar, S., Norain, K., Abdullah, W., Guarner, J., Goldsmith, C. S., Chua, K. B., Lam, S. K., Tan, C. T., Goh, K. J., Chong, H. T., Jusoh, R., Rollin, P. E., Ksiazek, T. G., Zaki, S. R.Group, t.N.V.P.W (2002a). Nipah virus infection. Pathology and pathogenesis of an emerging paramyxoviral zoonosis. *American Journal of Pathology*, *161*, 2153–2167.

Wong, K. T., Shieh, W. -J., Zaki, S. R., & Tan, C. T. (2002b). Nipah virus infection, an emerging paramyxoviral zoonosis. *Springer Seminars in Immunopathology*, *24*, 215–228.

Wong, K. T., Grosjean, I., Brisson, C., Blanquier, B., Fevre-Montange, M., Bernard, A., Loth, P., Georges-Courbot, M. -C., Chevallier, M., Akaoka, H., Marianneau, P., Lam, S. K., Wild, T. F., & Deubel, V. (2003). A golden hamster model for human acute nipah virus infection. *American J. Pathol.*, *163*, 2127–2137.

Wood, J. L., Leach, M., Waldman, L., Macgregor, H., Fooks, A. R., Jones, K. E., Restif, O., Dechmann, D., Hayman, D. T., Baker, K. S., Peel, A. J., Kamins, A. O., Fahr, J., Ntiamoa-Baidu, Y., Suu-Ire, R., Breiman, R. F., Epstein, J. H., Field, H. E., & Cunningham, A. A. (2012). A framework for the study of zoonotic disease emergence and its drivers: spillover of bat pathogens as a case study. *Philos. Trans. R. Soc. Lond. B. Biol. Sci.*, *367*, 2881–2892.

Yadav, P. D., Raut, C. G., Shete, A. M., Mishra, A. C., Towner, T. S., Nichol, S. T., & Mourya, D. T. (2012). Detection of Nipah virus RNA in fruit bat (*Pteropus giganteus*) from India. *Am. J. Trop. Med. Hyg.*, *87*, 576–578.

Yob, J. M., Field, H., Rashdi, A. M., Morrissy, C., van der Heide, B., Rota, P., bin Adzhar, A., White, J., Daniels, P., Jamaluddin, A., & Ksiazek, T. (2001). Nipah virus infection in bats (order Chiroptera) in peninsular Malaysia. *Emerg. Infect. Dis.*, *7*, 439–441.

Young, P., Halpin, K., Selleck, P. W., Field, H., Gravel, J. L., Kelly, M. A., & Mackenzie, J. (1996). Serologic evidence for the presence in *Pteropus* bats of a paramyxovirus related to equine morbillivirus. *Emerg. Infect. Dis.*, *2*, 239–240.

Zhu, Z., Dimitrov, A. S., Bossart, K. N., Crameri, G., Bishop, K. A., Choudhry, V., Mungall, B. A., Feng, Y. -R., Choudhary, A., Zhang, M. -Y., Feng, Y., Wang, L. -F., Xiao, X., Eaton, B. T., Broder, C. C., & Dimitrov, D. S. (2006a). Potent neutralization of Hendra and Nipah viruses by human monoclonal antibodies. *J. Virol.*, *80*, 891–899.

Zhu, Z., Dimitrov, A. S., Chakraborti, S., Dimitrova, D., Xiao, X., Broder, C. C., & Dimitrov, D. S. (2006b). Development of human monoclonal antibodies against diseases caused by emerging and biodefense-related viruses. *Expert Review of Anti-Infective Therapy*, *4*, 57–66.

Chapter 12

Synthesis

Philip R. Wakeley, Sarah North and Nicholas Johnson

Animal Health and Veterinary Laboratories Agency, Surrey, United Kingdom

INTRODUCTION

The examples in the preceding chapters give a comprehensive overview of the role of animals in the emergence and re-emergence of viral diseases. Every species on the planet, *Homo sapiens* included, hosts a range of viruses that are naturally transmitted between members of that species. However, opportunities arise for one species to encounter a virus from another species. Such opportunities lead to cross species transmission events and if the virus can infect the new host it has the opportunity to cause the emergence of a new disease. If that new species is human or livestock or, now increasingly, wildlife, there is a disease outbreak. The repeated exposure of humans to lentiviruses through hunting and butchery of bushmeat is a good example of this and has been responsible for the largest zoonotic disease outbreak in recent times, namely that of human immunodeficiency virus (HIV).

If that species is a domestic or companion animal, then the likelihood of humans being infected increases. This has been seen with the emergence of Nipah virus, through the pig, and for Hendra virus infection, through the horse, in southeast Asia and Australia, respectively. It is therefore possible to characterize the role of animals into discrete groups. The first is that of the **reservoir** or **vector**. This group potentially includes all animals, as any virus infecting an animal could theoretically adapt to infect humans. This in part depends on the binding of virus proteins to cellular receptors that can be highly selective, as seen for parvoviruses, or nonselective as observed for Nipah virus. However, most viruses appear to be strongly adapted to a particular species or genus and tend not to "jump" between species and cause disease. Bats harbor a multiplicity of coronaviruses (Tang et al., 2006) that are believed to be the progenitor of SARS coronavirus. Infection appears to be asymptomatic in bats with shedding of virus in fecal material being frequently observed. At some point a cross species transmission event occurred, leading to infection of animals that entered the wet markets of China.

The Role of Animals in Emerging Viral Diseases. http://dx.doi.org/10.1016/B978-0-12-405191-1.00012-0

A further transmission to humans occurred, leading to the outbreak of a respiratory disease that through air travel affected countries around the world (Hughes, 2004). In the Americas, bats harbor rabies virus and are becoming an increasing source of infection to humans (Banyard et al., 2011). In this case, bats are susceptible to disease but through predation by the vampire bat (*Desmodus rotundus*) or chance encounters with a wide variety of insectivorous bats, humans become infected. Dogs are also efficient transmitters of rabies (see Chapter 4) and it is through them that the virus has spread throughout the world. The virus is maintained within the dog population, although can be controlled through methodical management, such as animal movement restrictions, quarantine and vaccination against disease. Rodents are the source of a range of viruses that are capable of crossing the species barrier to humans (see Chapters 5 and 10). Transmission to humans occurs through exposure to excreta; control is limited to reducing the opportunities for exposure to particular rodents or exclusion of rodents from human premises. Finally, one of the largest reservoirs for emerging viruses is the bird population. Here, birds act as the host for influenza viruses that have caused some of the largest outbreaks of disease over the past hundred years.

The second role of animals is that of the **amplifying** or **bridge** host. These are those species that are not the natural host of a particular virus but, through infection and shedding of virus (amplification), provide the opportunity to infect humans. Nipah virus (Chapter 11) provides an excellent example of this. In late 1998 a new disease emerged in pig farms in Malaysia, followed shortly by cases of viral encephalitis in farm workers (Mohd Nor and Ong, 2000). The source of the virus was identified as fruit bats that roosted close to the farms and shed the virus in fecal matter. In the case of the Hendra virus outbreak in Australia (see Chapter 6), the source was also fruit bats and horses acted as the bridge host that led to the infection of veterinarians and others who handled the horses.

One final consideration is to see animals as the **victims** of virus emergence. Rather than this being a sentimental consideration, disease emergence can have a profound effect on particular species that in extreme cases could lead to species decline or extinction and influence the ecology of a particular environment. Outbreaks of rabies in a range of canid species have led to the possibility that rare species could be driven to extinction. Two examples from Africa illustrate this. The world's rarest canid, the Ethiopian wolf (*Canis simensis*), is restricted to the Ethiopian Highlands with populations limited to a few thousand. Recent outbreaks of rabies have caused dramatic disappearances of well-studied populations (Johnson et al., 2010) and it is only through interventions such as parenteral vaccinations that outbreaks have been controlled and further population declines prevented (Randall et al., 2006). Similarly, outbreaks of rabies and, more devastatingly, canine distemper virus in populations of the African wild dogs (*Lycaon pictus*) has led to the decline of this species across its whole range (Goller et al., 2010).

THE ROLE OF HISTORY IN VIRUS DISEASE EMERGENCE

A somewhat surprising observation from these chapters has been the role of historical events, often occurring hundreds of years in the past, influencing disease emergence in the present. Clearly human activities underpin this, but it is the emergence of disease long after the trigger event that is striking. Examples of this include the emergence of vampire bat rabies in Latin America, dog rabies in Africa, and the emergence of HIV and the current distribution of Lassa fever in West Africa. For the two examples for rabies, the key events are related to colonization by Europeans. In the New World, Europeans introduced domestic animals that provided a ready supply of prey for vampire bats that, in turn, is thought to have increased their numbers with larger, more numerous colonies. Europeans initially introduced dogs and horses to assist in the conquest and subjugation of the indigenous populations. Later introductions of domestic animals to supply food completed the radical change to the ecology of the New World. It is hypothesized that this supported an increase in the vampire bat population capable of sustaining rabies virus infection, leading to spillover transmission to human and livestock. While the initial colonization took place in the 15th and 16th centuries, rabies in vampire bats was only recognized in the early 20th century and continues to be a major problem for the livestock industry in Latin America and a public health hazard. In Africa, the translocation of domestic dogs by colonists arriving in the west and the south of the continent is widely suspected of introducing dog-associated strains of rabies (Smith et al., 1992). While there are ancestral populations of rabies variants in certain wildlife species such as the yellow mongoose (*Cynictis penicillata*) that likely predates colonization (Van Zyl et al., 2010), the virus strains associated with dogs are those responsible for the vast majority of human rabies cases.

Further colonial events such as economic exploitation of the human population and the environment are considered a major cause for the emergence of HIV in the human population of the Congo Basin (Chapter 9). While it is likely that human population growth and environmental destruction were inevitable without the "blame King Leopold II" explanation, Leopold's policies brought humans into contact with nonhuman primates. The exploitation of the Congo Basin by the royal private army, the Force Publique, began in the 1870s. This group terrorized the local population through mutilation and murder in pursuit of increased produce yields that in turn led to population disturbance, forced urbanization and accelerated the use of bush meat as a means of nutrition. The dating of the last common ancestors of HIV and SIV provides strong evidence for cross species transmission during this period. However, it was not until a further hundred years later that the virus achieved pandemic status. The importation of European breeds of livestock is also thought to have led to the rapid spread of diseases such as foot-and-mouth disease (Chapter 2) and rinderpest. While rinderpest has been eradicated, FMD continues to affect many regions of the world and is a major threat to the food chain.

For Lassa fever virus, the key historical event that could have caused the emergence of the virus in Sierra Leone is the disruption of the West African slave trade by the British in the 19th century. The slave trade itself was responsible for the translocation of numerous diseases to the New World from Africa that remain public health problems to the present day. There is evidence that mosquito-borne diseases such as yellow fever and dengue were translocated to the New World by the slave trade (Bryant et al., 2007). However, in the case of Lassa fever virus, the result of this intervention that freed many slaves caused a problem in relocation. While the British Empire repudiated slavery, many countries did not, maintaining the trade in humans from Africa to the Americas. Those that were freed were forced to remain in Sierra Leone because of the danger of recapture if they returned to their country of origin. If the "humans as vectors" hypothesis is correct, this population could have been responsible for the introduction of Lassa fever virus and the emergence of Lassa fever in Sierra Leone in the 20th century.

RETROSPECTIVE TESTING OR ANALYSIS TO DEFINE DISEASE EMERGENCE

Another common feature that has played a role in the investigation of emerging disease investigation has been the ability to use retrospective analysis of biological samples or data to pinpoint to time of emergence. Analysis of the piglet mortality index, a measure of litter deaths in the index farm where Nipah virus is believed to have been introduced has identified the likely time of cross-species transmission from Pteropid bats to livestock (Pulliam et al., 2012). Critically, this precedes the first human cases by a number of months and refutes the hypothesis that a particular environmental factor, the El Niño Southern Oscillation, triggered the emergence of the virus.

Another striking case is the retrospective identification of a cluster of HIV infections in Norway, a decade before the recognition of the clinical syndrome AIDS in the USA (Jonassen et al., 1997). The index case was a Norwegian male who worked on merchant shipping who traveled to numerous locations around the world between 1961 and 1965. On some of these, he visited ports on the west coast of Africa. He later returned, presumably after being infected, married and started a family. In 1966 he developed various disease symptoms, including persistant lymphadenopathy. His wife also developed recurrent mucocutaneous candidiasis and his daughter from the age of 2 developed a series of recurrent infections. All died in 1976. However, it was the ability to demonstrate HIV genomic RNA in formalin fixed samples using highly sensitive capture PCR that confirmed that these were perhaps the earliest cases of HIV in Europe. Further serological testing of archive samples in Africa has also identified early evidence for HIV infection as far back as the 1950s (Nahmias et al., 1986).

For the emergence of canine parvovirus (Chapter 3), clinical disease alerted veterinary authorities to the presence of a new disease in dogs. However, restrospective testing of canine serum samples allowed the presence of the causative agent to be detected and allowed the emergence of the disease to be traced through the 1970s (Parrish, 1990).

VIRUS DETECTION USING MOLECULAR TECHNIQUES

Rapid and accurate detection of pathogens in clinical or environmental samples is the cornerstone of disease diagnosis. It has also been instrumental in the detection of emerging viruses. Over the past 20 years there has been a move to enhance diagnostic capability using more conventional diagnostic tests such as ELISA, serology and virus isolation, by the introduction of molecular-based testing as an adjunct to these tests rather than replacing them. Molecular diagnostics involve the detection of specific nucleic acids and frequently the amplification of these sequences in order to allow the products to be easily detected and as such are not reliant on the viability of the virus. This is both advantageous in that generic tests can be applied where no virus isolation methodology exists (or no means to identify the virus either through a cytopathic effect or immunologically once propagated in cells—i.e., lack of monoclonal antibodies) but may also identify samples as containing dangerous pathogens when in fact the virus is nonviable and therefore presents no threat. A further benefit of these approaches is the generation of DNA samples from which the sequence of the virus under investigation can be derived and used to identify the disease-causing agent. This in turn drives the design of molecular tests, which has been greatly accelerated through advances in genome sequencing technologies that have provided the base information (the specific pathogen sequences) for the design of such molecular tests. The following sections focus on particular molecular tests that have been applied to the detection of new and emerging viruses.

The **polymerase chain reaction** (PCR) is the most commonly used and dominating technology applied to molecular testing both in the clinical and veterinary field. Its power lies in its simplicity of design and application. The specificity of the assay relies on the hybridization of short stretches of synthetic DNA (oligonucleotides) complementary to the target sequence and amplification of the sequences between these oligonucleotides using the enzymatic activity of *Taq* polymerase. PCR assays are easy to design as primer design software packages are freely available on the Internet. The speed of the test using PCR is limited by the processivity of *Taq* polymerase and also the requirement to heat and cool the reaction through cycles of denaturation of the double-stranded DNA (95°C), annealing of primers and elongation (68–72°C). Various enhancements have been made to the process and the machinery to generate so-called Fast PCR, but time to result still remains somewhat slow (some commercial companies quote 40 minutes as being fast) or when compared with the rapidity (but decreased sensitivity) of direct detection methods employing lateral flow devices, for instance. The addition of sequence-specific DNA probes into a PCR enables the real-time monitoring of the reaction. The traditional probe format is a "TaqMan" probe which involves labeling a DNA probe with a fluorescent reporter and a corresponding quencher. The close proximity of reporter and quencher causes quenching of the fluorescence. The probe hybridizes with its complementary DNA target and is enzymatically cleaved, causing separation of the fluorophore and quencher and results in an increase in the level of fluorescence, which can be detected after excitation with a laser. Different probe and amplification technologies have emerged and are briefly described below:

- A molecular beacon is a dual labeled fluorescent probe, complementary to the target sequence, which is held in a hairpin structure by complementary stem sequences. Upon hybridization of the molecular beacon with its complementary sequence, the hairpin structure relaxes and the fluorophore and quencher separate, resulting in emission of the fluorescence. Unlike the traditional TaqMan probe, the molecular beacon is not cleaved. Real-time PCR assays for the detection of swine vesicular disease virus (SVDV) and vesicular stomatitis virus (VSV) using molecular beacons (MBs) have been developed (Belák, 2005).

- A Scorpion probe is formed by the linkage of the primer and probe via a blocker. The probe part is held in a hairpin structure, similar to that of a molecular beacon, with a fluorescent reporter dye and quencher. The 3′ end of the probe (where the quencher is located) is linked via a blocker to the 5′ end of the PCR primer. During the amplification reaction the primer anneals to its complementary target sequence and the polymerase extends the primer. This results in the newly synthesized strand being attached to the probe. During a subsequent cycle the hairpin loop of the probe is denatured and hybridizes to a part of the newly produced PCR product. This results in the separation of the fluorophore from the quencher and causes light emission. Scorpion probes are thought to provide increased specificity and have been applied to the quantitation of HIV-1 in clinical samples (Saha et al., 2001).

- LATE (Linear After The Exponential) PCR is a form of asymmetric amplification developed by researchers at Brandeis University. LATE-PCR generates single-stranded amplicons after a short period of exponential amplification through the use of a limiting primer and an excess primer. The reaction switches from the production of double-stranded to single-stranded DNA when the limiting primer is exhausted, creating an abundance of single-stranded DNA which can be detected with probes over a temperature range. This enables the detection of low numbers of target organisms and the detection of a wide range of sequences with a single probe (Sanchez, 2004). LATE-PCR has successfully been utilized in the detection of foot and mouth virus (Pierce, 2010).

Isothermal amplification methods are now coming to prominence principally due to the ease with which these methods can be applied in that they do not require a programmable heating block being the heart of the PCR machine, but simply a constant source of heat. This has the advantage that the engineering challenge with respect to a programmable heating device is considerably simplified.

Nucleic Acid Sequence Based Amplification (NASBA) was one of the first methods to gain prominence (Guatelli et al., 1990). NASBA is an isothermal transcription based amplification system using three enzymatic processes essential in retroviral replication. These processes involve reverse transcriptase, e.g., avian myeloblastosis virus reverse transcriptase (AMV-RT), RNase H and DNA-dependent RNA polymerase (T7 RNA polymerase, most commonly). The method, using RNA as the starting material, is particularly suitable for the detection of genomic, ribosomal and messenger RNA. As for all the isothermal methods it does not require a thermocycler but just a continuous source of heat (41°C). The specificity of the method is conferred by a primer that hybridizes to the target. This primer also encodes a T7 RNA polymerase promoter sequence which comes into play during the exponential phase of the amplification process. Following annealing of the primer to target complementary DNA (cDNA) is generated under the action of AMV-RT, forming a DNA-RNA hybrid, the RNA being subsequently digested under the action of RNase H leaving just the cDNA. A second primer complementary to the cDNA then binds and extends a second strand of DNA leading to the formation of an intact T7 RNA promoter region. Once the promoter is formed T7 RNA polymerase is able to catalyze the production of very large numbers of RNA transcripts (complementary to the original RNA), which can themselves serve as starting material for further rounds of exponential amplification. NASBA has been applied to the detection of numerous viruses including those transmitted or amplified in animals prior to insect vector transmission. Influenza virus has been detected using NASBA from both birds (Lau et al., 2004) and more recently swine (Ge et al., 2010). NASBA has also been used to detect West Nile and St. Louis encephalitis virus (Lanciotti and Kerst, 2001), SARS coronavirus (Keightley et al., 2005), rabies virus (Wacharapluesadee et al., 2011) and human rhinoviruses (Sidoti et al., 2012).

An alternative isothermal amplification method that appears to offer many advantages over others available is **loop mediated isothermal amplification (LAMP).** First described by Notomi et al., (2000) and improved upon with respect to speed of reaction by Nagamine, Hase and Notomi (2002), this method employs the strand displacement activity of *Bst* polymerase (or equivalent) and four specific primers that hybridize to six different regions of the target DNA. Despite the apparent complexity of the molecules produced (cauliflower-like with multiple loops) they are essentially composed of repeats of the initial amplification molecule and as a consequence undergo DNA melting at a consistent and reproducible temperature in much the same way PCR products do. By incorporating thermostable reverse transcriptase (RT) able to operate at 62–65°C into the reaction mix, it is possible to generate RT-LAMP reactions that can be used to amplify viral RNA targets. The pace with which this technology is being adopted for diagnostic testing appears to be increasing, with diagnostic LAMP reactions for a number of viruses of significance to human health emerging over the past 5 years or so, e.g., Japanese encephalitis virus (Liu et al., 2012), hepatitis E virus (Zhang et al., 2012), West Nile virus (Shukla et al., 2012), rabies virus (Hayman et al., 2011), influenza virus including pandemic H1N1 (Nakauchi et al., 2011) and Rift Valley fever virus (Le Roux et al., 2009). However, despite the use of this technology in many detection assays a major limitation of the technology appears to be its apparent lack of tolerance to sequence variation in the primer binding sites. This is particularly important for detection of RNA viruses. Unlike PCR, which appears somewhat tolerant to a limited number of base changes in the primer/probe binding regions (particularly if these are not at the 3' termini of the primers), LAMP primers must bind with a high efficiency or the reaction fails to work. It is possible for particular assays to circumvent this problem by the mixing together of different primer sets compensating for sequence variation in the target region with apparently no effect on the performance of the assays when performed individually.

An ingenious strategy for DNA amplification has been adopted in a process known as **recombinase polymerase amplification** or **RPA**. This isothermal amplification method is similar to PCR in that only two opposing primers are employed but where heating is used to melt the template and subsequent products globally, RPA uses recombinase-primer complexes to scan the double-stranded DNA, facilitating strand exchange at cognate sites (specific site melting, in other words). RPA and assays combining reverse transcription and RPA have demonstrated the utility of this technology for the detection of biothreat agents such as Rift Valley fever virus, Ebola, Sudan virus and Marburg (Euler et al., 2013). Previously, methods have been described to detect Rift Valley fever (Euler et al., 2012) and HIV (Rohrman and Richards-Kortum, 2012) but it is anticipated that in time this technology will also become more prominent.

FUTURE PROSPECTS FOR VIRUS DETECTION

PCR will continue to dominate but has and will become increasingly more rapid so that it can compete with more rapid isothermal methods. Commercial manufacturers are now claiming "cycling times to result" of 25 minutes with some, such as Applied Biosystems producing specific FAST machines, recognizing that *Taq* polymerase is not working at a maximum rate in conventional machines. In well-equipped laboratories the rate-limiting step is shifting from the detection method used to the method of nucleic acid extraction.

Another limiting step is the transport of samples from the infected animal to the laboratory. One of the most exciting prospects for molecular testing is to take the test to the infected animal. The principle of **point of care** or **point of decision testing** has been established using RT-PCR methods and combined extraction/real-time PCR machines, such as the GeneXpert produced by Cepheid who have produced a range of assays for clinical use. However, the cost of these machines and consumables where they are most needed in the world may be prohibitive and this is clearly problematic for commercial companies producing such devices. Using isothermal amplification methods may circumvent problems of cost by using very low cost instruments or no instrument whatsoever for heating, as has been demonstrated for RT-LAMP HIV testing where a constant source of heat was provided by essentially a canister that relied on the exothermal energy derived from the reaction of calcium oxide with water (Curtis et al., 2012). Both RPA and LAMP assays can be detected in real time using intercalating dyes or more simply at the end of the reactions using tagged primers (FITC, biotin) on lateral flow devices (LFD), which may be of considerable benefit to laboratories where capital investment in such technologies is limited.

Of a particular concern with respect to point of care testing is not only the consideration of what is gained (speed of testing and reduced costs of transport) but also what is lost for those collecting epidemiological data or responsible for control of disease (centralized laboratories). This is a pressing issue, particularly in light of the availability of diagnostics via the internet.

Much effort has been put into the development of **microarrays** that can detect multiple virus families. Such arrays consist of thousands of oligonucleotides, targeting many known pathogens. Examples of these have been deployed in the detection of virus, bacterial and protozoan pathogens (Palacios et al., 2007). The use of pan-virus family polymerase chain reaction assays has revealed the existence of thousands of viruses in a range of hosts that previously have been poorly studied. In the wake of the SARS outbreak, the search for the reservoir host led to the discovery of a wide diversity of SARS-like coronaviruses being shed by bats. This was made possible by the application of sensitive pan-coronavirus primers to RNA extracted from throat and fecal samples in Chinese bats (Chu et al., 2006). A similar approach has led to the detection of coronaviruses in bat populations around the world (Carrington et al., 2008; Drexler et al., 2010). The major benefit of PCR is that it generates fragments of DNA that can be sequenced and used in phylogenetic analysis. Pan-paramyxovirus primers have been applied to the detection of viruses from this family from a range of vertebrate hosts and the phylogenies being created have led to the proposal that bats form the ancestral host for many viruses (Drexler et al., 2012).

An alternative route to the use of PCR is to directly sequence a virus from nucleic acid extracted from a sample. The development of mass sequencing and the application of algorithms that will construct contiguous sequences from large data sets have enabled the recovery of complete virus genomic sequences directly from clinical samples. A recent example of this approach detected a novel Rhabdo virus in patients suffering from acute hemorrhagic fever in Africa (Grard et al., 2012). A key benefit is that a virus can be characterized from a clinical sample or from a degraded sample. This raises a number of fundamental questions on the nature of viruses and disease. While a virus can be detected in a sample derived from a diseased individual, can it unequivocably be said to have caused that disease if "Kochs postulates" cannot be demonstrated? Also many viruses are detected that appear to have no role in disease within a particular host. The ability of these viruses to jump the species barrier and cause disease in a new host is difficult to assess.

CONCLUSIONS

The consequences of disease emergence are clear in human morbidity and death. In addition is the economic cost, highlighted in the foreword to this book. Disease emergence can also lead to changes in human behavior and also innovation through the development of vaccines and therapeutics. Combating disease emergence, particularly raising awareness among susceptible populations and detecting the causative agent, is critical in responding to such events. Due to the nature of diseases of animal origin it is unlikely that eradication will ever be an option (see Box 12.1) as there will always remain a reservoir of pathogen (Dowdle, 1999). Therefore, rapid detection and response will be the most effective approaches to controlling future disease emergence.

One area that has received particular attention in recent years has been the development of the One Health approach to responding to disease emergence (Hayman et al., 2012). The One Health concept is a worldwide strategy for expanding interdisciplinary collaborations and communications in all aspects

---

BOX 12.1 Key terms for progressive reduction of disease (Dowdle, 1999).

Control
The reduction of incidence of a disease to an arbitrary level where it is no longer a public health priority.

Elimination
The interruption of transmission of the pathogen when disease incidence becomes zero in a population within a defined geographic area.

Eradication
The interruption of transmission of a pathogen worldwide and the reduction of disease incidence to zero.

Extinction
The infectious agent no longer exists in nature or the laboratory.

Biological criteria for disease eradication
- Humans are the sole pathogen reservoir
- An accurate diagnostic test exists
- An effective, practical intervention is available at a reasonable cost.

of health care for humans, animals and the environment (Narrod et al., 2012). A key observation from this that will be developed below is that intervention early in areas such as wildlife and domestic livestock will yield benefits for human health. The challenge is to persuade policy makers that this is the case and that limited resources should be targeted toward achieving this.

The origins of the One Health approach are found in the observations of Robert Virchow, a nineteenth-century physician who stated that "between animal and human medicine there is no dividing line, nor should there be" (Conrad et al., 2009). This then developed into the concept of "One Medicine," by the veterinary epidemiologist, Calvin Swabe, who emphasized the need for clinicians and veterinarians to coordinate efforts at disease control and reverse the trend of increasing specialization and division of resources. The One Medicine approach to disease control and management has evolved into One Health, which encourages all disciplines to focus efforts on the interface between humans, wildlife and domestic animals and can be applied to the control of all zoonotic pathogens including viruses, bacteria, protozoa, spongiform encephalopathies and fungi. This has been defined as "the collective efforts of multiple disciplines working locally, nationally and globally to attain optimal health for people, animals and environment." The key benefits to this approach are reduction in human and animal morbidity and mortality, reduction in costs associated with responding to disease outbreaks or persistence, and innovations in animal husbandry and human health (Zinsstag et al., 2012). The obstacles to introducing a One Health approach are weak leadership, institutional resistance, lack of finance to support linkage between human and veterinary medicine and a poor understanding of ecology. Therefore, a key challenge to introducing a One Health approach to disease management is to demonstrate its efficacy and the potential benefits. For this the most obvious one is to demonstrate cost benefit. Narrod and co-workers (2012) have developed modified risk models that attempt to calculate the economic impact of zoonotic diseases and highlight the benefits of early intervention before the disease reaches the human population. A good example of this is rabies control. The majority of governments target prevention of human deaths by reactive vaccination, commonly referred to as post-exposure prophylaxis (PEP). While highly effective at preventing human deaths from rabies, it does not prevent the persistence of rabies within animal reservoirs, mainly the domestic dog (Lembo, 2012). By maintaining PEP, but augmenting this with measures that control dog rabies, there are clear long-term financial cost savings and the possibility of eliminating rabies at the local level (Zinsstag et al., 2009).

One final consideration is what viruses will emerge in the future. In recent years the spotlight has been on bats as a source of disease and many putative viruses have been detected within bat species from around the world (Drexler et al., 2012) using the technologies described above. Whether these will be the precursors of future disease emergence is unclear. However, future diseases will probably be caused by viruses that are similar to the ones we are aware of.

Reassortment of the influenza genome and subtle changes to virus coat proteins, such as have occurred for canine parvovirus, can rapidly change an apparently harmless virus to one with radically different infectious properties. Strategies that can detect such changes in virus virulence may offer the best hope of identifying emerging viruses that will impact on human populations.

ACKNOWLEDGMENTS

This work has been supported through the project "Anticipating the global onset of novel epidemics (ANTIGONE)" funded by the European Union project number 278976.

REFERENCES

Banyard, A. C., Hayman, D., Johnson, N., McElhinney, L. M., & Fooks, A. R. (2011). Bats and Lyssaviruses. *Advances in Virus Research*, *79*, 239–289.

Belák, S. (2005). The molecular diagnosis of porcine viral diseases: a review. *Acta. Vet. Hung.*, *53*, 113–124.

Bryant, J. E., Holmes, E. C., & Barrett, A. D. T. (2007). Out of Africa: A molecular perspective on the introduction of yellow fever virus into the Americas. *PLoS Pathogens*, *3*, e75.

Carrington, C. V., Foster, J. E., Zhu, H. C., Zhang, J. X., Smith, G. J., Thompson, N., Auguste, A. J., Ramkissoon, V., Adesiyun, A. A., & Guan, Y. (2008). Detection and phylogenetic analysis of group 1 coronaviruses in South American bats. *Emerg. Infect. Dis.*, *14*, 1890–1893.

Chu, D. K. W., Poon, L. M., Chan, K. H., Chen, Y., Guan, Y., Yuen, K. Y., & Peiris, J. S. M. (2006). Coronaviruses in bent-winged bats (*Miniopterus* spp.). *J. Gen. Virol.*, *87*, 2461–2466.

Conrad, P. A., Mazet, J. A., Clifford, P. A., Scott, C., & Wilkes, M. (2009). Evolution of a transdisciplinary "One Medicine – One Health" approach to global health education at the University of California, Davis. *Preventative Veterinary Medicine*, *92*, 268–274.

Curtis, K. A., Rudolph, D. L., Nejad, I., Singleton, J., Beddoe, A., Weigl, B., LaBarre, P., & Owen, S. M. (2012). Isothermal amplification using a chemical heating device for point-of-care detection of HIV-1. *PLoS ONE*, *7*, e31432.

Dowdle, W. (1999). The principles of disease elimination and eradication. *Morbidity and Mortality Weekly Report*, *48*(SU01), 23–27.

Drexler, J. F., Gloza-Rausch, F., Glende, J., Corman, V. M., Muth, D., Goettsche, M., Seebens, A., Niedrig, M., Pfefferle, S., Yordanov, S., Zhelyakov, L., Hermanns, U., Vallo, P., Lukashev, A., Mülller, M. A., Deng, H., Herrler, G., & Drosten, C. (2010). Genomic characterization of severe acute respiratory syndrome-related coronaviruses in European bats and classification of coronavirues based on partial RNA-dependent RNA polymerase gene sequences. *J. Virol.*, *84*, 11336–11349.

Drexler, J. F., Corman, M. V., Müller, M. A., Maganga, G. D., Vallo, P., Binger, T., Gloza-Rausch, F., Rasche, A., Yordanov, S., Seebens, A., Oppong, S., Sarkodie, Y. A., Pogombo, C., Lukashev, A. N., Schmidt-Chanasit, J., Stöcker, A., Carneiro, A. J. B., Erbar, S., Maisner, A., Fronhoffs, F., Buettner, R., Kalko, K., Kruppa, T., Franke, C. R., Kallies, R., Yandoko, E. R. N., Herrler, G., Reusken, C., Hassanin, A., Krüger, D. H., Matthee, S., Ulrich, R. G., Leroy, E. M., & Drosten, C. (2012). Bats host major mammalian paramyxoviruses. *Nature Communications*, *3*, 796.

Euler, M., Wang, Y. J., Nentwich, O., Piepenburg, O., Hufert, F. T., & Weidmann, M. (2012). Recombinase polymerase amplification assay for rapid detection of Rift Valley fever virus. *J. Clin. Virol.*, *54*, 308–312.

Euler, M., Wang, Y. J., Heidenreich, D., Patel, P., Strohmeier, O., Hakenberg, S., Niedrig, M., Hufert, F. T., & Weidmann, M. (2013). Development of a panel of Recombinase Polymerase Amplification assays for detection of biothreat agents. *J. Clin. Microbiol.*, *51*, 1110–1117.

Ge, Y. Y., Cui, L. B., Qi, X., Shan, J., Shan, Y. F., Qi, Y. H., Wu, B., Wang, H., & Shi, Z. Y. (2010). Detection of novel swine origin influenza A virus (H1N1) by real-time nucleic acid sequence-based amplification. *J. Virol. Methods*, *163*, 495–497.

Goller, K. V., Fyumagwa, R. D., Nikolin, V., East, M. L., Kilewo, M., Speck, S., Muller, T., Mazke, M., & Wilbbelt, G. (2010). Fatal canine distemper infection in a pack of African wild dogs in the Serengeti ecosystem, Tanzania. *Vet. Microbiol.*, *146*, 245–252.

Grard, G., Fair, J. N., Lee, D., Slikas, E., Steffen, I., Muyembe, J., Sittler, T., Veeraraghavan, Ruby, G. J., Wang, C., Makuwa, M., Mulembakani, P., Tesh, R. B., Mazet, J., Rimoin, A. W., Taylor, T., Schneider, B. S., Simmons, G., Delwart, E., Wolfe, N. D., Chiu, C. Y., & Leroy, E. M. (2012). A novel Rhabdovirus associated with acute hemorrhagic fever in Central Africa. *PLOS Pathogens*, e1002942.

Guatelli, J. C., Whitfield, K. M., Kwoh, D. Y., Barringer, K. J., Richman, D. D., & Gingeras, T. R. (1990). Isothermal, in vitro amplification of nucleic acids by a multienzyme reaction modelled after retroviral replication. *Proc. Natl. Acad. Sci. USA*, *87*, 1874–1878.

Hayman, D. T. S., Johnson, N., Horton, D. L., Hedge, J., Wakeley, P. R., Banyard, A. C., Zhang, S. F., Alhassan, A., & Fooks, A. R. (2011). Evolutionary history of rabies in Ghana. *PLOS Neglected Tropical Diseases*, *5*, e1001.

Hayman, D. T., Gurley, E. S., Pulliam, J. R., & Field, H. E. (2012). The application of One Health approaches to Henipavirus research. *Curr. Top. Microbiol. Immunol.* In press.

Hughes, J. M. (2004). SARS: an emerging global microbial threat. *Transactions of the American Clinical and Climatological Association*, *115*, 361–374.

Johnson, N., Mansfield, K. L., Marston, D. A., Wilson, C., Goddard, T., Seldon, D., Hemson, G., Edea, L., van Kesteren, F., Shiferaw, F., Stewart, A. E., Sillero-Zubiri, C., & Fooks, A. R. (2010). A new outbreak of rabies in rare Ethiopean Wolves (*Canis simensis*). *Arch. Virol.*, *155*, 1175–1177.

Jonassen, T. Ø., Stene-Johansen, K., Berg, E. S., Hungnes, O., Lindboe, F., Frøland, S. S., & Grinde, S. (1997). Sequence analysis of HIV-1 group O from Norwegian patients infected in the 1960s. *Virology*, *231*, 43–47.

Keightley, M. C., Sillekens, P., Schippers, W., Rinaldo, C., & St George, K. (2005). Real-time NASBA detection of SARS-associated coronavirus and comparison with real-time reverse transcription-PCR. *J. Med. Virol.*, *77*, 602–608.

Lanciotti, R. S., & Kerst, A. J. (2001). Nucleic acid sequence-based amplification assays for rapid detection of West Nile and St. Louis encephalitis viruses. *J. Clin. Microbiol.*, *39*, 4506–4513.

Lau, L. T., Banks, J., Aherne, R., Brown, I. H., Dillon, N., Collins, R. A., Chan, K. Y., Fung, Y. W. W., & Xing, J. Y. (2004). Nucleic acid sequence-based amplification methods for detect avian influenza virus. *Biochem. Biophys. Res. Commun.*, *313*, 336–342.

Lembo, T. (2012). The blueprint for rabies prevention and control: A novel operational toolkit for rabies elimination. *PLoS Neglected Tropical Diseases*, *6*, e1388.

Le Roux, C. A., Kubo, T., Grobbelaar, A. A., van Vuren, P. J., Weyer, J., Nel, L. H., Swanepoel, R., Morita, K., & Paweska, J. T. (2009). Development and evaluation of a real-time reverse transcription-loop-mediated isothermal amplification assay for rapid detection of Rift Valley fever virus in clinical specimens. *J. Clin. Microbiol.*, *47*, 645–651.

Liu, H., Liu, Z. J., Jing, J., Ren, J. Q., Liu, Y. Y., Guo, H. H., Fan, M., Lu, H. J., & Jin, N. Y. (2012). Reverse transcription loop-mediated isothermal amplification for rapid detection of Japanese encephalitis virus in swine and mosquitoes. *Vector Borne Zoonotic Dis.*, *12*, 1042–1052.

Mohd Nor, M. N., & Ong, B. L. (2000). Nipah virus infection of pigs in peninsular Malaysia. *Rev. sci. tech. Off. Int. Epiz.*, *19*, 160–165.

Nagamine, K., Hase, T., & Notomi, T. (2002). Accelerated reaction by loop-mediated isothermal amplification using loop primers. *Molecular and Cellular Probes*, *16*, 223–229.

Nahmias, A. J., Weiss, J., Yao, X., Lee, F., Kodsi, R., Schanfield, M., Matthews, T., Bolognesi, D., Durack, D., Motulsky, A., Kanki, P., & Essex, M. (1986). Evidence for human infection with an HTLV III/LAV-like virus in Central Africa. *Lancet*, *i*, 1279–1280.

Nakauchi, M., Yoshikawa, T., Nakai, H., Sugata, K., Yoshikawa, A., Asano, Y., Ihira, M., Tashiro, M., & Kageyama, T. (2011). Evaluation of reverse transcription loop-mediated isothermal amplification assays for rapid diagnosis of pandemic influenza A/H1N1 2009 virus. *J. Med. Virol.*, *83*, 10–15.

Narrod, C., Zinsstag, J., & Tiongco, M. (2012). A One Health framework for estimating the economic costs of zoonotic disease on society. *EcoHealth*, *9*, 150–162.

Notomi, T., Okayama, H., Masubuchi, H., Yonekawa, T., Watanabe, K., Amino, N., & Hase, T. (2000). Loop-mediated isothermal amplification of DNA. *Nucleic Acids Research*, *28*, e63.

Palacios, G., Quan, P. -L., Jabado, O. J., Conlan, S., Hirschberg, D. L., Liu, Y., Zhai, J., Renwick, N., Hui, J., Hegyi, H., Grolla, A., Strong, J. E., Towner, J. S., Geisbert, T. W., Jahrling, P. B., Büchen-Osmond, Ellerbrook, H., Sanchez-Seco, M. P., Lussier, Y., Formenty, P., Nichol, S. T., Feldmann, H., Briese, T., & Lipkin, W. I. (2007). Panmicrobial oligonucleotide array for diagnosis of infectious diseases. *Emerg. Infect. Dis.*, *13*, 73–81.

Parrish, C. R. (1990). Emergence, natural history, and variation of canine, mink, and feline parvoviruses. *Adv. Virus Res.*, *38*, 403–450.

Pierce, K. E., Mistry, R., Reid, S. M., Bharya, S., Dukes, J. P., Harshorn, C., King, D. P., & Wangh, L. J. (2010). Design and optimization of a novel reverse transcription linear-after-the-exponential PCR for the detection of foot-and-mouth disease virus. *J. App. Microbiol.*, *109*, 180–189.

Pulliam, J. R., Epstein, J. H., Duschoff, J., Rahman, S. A., Bunning, M., Jamalludin, A. A., Hyatt, A. D., Field, H. E., Dobson, A. P., Daszak, P.Henipavirus Ecology Research Group (HERG) (2012). Agricultural intensification, priming for persistence and the emergence of Nipah virus: a lethal bat-borne zoonosis. *J. R. Soc. Interface*, *9*, 89–101.

Randall, D. A., Marino, J., Haydon, D. T., Sillero-Zuberi, C., Knobel, D. L., Tallents, L. T., Macdonald, D. W., & Laurenson, M. K. (2006). An integrated disease management strategy for the control of rabies in Ethiopian wolves. *Biological Conservation*, *131*, 325–337.

Rohrman, B. A., & Richards-Kortum, R. R. (2012). A paper and plastic device for performing recombinase polymerase amplification of HIV DNA. *Lab Chip*, *12*, 3082–3088.

Saha, B. K., Tian, B., & Bucy, R. P. (2001). Quantitation of HIV-1 by real-time PCR with a unique fluorigenic probe. *J. Virol. Methods*, *93*, 33–42.

Sanchez, J. A., Pierce, K. E., Rice, J. E., & Wangh, L. J. (2004). Linear-After-The-Exponential (LATE)-PCR.: An advanced method of asymmetric PCR and its uses in quantitative real-time analysis. *Proc. Natl. Acad. Sci. USA*, *101*, 1933–1938.

Shukla, J., Saxena, D., Rathinam, S., Lalitha, P., Joseph, C. R., Sharma, S., Soni, M., Rao, P. V. L., & Parida, M. (2012). Molecular detection and characterization of West Nile virus associated with multifocal retinitis in patients from southern India. *International J. Infect. Dis.*, *16*, E53–E59.

Sidoti, F., Bergallo, M., Terlizzi, M. E., Alessio, E. P., Astegiano, S., Gasparini, G., & Cavallo, R. (2012). Development of a Quantitative Real-Time Nucleic Acid Sequence-Based Amplification Assay with an Internal Control Using Molecular Beacon Probes for Selective and Sensitive Detection of Human Rhinovirus Serotypes. *Molecular Biotechnology*, *50*, 221–228.

Smith, J. S., Orciari, L. A., Yager, P. A., Seidel, H. D., & Warner, C. K. (1992). Epidemiologic and historical relationships among 87 rabies virus isolates as determined by limited sequence analysis. *J. Infect. Dis.*, *166*, 296–307.

Tang, X. C., Zhang, J. X., Zhang, S. Y., Wang, P., Fan, X. H., Li, L. F., Dong, B. Q., Liu, W., Cheung, C. L., Xu, K. M., Song, W. J., Vijaykrishna, D., Poon, L. L., Peiris, J. S., Smith, G. J., Chen, H., & Guan, Y. (2006). Prevalence and genetic diversity of coronaviruses in bats from China. *J. Virol.*, *80*, 7481–7490.

Van Zyl, N., Markotter, W., & Nel, L. (2010). Evolutionary history of African mongoose rabies. *Virus Res.*, *150*, 93–102.

Wacharapluesadee, S., Phumesin, P., Supavonwong, P., Khawplod, P., Intarut, N., & Hemachudha, T. (2011). Comparative detection of rabies RNA by NASBA, real-time PCR and conventional PCR. *J. Virol. Methods*, *175*, 278–282.

Zhang, L. Q., Zhao, F. R., Liu, Z. G., Kong, W. L., Wang, H., Ouyang, Y., Liang, H. B., Zhang, C. Y., Qi, H. T., Huang, C. L., Guo, S. H., & Zhang, G. H. (2012). Simple and rapid detection of swine hepatitis E virus by reverse transcription loop-mediated isothermal amplification. *Arch. Virol.*, *157*, 2383–2388.

Zinsstag, J., Dürr, S., Penny, M. A., Mindekem, R., Roth, F., Menendez Gonzalez, S., Naissengar, S., & Hattendorf, J. (2009). Transmission dynamics and economics of rabies control in dogs and humans in an African city. *Proceedings of the National Academy of Sciences, USA*, *106*, 14996–15001.

Zinsstag, J., Mackenzie, J. S., Jeggo, M., Heymann, D. L., Patz, J. A., & Daszak, P. (2012). Mainstreaming One Health. *EcoHealth*, *9*, 107–110.

Note: Page numbers with "f" denote figures; "t" tables; and "b" boxes.

Printed and bound by CPI Group (UK) Ltd, Croydon, CR0 4YY

03/10/2024

01040422-0011